P9-DTS-569

ALLE · ZEIT · WACH · 1842

To Sally
With many thanks
Dov

Dov Bahat

Tectono-fractography

With 197 Figures in 299 Parts

Springer-Verlag
Berlin Heidelberg New York
London Paris Tokyo
Hong Kong Barcelona
Budapest

Professor Dov Bahat
Department of Geology and Mineralogy
Ben Gurion University of the Negev
Beer Sheva 84105
Israel

ISBN 3-540-53281-1 Springer-Verlag Berlin Heidelberg New York
ISBN 0-387-53281-1 Springer-Verlag New York Berlin Heidelberg

Printed in Germany

Typesetting: K+V Fotosatz, Beerfelden; Printing: Beltz, Hemsbach; Binding: Schäffer, Grünstadt
32/3145-543210 – Printed on acid-free paper

Preface

"... he who repeats a thing in the name of him who said it brings deliverance to the world ..."
Mishnah, Sayings of the Fathers 6; 6

Main Objectives

The present book intends to fulfill a number of purposes, which are arranged under the following scheme:
1. A topical review of main subjects in fractography, that branch of science which analyses fracture surface morphology and related features and their causes and mechanisms in technological materials. Among the materials that bear significant affinities to rock are inorganic glass, ceramics, metals and polymer glass.
2. A historical review of the main studies published to date on rock fractography.

In both these fields of review, one is confronted by the similarities between small-scale (micrometre) and large-scale (tens of metres) fracture surface morphologies. The similarities, on the one hand, and the differences on the other must surely promote further development of fractographical approaches in structural geology, where extrapolation from microfractography to large-scale fractography is virtually a directive.

As geologists become more familiar with the fractography of rocks, they undoubtedly will become aware of the great power of this descriptive discipline as a tool, in both qualitative and quantitative analysis. Rock fractography must yet be routinely applied in the structural analysis of rock formations in which fracture morphology is sufficiently prominent or extensive.

These applications would naturally deepen the involvement of field workers with rather complicated mechanical issues related to the genesis of fractures, and promote the interpretation of problems in structural geology by fractographical approaches, to which the term tectonofractography may be applied.

The detailed treatment of some issues in fractography (Chap. 2) may also be of interest to the non-geologist. Material scientists and prehistoric archaeologists might find here potentially useful applications in some of their specific fields.

The selection of topics for this book and their assignation to each chapter was not a simple matter. One was faced, for instance, with the dilemma of where to place Section 5.9.3 on the interaction mechanisms of cracks. Should it be in Chapter 1, which presents basic concepts of fracturing, or rather in Chapter 5, where it is relevant to the discussion on fault interaction? One hopes that the reader appreciates such difficulties, even where he would possibly have favoured a different alternative.

This book, as noted, purports to present a state-of-the-art overview of the field of tectonofractography. As such, it is bound to draw fire from the initiated, as well as from outer circles. Nevertheless, reasoned criticism would be welcomed, and any alert response may help in advancing the subject.

The Study of Joints

The ubiquity of vertical joints in sedimentary rocks that occur through a wide spectrum of tectonic environments has puzzled investigators since the early days of geology (e.g. Hall 1843). The orientation of joints was correlated with strikes of latter-day faulting and folding (Sheldon 1912), as well as with old basement fractures (Barton 1933). Models of residual elastic strain (Friedman and Logan 1970a) and contemporary stress (Engelder 1982) were also compared to regional joint patterns. In addition to regional investigations, studies were devoted to mesoscale aspects of jointing, such as the surface morphology of fractures (Woodworth 1895; Hodgson 1961a). Joint-producing stress systems (Becker 1893; Parker 1942), the role of pore pressure (Secor 1965) and various engineering aspects (Deere 1963; Barton 1973; Wheeler and Dixon 1980) were investigated as well. Segall and Pollard (1983) made a detailed study of stress and strain close to joint tips in granite.

Pollard and Aydin (1988) note that 10933 citations were listed under joint, joints and jointing in GEOREF for the period 1785 to 1987. Price (1966), who investigated structural aspects of fractures, observed that "despite the fact that joints are so common and have been studied widely, they are perhaps the most difficult of all structures to analyse" (p. 110). It seems that this statement still holds true for today. There are to date very few in-depth applications of joint characterization and classification in the various studies that are devoted to tectonic analysis. Nickelsen and Hough (1967), Price (1974) and Engelder (1985) are three notable exceptions.

Applied Aspects of Joint Investigation

Most important is the practical side of studying joints. According to Brown (1989), "Fractures are well known for their effects on the mechanical and transport properties of rock. Mechanical properties, such as bulk elastic constants and shear strength, are strongly affected by the presence of fractures". Brown adds that "Fractures also control the hydraulic conductivity of crystalline and tight sedimentary rock". Pollard and Aydin (1988) note that joints "influence mineral deposition by guiding ore-forming fluids, and they provide fracture permeability for water, magma, geothermal fluids, oil and gas. Because joints may significantly affect rock deformability and fluid transport, they are carefully considered by engineering geologists in the design of large structures, including highways, bridges, dams, power plants, tunnels, and nuclear-waste repositories". Pollard and Aydin stress that knowledge of the spacing, orientation, aperture (fracture opening), and connectivity of joints at such sites is crucial for the construction of nuclear waste repositories and for prediction of the long-term behavior of ground-water flow.

Fractography and Tectonofractography

Kerkhof (1973) points out that fractography has a considerable resolving power and suggests an analogy between fractography and photography: "While in photography the light fixes the picture of an event, in fractography it is the fracture which plays the corresponding part". This science is a very versatile field. Scientistis who investigate glass, single crystals, ceramics, metallurgical and polymeric materials as well as rocks can better understand the fracturing processes by using fractographical criteria. Engineers might wish to see in it a tool for diagnosing material failure. The analysis of fracture surface morphology (or markings) may provide answers to many specific questions, such as:

1. What are the various morphological features on the fracture surface, and what are their characteristics?
2. Can discrimination of these characteristics lead to a knowledge of
   a) fracture origin,
   b) modes and directions of fracture propagation,
   c) stress configurations and magnitudes,
   d) failure modes (fatigue, brittle, ductile, impact, thermal influences),
   e) properties of initial failure flaw, and critical failure flaw,
   f) establish whether the surface features reflect a single or multistage fracture process.
3. Analysis of fracture mechanics, the determination of fracture toughness and related parameters.
4. Correlation between the surface features of a particular fracture and conditions affecting the medium which contains the fracture.

Tectonofractography is a new branch of tectonics and tectonophysics, which applies fractographical analysis to rock fractures and to regional fracture systems (joint sets) with the objective of identifying the tectonic processes that produced the fractures and of determining the mechanical conditions involved. What appears microscopic in the laboratory to most other scientists, appears reaonably large to the geologist who examines outcrops in the field. The geologist is in a position to combine the fractographic information with rock mechanics and structural information. This provides the basis for tectonofractography. An advantage of the tectonofractographic method is that a structural interpretation based on fracture marking analysis is mostly unequivocal. For example, when a geologist observes two joints cutting each other he has difficulties in determining the sequence of events. However, when a fracture marking on a joint surface is fractured, the information can be conclusively time-ordered.

Terminology

A uniform nomenclature of features characteristic to the morphologies of fracture surfaces is advocated in Section 2.2. However, a number of basic terms are somewhat interchangeably used.

Of the four terms flaw, crack, fracture and joint, fracture is the broadest one. The term fracture is applied here to a relatively smooth break in the material that could initiate in a heterogeneus field, but where tension was the dominant stress. The term is used for features of approximately planar dimension that vary in size from millimetres to tens of metres.

"Crack" in the present text is almost synonymous with "fracture". It is generally applied to small fractures.

"Flaw" substitutes occasionally for "microcrack". "Joint" refers to a natural fracture in a rock that developed in a single event. A fracture surface may involve more than one distinct joint (e.g. Fig. 4.12).

"Cleavage" is a term favoured in microscale descriptions. It refers to a smooth break that differs from the above structures by a particular affinity to parallel lattice planes determined by weak interatomic bonds.

The penality for writing a book in a rapidly developing field of research is that it soon becomes redundant. Yet in the broader design one works to achieve this redundancy. Only when this work is surpassed has it achieved its purpose.

"Fracture Province" is defined as an area within which fractures are related in space, time and style, presumably of a common genetic origin.

Acknowledgements

I wish to thank all those who contributed to the preparation of this book. These include Reginald Shagam, who introduced me to rock fractography, and previous and current students of the Department of Geology and Mineralogy at The Ben Gurion University, who accompanied me on field trips: Avi Avigour, Ron Bogoch, Avraham Dodic, Dorit Korngreen, Gideon Leonard, Yuval Levi, Benyamin Rophe, Beni Sagiv and Yuval Winter. I am also pleased to thank my companions on various excursions: Alexa Yelin (Sinai), Michel Coulon (France), Bijorn T. Larsen (Norway), Peter Bonne (England) and Yaacov Arkin (Negev).

Artur Axselrod, Dan Bahat, David Elazar, Helena Pascal, Eli Shimshilashvilli and Hezi Y. Tehori, provided valuable technical assistance. Eli Ascher, plant engineer of former Tempo Inc. and Tamaz Inc. glass bottle factory at Yeruham, provided equipment, many samples and technical help in my experiments. Members of my department were an unceasing source of encouragement and help. Chaim Benjamini was always supportive. Rina Jaget made fracture surface morphology drawings. Adi Savransky was my editorial assistant. Ruth Massil typed the text. Michel Coulon made the photograph shown in Fig. 2.3 available to me.

A. Rabinovitch has been my debating companion for many years, and also cooperated with Gideon Leonard and myself in glass experiments. The collaboration, agreements and disagreements with Terry Engelder on the subject of jointing have been educational to me. M. J. Kovac and G. L. Smay from American Glass Research Inc. broadened my knowledge on the fractography of glass bottles.

I am pleased to record special thanks to scientists who critically reviewed chapters in draft: Chapter 1, Leslie Banks-Sills, A. Rabinovitch and K. Schulgasser; Chapter 2, B. Sina; Chapters 3 to 5, I. Perath. B. Sina also wrote most of Section 2.1.2.6 and supplied illustrations for that section.

Financial assistance from the Ministry of Energy and Infrastructure, the Earth Science Research Administration, is gratefully acknowledged.

Final thanks go to Hanna, Dan, Zvi and Jonathan, my life companions, who coped with my occasional autistic temper during work on the book and shared with me the stress and strain.

DOV BAHAT April 1991

Nomenclature

Most quantities in the text are included in this list of symbols. The wish to retain conventional notations resulted unavoidably in several cases in the use of the same symbols for different quantities or in using two letters for the same entity

$2c$ or l; $2b$	internal crack length; internal crack width
$2c_i$ or l_o; a_i	initial flaw length; initial edge flaw depth
$2c_{cr}$; a_{cr}	critical flaw length; critical edge flaw depth
r_m, r, r_b	mirror radii for the boundaries of mirror-mist, mist-hackle and initiation of macroscopic crack branching
b_o; r_o	equilibrium atomic spacing; mean ionic distance
r, θ	radius and angle in polar coordinates
θ,α	crack angle and cone angle in Hertzian cone
Ω	angle produced by the rotation of principal stress
ϕ	angle between the primary crack and stress direction
α	half branching angle or angle of dilation in en echelon array or half angle produced by conjugate fractures
γ; θ	deviation angle of a branching crack due to fracture interaction; deviating angle of parent crack or angle between primary crack and initial secondary crack
ω	twist angle between the joint shoulder and en echelon crack
β	angle of internal friction
α; β	deviation angle of various planes from the parent joint in the fringe of a discoid; the kink angle between two adjacent planes in the fringe
m, i	median and interface angles which characterize the plume
ϱ	radius of curvature at the tip of the crack
σ, σ_L; σ_p	normal stress; uniform applied tension; proof stress
σ_{xx}, σ_{yy}, σ_{xy}, σ_{yz}	rectangular Cartesian stress components

σ_{rr}, $\sigma_{r\theta}$, $\sigma_{\theta\theta}$	polar stress components
$\sigma_1 > \sigma_2 > \sigma_3$; $\bar{\sigma}_1 > \bar{\sigma}_2 > \bar{\sigma}_3$	principal stresses; effective principal stresses
σ_{th}; σ_f	theoretical strength; fracture stress at the crack tip
σ_{ys}; e_{ys}; e	yield stress; uniaxial tensile yield strain; tensile strain
σ_{cir}, σ_{ax}; σ_t	circumferential stress and axial stress in the wall of a cylindrical bottle; thermal stress
S; E; υ	tensile strength of the rock; Young's modulus; Poisson ratio
P; d; g; h	pore pressure; rock density; acceleration of gravity; depth
K_I, K_{II}, K_{III}	stress intensity factors for three modes
K_o; K_{Ii}	the (fatigue) endurance limit; initial stress intensity factor
K_c; K_{Ic}; K_{1c}; K_{ID}; K_b	fracture toughness; mode I static plane strain fracture toughness; mode I static plane stress fracture toughness; dynamic fracture toughness; K at crack branching
ΔK	difference between maximum and minimum K
A_m; A	mirror-mist constant; mist-hackle constant
G; G_I; G_c; G_b	the strain energy release rate; mode I strain energy release rate; the critical strain energy release rate; G at branching
γ, γ_p; γ_m	fracture surface energy per unit area; γ for crack propagation before branching; γ at mist initiation
U_s; U_e; U; ΔU	surface energy; stored elastic strain energy; total crack energy; surplus of energy released due to crack joining
r_y; $\bar{\varrho}$	radius of circular plastic zone at the crack tip; plastic zone length
v_c; v_c^*; δ	half crack tip displacement; half critical crack tip displacement at crack extension; crack opening displacement
da/d_N	increment of crack growth per cycle
t_f; t_{min}; t_o	time to failure; minimum time to failure; time to failure under constant applied stress σ_a
σ/σ_N	ratio of stress in room temperature distilled water to strength in liquid nitrogen temperature
$t/t_{0.5}$	ratio of time to failure at stress σ to breaking time for stresses equal to half σ_N
V or V_b; V_{th}; V_T	crack (fracture) velocity; theoretical crack velocity; terminal fracture velocity
V_o; V_{max}	crack growth velocity at the stress corrosion limit; maximum measured crack velocity

V_t; V_l	transverse (shear, secondary) wave velocity; longitudinal (dilatation, primary) wave velocity
v_m; Δ_v^+	molar volume; activation volume for the chemical reaction
ΔE^+	activation energy for the chemical reaction
v_{crit}	critical volume of a rock
n; $\bar{n}$, $\acute{n}$	order of chemical reaction; stress corrosion susceptibility; branching stages
X_p; X_t	partial pressure; boundary layer thickness
D_{H_2O}; T; R	diffusivity of water; absolute temperature; gas constant or stress ratio
Y, Z, Q	geometric factors of the flaw
P; D; t	internal pressure; inside diameter of the bottle; wall thickness of the bottle
W_{mist}; W; w; W_v	width of the mist region; the width of en echelon crack; part of W which is not overlapped; plume waviness
FSM; SLJ; MLJ	fracture surface markings, single-layer joints; multi-layer joints
m.a.	million years

Contents

Chapter 1 Introduction to Fracture 1

1.1 Basic Concepts of Elastic Fracture 1
1.1.1 The Stress Concentration Factor 1
1.1.2 The Griffith Energy-Balance Concept 3
1.1.3 Obreimoff's Experiment 7
1.1.4 Fracture Mechanics 9
1.1.5 The Maximum $\sigma_{\theta\theta}$ Criterion 14
1.1.6 Fracture in Compression 16
1.1.7 Experimental 17
1.2 Plastic Zones Ahead of the Crack 22
1.2.1 Introduction 22
1.2.2 Various Manifestations of the Plastic Zone 22
1.2.3 The Size of the Plastic Zone in Silicate Glasses ... 25
1.2.4 Plane Strain and Plane Stress 26
1.2.5 Secondary Cracks in the Plastic Zone 29
1.2.6 The Damage Zone 30
1.3 Atomistic Concepts of the Crack Tip 31
1.3.1 Thomson's Three Prototypes 31
1.3.2 The LRT Atomistic Surface Force Model 33
1.3.3 The Dissociative Chemisorption Model 34
1.4 Kinetic Processes in Fracture 36
1.4.1 Subcritical Crack Growth 36
1.4.2 Failure Prediction 44
1.4.3 Supercritical Crack Velocities 46
1.4.4 Fracture Branching 49
1.5 Microstructural Aspects of Fracture in Polycrystalline (Grainy) Materials 51
1.5.1 General 51
1.5.2 Dependence of Mechanical Properties on Microstructure 52
1.5.3 Crack Shielding 53
1.5.4 Fracture in Concrete 54
1.5.5 Limitations and Deviations from Simple Microstructure-Strength Relationship 56
1.6 Fracture in Rocks 57
1.6.1 Joint Initiation Stage 58

1.6.2 Joint Propagation 59
1.6.3 Joint Arrest 61

Chapter 2 Fractography in Technical Materials 63

2.1 Fracture Surface Morphology – Basic Geometry 63
2.1.1 Introduction 63
2.1.2 Fracture Categories 67
2.1.3 The Quantitative Mirror Plane 101
2.1.4 Crack Branching 111
2.2 Terminology 118
2.3 Applied Fractography 119
2.3.1 Fractography as a Tool of Fracture Diagnosis in Glass Bottles 120
2.3.2 The Fractography of Metal Failures 132
2.3.3 Fractography in Polymethylmethacrylate 135

Chapter 3 Rock Fractography 139

3.1 Fracture Markings on Joint Surfaces 139
3.1.1 Early Studies 139
3.1.2 Plumes and Related Structures 142
3.1.3 Rib Markings 159
3.1.4 Combined Markings of Plumes and Ribs 166
3.1.5 Affinities of Specific Joint Markings to Certain Joint Directions 167
3.1.6 The Fringe 170
3.1.7 Discoid Radial and Ring Joints 183
3.1.8 Fracture Mechanisms 189
3.1.9 Fracture Markings in Thermally Deformed Rocks and in Granite 195
3.2 Induced Fracturing in Rocks 202
3.2.1 Controlled Laboratory Conditions 202
3.2.2 Fractography Induced by Coring 205
3.2.3 Markings Induced by Quarryin 209

Chapter 4 Characterization and Classification of Fracture Surface Morphology in Geologic Formations 211

4.1 Qualitative Characterization of Fracture Surface Markings 211
4.1.1 Descriptive Parameters of Fracture Surface Markings 211
4.2 Quantitative Characterization of Fracture Surface Markings 223
4.2.1 Measureable Parameters of Fracture Surface Markings 223
4.3 Classification of Fracture Surface Markings 237

Chapter 5 Tectonofractography 239

5.1 Application of Joint Surface Morphology in Tectonics .. 239
5.1.1 Assumptions 239
5.1.2 Problems Involved in the Transition from Material Science to Tectonophysics 240
5.2 Burial Jointing 240
5.2.1 Early Burial Joints in Lower Eocene Chalks Near Beer Sheva 241
5.2.2 Jointing in Middle Eocene Chalks South of Beer Sheva 245
5.2.3 Burial Joints and Syntectonic Joints in the Appalachian Plateau, U.S.A. 248
5.2.4 Common Features of Burial Joints 255
5.3 Syntectonic Jointing 258
5.3.1 Syntectonic Jointing in Santonian Chalks, Israel .. 259
5.3.2 Upper Cretaceous Chalks in East France 265
5.3.3 Upper Cretaceous Chalks in South England 267
5.3.4 Upper Palaeozoic Fractures in Shales of the Appalachian Plateau, New York 274
5.3.5 Jointing in the Entrada Sandstone, Utah 274
5.3.6 Syntectonic Jointing in Association with Fault Termination 275
5.3.7 Syntectonic Joints Related to Unfolding 277
5.3.8 Syntectonic Discoidal Fracturing in Flint Clays ... 279
5.3.9 Characteristic Features of Syntectonic Joints 279
5.4 Uplift Jointing 282
5.4.1 Burial Joints and Uplift Joints in Lower Eocene Chalks Around Beer Sheva 282
5.4.2 Uplift Joints in Middle Eocene Chalks and Limestones from the Northern Negev (Southern Israel) 285
5.4.3 Unloading and Release Joints from the Appalachian Plateau, U.S.A. 287
5.4.4 Sequential Formation of Uplift Joints 290
5.4.5 Vertical Propagation of Uplift Joints 292
5.4.6 Orthogonal Joint Sets 297
5.4.7 Fracture Markings on Horizontal Surfaces 297
5.4.8 Common Features of Uplift Joints 298
5.5 Post Uplift Jointing 301
5.6 Comparison of Jointing During the Three Major Tectonic Phases: a Summary 302
5.6.1 Spatial Characters of the Joint Plane 302
5.6.2 Fracture Surface Markings 310
5.7 Propagation, Opening, Mineralization and Joint Intensity of the Various Joint Types 313

5.7.1 Propagation, Opening and Mineralization 313
5.7.2 Joint Intensity 316
5.8 Approximate Maximum Depths at Which Various Joint Types Develop 316
5.9 Fracture Interaction 317
5.9.1 Extents of Joint Interaction 318
5.9.2 Conditions of Joint Interaction 320
5.9.3 Interaction Mechanisms of Experimental Cracks and Faults 322
5.9.4 Extensional Branching of Faults 323

References 325

Subject Index 341

Chapter 1 Introduction to Fracture

1.1 Basic Concepts of Elastic Fracture

The following chapter has two objectives; (1) to introduce the reader to some basc concepts of fracture; (2) to provide theoretic background for subjects that are treated in the following chapters. The curriculum of this chapter is primarily aimed at the student of natural fracturing (jointing) in rocks.

Rock fractography requires a knowledge of basic concepts of elastic fracturing. However, since rocks are generally grainy, multi-phase materials, deviations from elastic models are unavoidable, and allowances must be made for influences of the "plastic zone" and the "damage zone" at the crack tip (Sect. 1.2). Microstructural aspects of fracturing in polycrystalline materials (Sect. 1.5) are essential items in the analysis of rock fracturing (Sect. 1.6). The wish to link models of rock fracturing to ultrascale (intermolecular) processes will also be always at the back of most investigators' minds (Sect. 1.3). The graph which relates the fracture tensile stress intensity with the fracture velocity (Sect. 1.4) has two parts: the subcritical and the supercritical regimes which are below and above K_{Ic} respectively (Sect. 1.4.1.3). Alternative methods may be applied to calculate the time to failure (Sect. 1.4.2). Fracture branching and fracture arrest are still the two sides of the same mysterious coin in the regime of supercritical fracturing (Sects 1.4.3 and 1.4.4).

Emphasis on the various topics is selective and to some extent different from other texts written either for engineers (e.g. Knott 1973; Broek 1982; Hertzberg 1976) or physicists (e.g. Lawn and Wilshaw 1975a; Thomson 1986). For instance, Section 1.1.2 on the Griffith energy-balance concept draws heavily on studies of concrete (a grainy, rock-simulative material generally unfamiliar to geologists), whereas concepts related to plastic fracture mechanics, such as the J-integral and the COD, are only briefly mentioned (Sects. 1.2.1, 1.2.2).

1.1.1 The Stress Concentration Factor

Scientists have concentrated on two basic parameters in reasoning their fracture theories: the crack tip shape and the crack size. A pioneering analysis of the dependence of fracture on crack shape is offered by Inglis (1913). He considers an elliptical hole under a uniform applied tension. Using the theory of linear elasticity, Inglis shows that a remote stress may be magnified many times at the highly curved end of an ellipse, an effect which is termed stress concen-

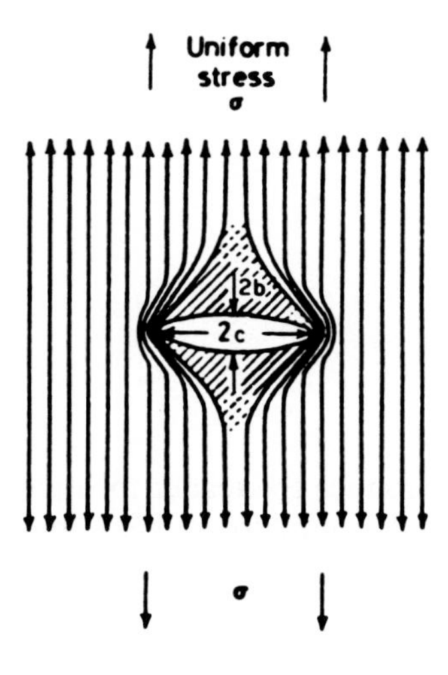

Fig. 1.1. Perturbation of lines of force by an elliptical hole of length *2c* and width *2b* in a plate. (After Knott 1973)

tration. Accordingly, Inglis recognizes that a small narrow flaw can be regarded as a potential of weakness.

Consider a plate containing a central elliptical hole, which is subjected to a uniform tensile stress (Fig. 1.1). Regard the stress as being transmitted from one end of the plate to the other by means of "lines of force" (like magnetic lines of force). At the ends of the plate remote from the hole the lines are spaced uniformly. The stress field close to the hole is said to be perturbed, and, because the lines are supposed to behave like elastic strings (they may be regarded as the sequences of atomic bonds) and, therefore, try to minimize their lengths, they cluster together near the ends of the long axis to give a decrease in the local spacing and, therefore, an increase in the local tension (Knott 1973, p. 12). At the end of the short axis the effect is reversed and there are regions of compression. Conversely, if a circular hole is subjected to a remote uniaxial compression σ, the hole will have local compression of 3σ at the two side poles and tension of magnitude σ near the top and bottom poles.

Inglis's basic equation for the tangential stress σ_{yy} at the tip of an elliptical crack whose length 2c is normal to the applied tension σ_L is (Fig. 1.1):

$$\sigma_{yy}(C,O) = \sigma_L(1+2c/b) = \sigma_L[1+2(c/\varrho)^{1/2}] \ , \tag{1.1}$$

where 2b is the crack width, ϱ is the radius of curvature at the tip of the elliptical crack (Fig. 1.2) and $\varrho = b^2/c$. The greatest stress concentration occurs at the point C, the point of maximum curvature along X. When $b \ll c$, Eq. (1.1) reduces to

$$\sigma_{yy}(C,O)/\sigma_L \approx 2c/b = 2(c/\varrho)^{1/2} \ . \tag{1.2a}$$

For a tip of a deep thin crack, $\sigma_{yy}(C,O)$ may be replaced in Eq. (1.2a) by the theoretical cohesive strength σ_{th} and Eq. (1.2a) becomes

$$\sigma_{th}/\sigma_L \approx 2(c/\varrho)^{1/2} \ . \tag{1.2b}$$

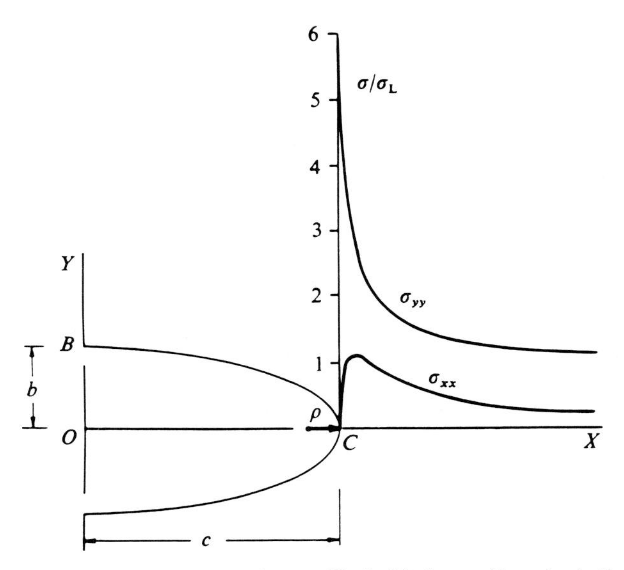

Fig. 1.2. Stress concentration at elliptical hole, $c = 3b$ and ϱ is the radius of curvature at C. Note that concentration localized within $\approx c$ from tip, with high stress gradients within $\approx \varrho$ from tip. (Lawn and Wilshaw 1975a)

The ratio $2c/b$ is commonly referred to as the elastic stress concentration factor.

Figure 1.2 shows that σ_{yy} drops rapidly from the end of the ellipse along X to σ_L and σ_{xx} has a maximum at about a ϱ distance from the tip.

1.1.2 The Griffith Energy-Balance Concept

Griffith (1920, 1924) expands on Inglis's crack. He views the crack as a reversible thermodynamic system within the framework of the theory of elasticity and investigates the total energy of a two-dimensional thin elliptically shaped crack under remote applied stress (Fig. 1.3). The energy of a fracture surface is higher than the energy at the interior of the material due to the greater unsatisfied atomic bonding at the surface compared to the interior. Therefore, a creation of a new surface requires an additional supply of energy to the system. That is, the crack extension involves an increase of the surface energy U_s which is given per unit thickness of the plate by

$$\Delta U_s = 4\Delta c\gamma \ , \tag{1.3}$$

with γ the surface free energy per unit area. This is balanced by the decrease in the stored elastic strain energy U_e of a semicircular region around the crack (Holloway 1973), given for plane stress conditions per unit thickness of the plate (when $\varrho \to 0$) as

$$U_e = \pi c^2 \cdot \sigma_L^2/E \ , \tag{1.4}$$

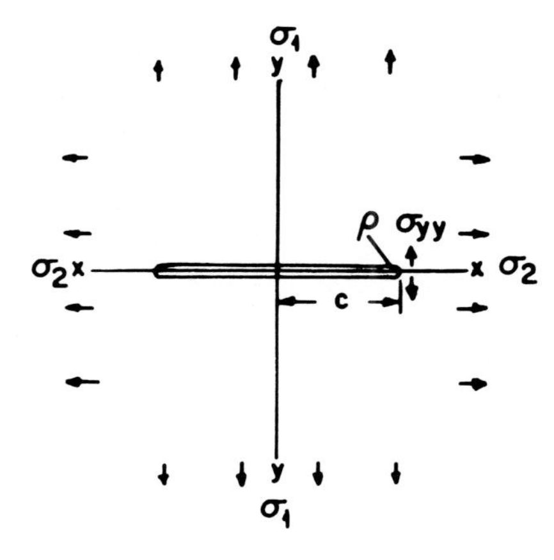

Fig. 1.3. Griffith's elliptic hole (1924) of the larger semi-axis subjected to remote uniform tension σ_1 where σ_2 "has no influence on the rupture stress", σ_{yy} is the maximum stress at the corners of the crack and ϱ as in Fig. 1.2, provided ϱ is small compared to c

where E is the modulus of elasticity. A formal statement of the Griffith energy-balance concept is given by the equilibrium condition that

$$dU_s/dc = dU_e/dc. \tag{1.5}$$

Differentiating Eqs. (1.3) and (1.4) yields $d/dc(4\gamma c) = 4\gamma$ and $d/dc(\pi c^2 \sigma_L^2/E) = 2\pi\sigma_L^2 c/E$, respectively, and equating these results derives the well-known Griffith equation

$$\sigma_L = (2E\gamma/\pi c)^{1/2}\ , \tag{1.6}$$

which predicts that the critical stress which results in the catastrophic growth of a crack 2c long is given by $(2E\gamma/\pi c)^{1/2}$ and, conversely, when σ_L is applied

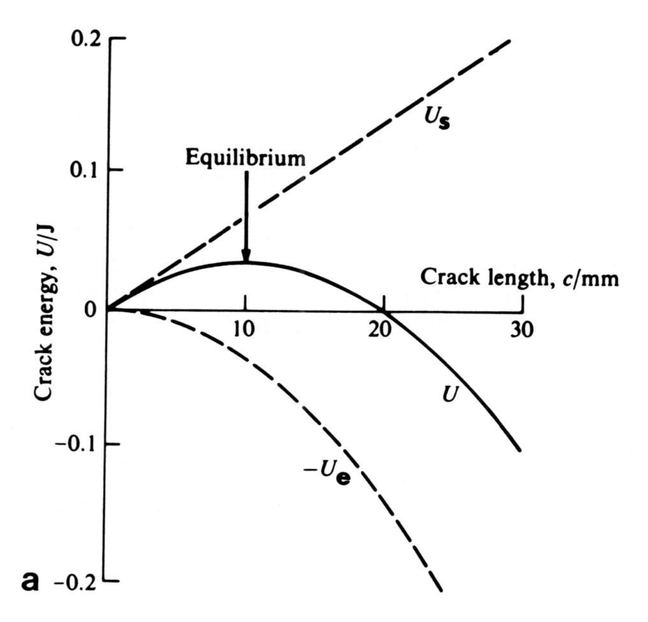

Fig. 1.4a

the critical crack length is $2E\gamma/\pi\sigma_L^2$. E and γ contribute to the strength of the material, and (Griffith 1920): the breaking load "is inversely proportional to the square root of the length of the crack". Griffith measured the surface tension of molten glass in the 1110–745 °C range and extrapolated it to 15 °C assuming an approximate linear γ function of temperature. Equation (1.6) is useful for a great range of crack and specimen geometries (Paris and Sih 1965).

Plots of the total crack energy U, which is given by $-\pi c^2 \sigma_L^2/E + 4c\gamma$ versus crack length, illustrate quite well (Fig. 1.4a) the U maximum at equilibrium (Knott 1973, Fig. 4.7; Lawn and Wilshaw 1975a, Fig. 1.5). This maximum

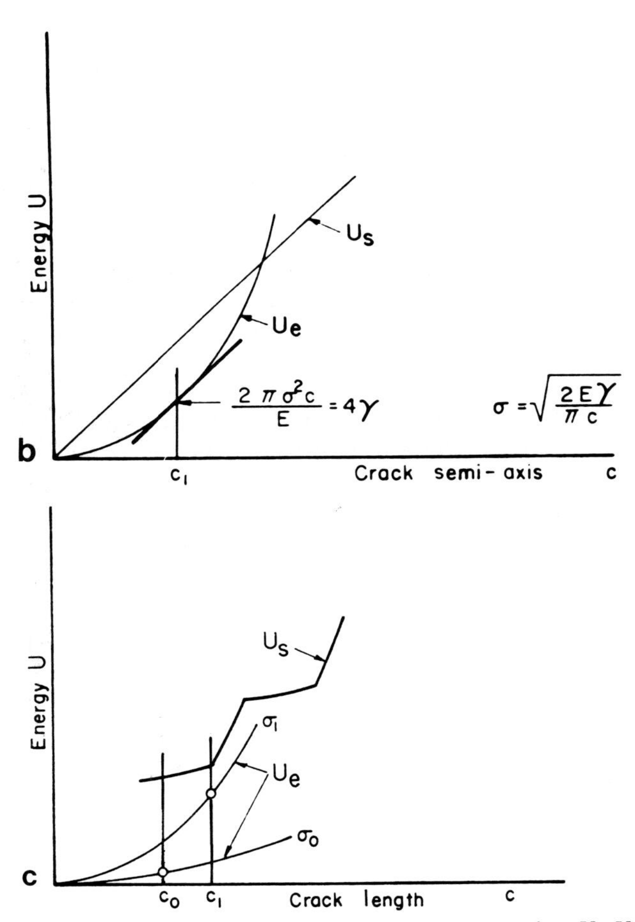

Fig. 1.4a–c. Energetics of Griffith crack in uniform tension. U_s, U and U_e are the surface energy, total crack energy and the stored elastic strain energy, respectively. **a** Data for glass from Griffith's paper: $\gamma = 1.75$ J/m^2, E = 6.2×10^{10} N/m^2, $\sigma_L = 2.63 \times 10^6$ N/m^2 (selected to give equilibrium at c = 10 mm). (Lawn and Wilshaw 1975a). **b** Schematic crack propagation in a uniform material, see further explanation in text. (After Glucklich 1963). **c** Schematic crack propagation in a biphase material. (After Glucklich 1963)

resembles the maximum in free energy plotted against the radius of a nucleated new crystal. Beyond a critical radius size, crystal growth occurs spontaneously (e.g. Carmichael et al. 1974, Fig. 4.14).

Whereas U_s is described by a straight line, indicating that its rate increase is crack length-independent and is given by the slope (4γ), the U_e release follows a parabolic curve [Eq. (1.4)], and the rate of U_e release is given by $2\pi\sigma_L^2 c/E$ (Fig. 1.4b). Equilibrium occurs at c_1 where the slopes of the two curves are equal. At $c > c_1$ the rate of U_e release is greater than the rate of U_s requirement and spontaneous fracture propagation initiates.

Studies of fracture in concrete (Kaplan 1961; Glucklich 1963; Soroka 1977) reveal much greater U_s values than expected. Apparently, in these grainy heterogeneous materials which consist of more than one phase, fracture propagation is different from that of glass. In glass, a single crack that propagates beyond c_1 (Fig. 1.4b) may lead to a failure. In concrete, on the other hand, fracture is characterized by the growth of many microcracks that develop at various zones of stress concentration and U_s is derived by the sum of surface energies of all the microcracks. In addition, when tension occurs in two-phase material the initial crack of length c_o starts to grow stably due to the increasing stress σ_o (Fig. 1.4c). The stress will increase faster than the external load, due to the decrease of the sample cross-section exposed to tension. This occurs as long as the crack propagates in the phase of lower toughness. As soon as the crack reaches an area (a grain) of the more resistant phase, U_s significantly increases. This occurs due to both the increase of the cracked zone at the crack tip (see damage zone in Sect. 1.2.6) when the crack propagates around the rigid grain and when the crack cuts into the grain. It requires a further increase of stress from σ_o to σ_1. U_s is characterized by a curve of increasing

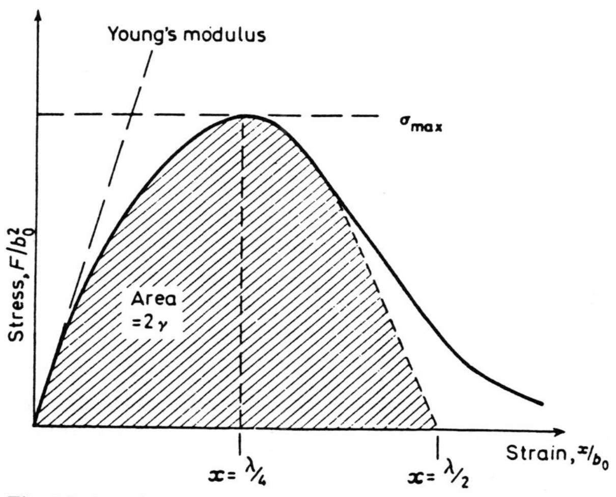

Fig. 1.5. Atomic stress-strain curve where area is given as 2γ and F is the force acting between pairs of atoms, see more notation in text. (Knott 1973)

slope with stepwise jumps. Hence a dependence of the material strength on two types of crack arresters.

In determining the theoretical cohesive strength σ_{th} [Eq. (1.2 b)] on an atomistic level, fracture is considered to occur when the bonds between atomic planes are stressed beyond a certain maximum (Fig. 1.5). The stress displacement curve is approximated by a sine curve of a λ wavelength (Orowan 1949):

$$\sigma_f = \sigma_{th} \sin(2\pi x/\lambda) \;, \qquad (1.7)$$

where σ_f is the applied stress that would lead to fracture (the fracture stress) and x is a small displacement from equilibrium. A mathematical computation for an elastic material results in

$$\sigma_{th} = (E\gamma/b_o)^{1/2} \;, \qquad (1.8)$$

where b_o is the equilibrium atomic spacing and 2γ is the area under the stress-displacement curve. For typical glass parameters where $E = 7 \times 10^4$ MPa, $b_o = 30$ nm, $\gamma = 5$ J/m^2, $\sigma_{th} \sim E/7$ or 1×10^4 MPa (Freiman 1980). Strengths of this order of magnitude have been measured in pristine glass fibers (Procter et al. 1967).

Griffith (1920) points out that the theoretical strength of isotropic solids which should reflect bond strength according to the molecular theory is actually reduced by microcracks (now known as "Griffith flaws") on the surface (Fig. 1.6a, b). The presence of these flaws explains why the strength of glasses are often much lower than σ_{th} (Sect. 1.4.1).

Leonardo da Vinci found an inverse relationship between the length of iron wires and the breaking load for constant diameter wires (Irwin 1964). Griffith (1920) observes that in glass fibres σ_f increases as thickness decreases (Fig. 1.6c) and at extremely small thicknesses the theoretical strength is approached. A combination of Eqs. (1.8) and (1.6) gives

$$\sigma_f/\sigma_{th} \sim (b_o/c)^{1/2} \;. \qquad (1.9)$$

Substituting values for bulk glass leads to a value of an inherent crack size of about 0.002 cm (Kanninen and Popelar 1985). These authors observe that the great discrepancy between the theoretical strength of many materials and their actual reduced strength due to flaws in them corresponds to the analogous discrepancy between th theoretical shear strength of a crystalline solid and the observed values. This "led to the identification of the dislocation as the fundamental element in metal plasticity in the mid 1930 s".

1.1.3 Obreimoff's Experiment

Obreimoff's experiment (1930) is of special interest. In this, a glass wedge is inserted to peel off a cleaved flake from mica (muscovite) (Fig. 1.7). Fracture propagates along the cleavage plane no faster than the wedging, and the system suffers from no displacement as the crack grows. These conditions are comparable to a "fixed grips" configuration (Sect. 1.1.7.1). Obreimoff discovered that on a gradual removal of the glass wedge in vacuum a reverse "healing"

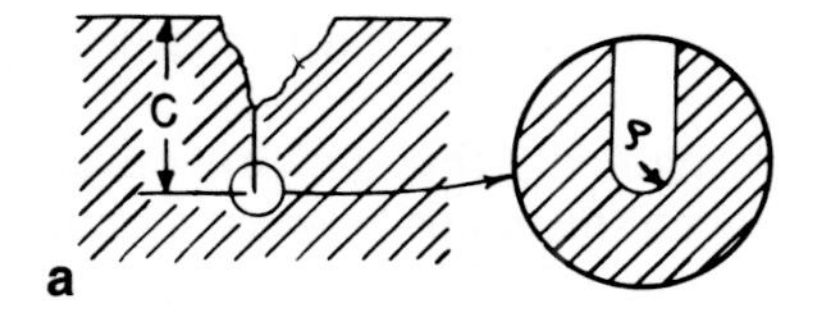

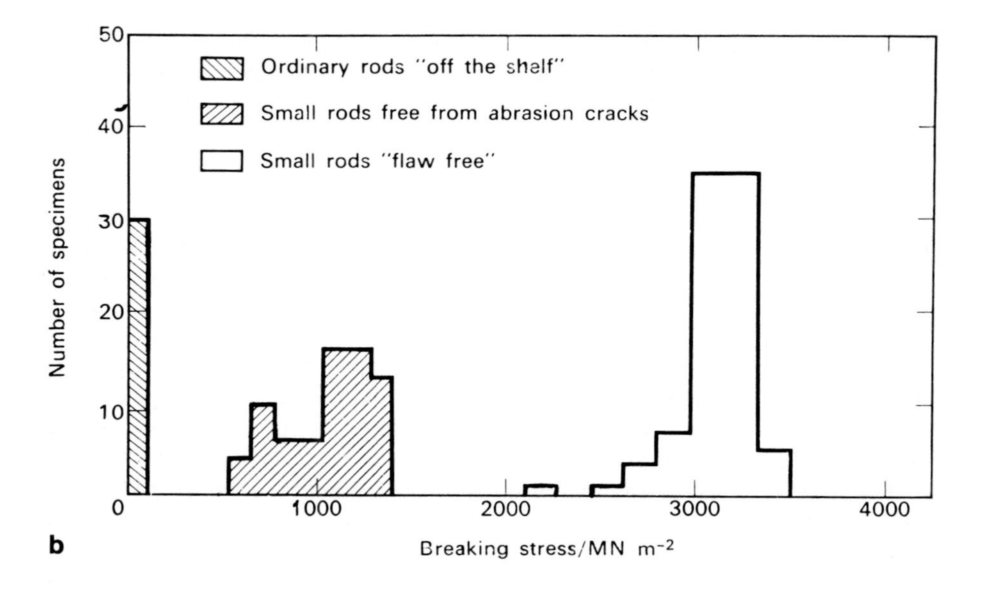

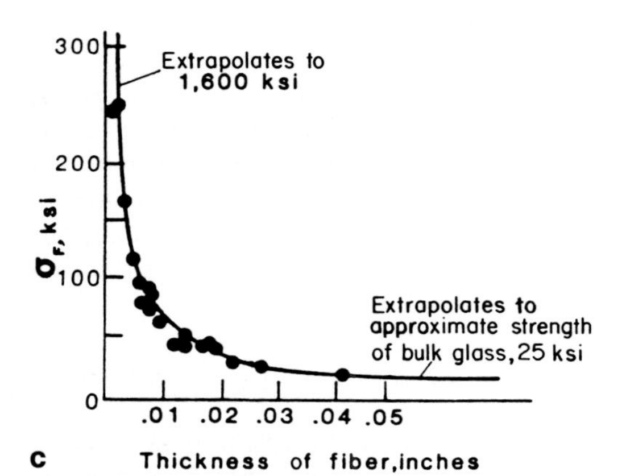

Fig. 1.6a–c. Cracks and strength of glass. **a** Schematic representation of abrasion flaw in glass surface. Glass is partly removed at the surface, but the crack effectively continues to further depths. (Mould 1967). **b** Comparison of the breaking strengths of ordinary untouched and flaw free glass rods. (Holloway 1973). **c** Results of Griffith's experiments on glass fibres. (Kanninen and Popelar 1985)

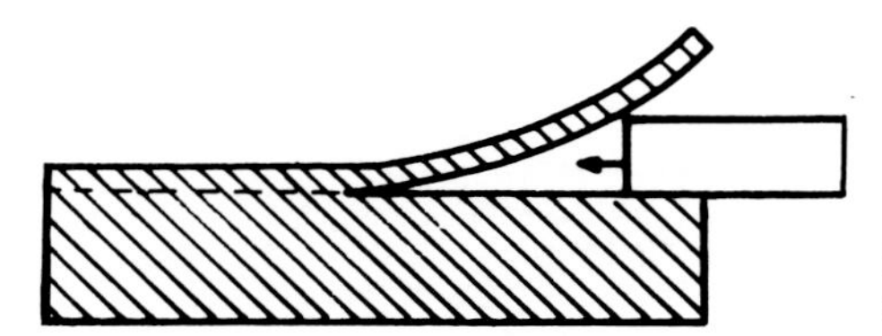

Fig. 1.7. Obreimoff's wedging experiment on mica

took place and the crack separating the flakes practically disappeared. This reversibility did not occur in normal atmosphere, which elucidated the importance of chemical reaction on the flake surfaces (see further recent elaborations by Lawn et al. 1987 and Section 1.3.2).

1.1.4 Fracture Mechanics

Fracture mechanics stems from the quantitative relationship between strength and crack size (Griffith 1920). It is, however, mostly identified with Irwin's introduction of the concept of the stress intensity factor K in 1948 and in other papers.

Fracture mechanics assumes that cracks are present on the surface of the brittle article under service conditions. It expands from engineering needs to define the critical conditions for instability regardless of the mechanisms that lead to fracture. This concept developed in response to previous inadequacies of designs based on strength measurements and unsatisfying safety factors that ended in catastrophic failures of large mobile structures like aircrafts and ships.

Extended investigations on the stress intensity factor (e.g. Sih et al. 1962) have provided an elaborate list of K formulae which are applicable to various sample geometries and crack conditions. For example, for a large body subjected to a remote tensile stress σ normal to an internal crack of length $2c$

$$K = \sigma\sqrt{\pi c}\ . \tag{1.10}$$

If, however, in the same sample the identical stress is applied on an edge crack of length (depth) c the formula is

$$K = 1.12\sigma\sqrt{\pi c}\ . \tag{1.11}$$

1.1.4.1 Three Modes of Crack Propagation

It is useful to distinguish three basic modes of distortion around the crack tip (Irwin 1960). These modes are defined with reference to a frame, often the Cartesian coordinate system (Fig. 1.8). Mode I (tensile or opening mode) corresponds to perpendicular separation of the crack walls under the action of tensile stresses; mode II (sliding mode) reflects shearing of the crack walls in the plane of the crack parallel to the direction of the crack (normal to the crack front); mode III (tearing mode) is the shearing in the plane of the crack parallel to the crack front. When the xy plane is used for reference, modes I and II

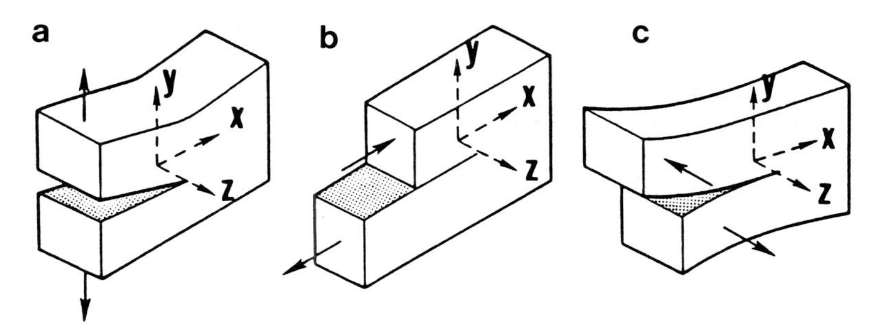

Fig. 1.8 a – c. Three basic loading modes for a cracked body. **a** Mode I, opening. **b** Mode II, sliding. **c** Mode III, tearing. (After Kanninen and Popelar 1985)

are termed plane strain distortions where points on the crack surface are displaced in the reference plane perpendicular and parallel to the crack surface, respectively. No displacement occurs along the z direction in modes I and II. In mode III points on the crack plane are displaced along the z axis, normal to the reference plane and this mode is termed anti-plane strain.

Irwin (1957) and Williams (1957) solved the biharmonic equation (see also earlier work of Westergaard 1939) for the elastic stress field near a sharp crack of a slit shape which has a single curved or straight border as resolved for rectangular and polar coordinates (Fig. 1.9). The solutions for the rectangular and polar coordinates are published in various text books. The origin of the polar reference frame coincides with the crack tip, r and θ are the usual polar coordinates, σ_{rr} is the radial stress component. $\sigma_{\theta\theta}$ is the tangential stress component, and $\sigma_{r\theta}$ is the shear stress component. All resolved in the stress field near the crack tip.

K_I, K_{II} and K_{III}, known as the stress intensity factors, are the central terms in Irwins analysis. They appear in equations that represent modes I, II and III,

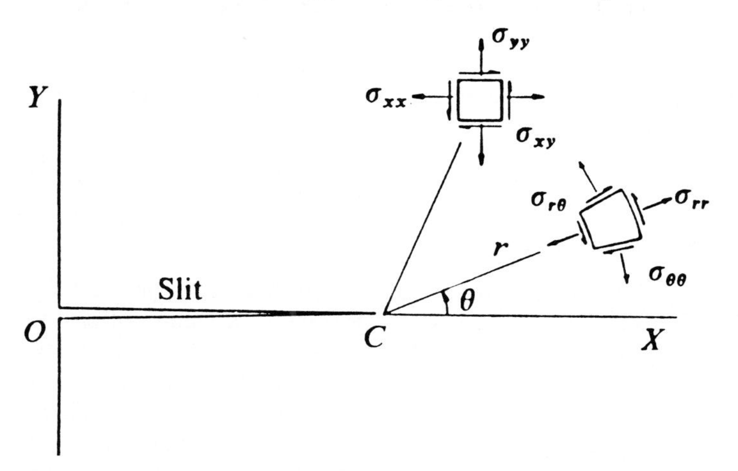

Fig. 1.9. Stresses at the tip of a crack showing rectangular and polar coordinate components (after Lawn and Wilshaw 1975 a)

respectively. Solutions for mode I are given below. More elaboration on these relationships is presented by Lawn and Wilshaw (1975a).

$$\begin{Bmatrix} \sigma_{xx} \\ \sigma_{yy} \\ \sigma_{xy} \end{Bmatrix} = K_I/(2\pi r)^{1/2} \begin{Bmatrix} \cos(\theta/2)[1-\sin(\theta/2)\sin(3\theta/2)] \\ \cos(\theta/2)[1+\sin(\theta/2)\sin(3\theta/2)] \\ \sin(\theta/2)\cos(\theta/2)\cos(3\theta/2) \end{Bmatrix} \quad (1.12)$$

$$\begin{Bmatrix} \sigma_{rr} \\ \sigma_{\theta\theta} \\ \sigma_{r\theta} \end{Bmatrix} = K_I/(2\pi r)^{1/2} \begin{Bmatrix} \cos(\theta/2)[1+\sin^2(\theta/2)] \\ \cos^3(\theta/2) \\ \sin(\theta/2)\cos^2(\theta/2) \end{Bmatrix} \quad (1.13)$$

The stress intensity factor K is the amplitude of the inverse square root stress singularity at the crack tip transmitted from a remote stress held for each mode and has units of stress x $(\text{length})^{1/2}$ or force x $(\text{length})^{-3/2}$. It is a constant independent of r and θ which depends on the loading and crack size, as well as on the shape and manner of sample loading and geometrical setting of the crack in the sample. The analysis is valid provided r is small compared to the dimensions of the crack length. Equations (1.12) and (1.13) do not contain terms other than $1/r^{1/2}$ because the higher-order terms are negligible.

The elastic stress concentration factor (Sect. 1.1.1) differs from the stress intensity factor K in that the former is dimensionless, being dependent on geometry and independent of load.

1.1.4.2 Fracture Toughness

Modern engineering needs introduced the term fracture toughness K_c as a simple relationship

$$K_c = (2E\gamma)^{1/2} \qquad \text{(plane stress)} \quad (1.14)$$

$$K_c = [2E\gamma/(1-\upsilon^2)]^{1/2} \qquad \text{(plane strain)}\ . \quad (1.15)$$

These equations help to define structural materials which are tough as well as strong when υ is the Poisson's ratio. E values give an idea on the resistance to yield stress and macroscopic plastic deformation and γ is a measure of the resistance to the propagation of the existing cracks in glass (Holloway 1973).

K_c is usually equated with K_{Ic}. K_{Ic} is arbitrarily taken as the K_I needed to drive the crack at a certain velocity at the onset of unstable growth in glass. This velocity is $\gtrsim 10^{-1}$ m/s according to Wiederhorn et al. (1974a), but other authors (e.g. Kerkhof 1973; Carlsson et al. 1973; Freiman et al. 1974) suggest other transitional velocities.

Fracture toughness is a reproducible material property under a given environment and experimental procedure, so that it is a useful guide for given service conditions. For instance (Kanninen and Popelar 1985), the critical length of the internal crack for spontaneous growth (Fig. 1.4a, b) is

$$2c_{cr} = 2/\pi (K_{Ic}/\sigma)^2 \quad (1.16)$$

and the critical length of the edge crack will be

$$c_{cr} = 1/1.25\,\pi (K_{Ic}/\sigma)^2\ . \quad (1.17)$$

It is clear than that the critical edge crack is considerably smaller than the critical internal crack and therefore the former crack is more risky than the latter (see more on c_{cr} in Fig. 2.31).

A useful relationship of this dependence is:

$$K_{Ic} = (1.2\pi/Q)^{1/2}\sigma_f c_{cr}{}^{1/2} , \tag{1.18}$$

where Q is a modifying geometrical factor ranging in value from 1.0 for a long shallow flaw to 2.46 for a semi-circular flaw (Randall 1966), 1.2 is the coefficient of surface flaw.

The great practical success of fracture mechanics stems from the ability to predict the behavior of a large structure on the basis of laboratory scale results when incorporated to a numerical analysis. K_c does depend, however, on the dimensions of the test specimen (Sect. 1.2.4).

1.1.4.3 The Strain Energy-Release Rate G

An important manifestation of the Griffith energy balance concept is the strain energy-release rate per unit width of crack front. Here the rate is with respect to crack length not time. G (in honor of Griffith) at constant displacement (Sect. 1.1.7.1) is defined by:

$$G = -dU_e/dc . \tag{1.19}$$

where U_e is the stored elastic strain energy [Eq. (1.4)] and c is the crack half length. G depends on the crack geometry and loading. G is also termed the crack extension force. The term driving force is often used.

An expression of the Griffith equilibrium requirement is that a crack would spontaneously propagate when the strain-energy released per increment of crack extension would equal the maximum rate of energy required for creating the new crack surfaces:

$$G_c = \pi c_{cr}\sigma^2/E = 2\gamma , \tag{1.20}$$

where G_c is the critical strain energy release rate which occurs at $G \sim 8$ J/m^2 for soda-lime glass (Table 1.1).

The actual fracture surface energy γ usually deviates from the specific surface free energy [Eq. (1.3)] due mainly to irreversible plastic deformation (Sect. 1.2.2), a damage zone or surface chemical reactions at the crack tip. Orowan (1948) considers an increase of some three orders of magnitude in γ above the specific surface free energy due to plastic deformation at the crack tip in a metal. A reduction of γ by about 3 in the cleavage of muscovite under room conditions compared to vacuum conditions has been observed by Obreimoff (1930; Sect. 1.1.3). Roesler (1956) found for glass a range of γ from 10 J/m^2 for a few seconds load duration to 1.8 J/m^2 for long loadings, suggesting γ to be 4.1 J/m^2. Hence, a slow crack growth affected by a surface chemical reaction may be responsible for a considerable range of γ. Generally, γ increases linearly in glass with increasing Young's modulus (Wiederhorn 1969) (see also Sect. 1.5.5).

Table 1.1. Fracture toughness values. (Holloway 1986)

Materials	G, ($J\ m^{-2}$)
Fused silica	9
Soda lime silica glass	8
Borosilicate glass	9
Pyroceram	50
Mullite	23
Silicon carbide	40
Glassy carbon	17
Al_2O_3	14
Polystyrene	180
Polymethylmethacrylate	400
Polycarbonate	1000
Epoxy resin	100
Epoxy resin + 15 vol% SiO_2 flour	500
Cast iron	2000
Mild steel (at −196 °C)	200

Fracture propagation in glass is controlled by linear elastic conditions and is described by a linear growth (at the specimen scale) of U_s (Fig. 1.4b). In heterogeneous grainy materials (like concrete, ceramics and rocks, see Sect. 1.5.4) U_s rises faster than U_e (Fig. 1.4c). Following increase in σ from σ_o through σ_4 (Fig. 1.10), at c_5 the slope does not increase any more when no further microcrack growth occurs. At that stage the rate of the release equals the rate of U_s increase. This occurs at G_c when the two curves have identical slopes and spontaneous crack growth starts.

G is often related to the plane stress K by (Lawn and Wilshaw 1975a):

$$G = K_I^2/E + K_{II}^2/E + K_{III}^2(1+\upsilon)/E \tag{1.21}$$

and to the plane-strain stress intensity factors by:

$$G = K_I^2(1-\upsilon^2)/E + K_{II}^2(1-\upsilon^2)/E + K_{III}^2(1+\upsilon^2)/E \ . \tag{1.22}$$

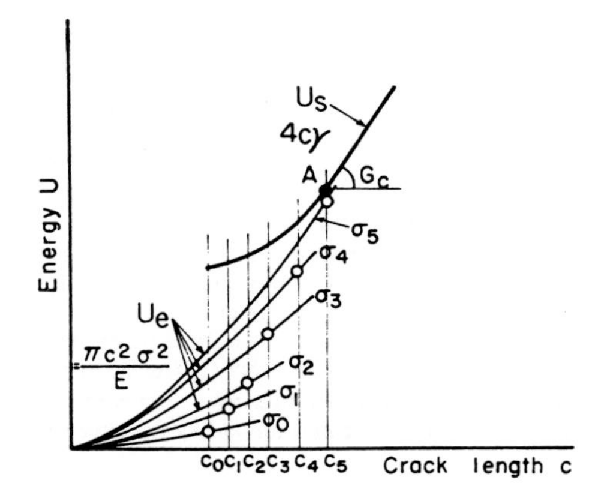

Fig. 1.10. Schematic propagation of a Griffith crack under uniform tension in a biphase material showing the critical strain energy release rate G_c. (After Glucklich 1963)

The strain energy release rates for different modes are additive.

The Irwin's (1957) criterion for crack propagations is $G = G_c$, and G_c is related to K_c as

$$G_c = K_c^2/E \quad \text{(plane stress)} \tag{1.23}$$

$$G_c = K_c^2(1-\upsilon^2)/E \quad \text{(plane strain)} , \tag{1.24}$$

whereas K is a local crack-tip parameter, G is a global one.

1.1.5 The Maximum $\sigma_{\theta\theta}$ Criterion

An important theory of fracture is known as the maximum tangential tensile stress $\sigma_{\theta\theta}$ or maximum $\sigma_{\theta\theta}$ criterion proposed by Erdogan and Sih (1963). The basic idea is that from all the stresses around the crack tip the most effective one that controls crack propagation is the maximum $\sigma_{\theta\theta}$ (Fig. 1.9). When exposed to static mode I, $\sigma_{\theta\theta}$ maximum is at $\theta = 0$ and it corresponds to σ_{yy}. Therefore, crack propagation occurs at the continuation of and parallel to the parent fracture. Under dynamic conditions $\sigma_{\theta\theta}$ maximum occurs at $\theta \neq 0$ (Yoffe 1951) (Sect. 1.4.4.1). A good experimental way to demonstrate the role of the maximum $\sigma_{\theta\theta}$ theory is by applying mixed mode loading on a plate.

Erdogan and Sih (1963) show the behavior of a central crack of length 2c in a plate under uniform applied tension inclined at an angle ϕ (Fig. 1.11 a). The results of their theoretical and experimental studies is a curve of ϕ versus θ (Fig. 1.11 b). θ varies from 0° under pure mode I ($\phi = 90°$) to $\theta = 70.5°$ under pure mode II ($\phi = 0°$)). The value of the angle θ produced between the plane of the initial crack and the first increment of the secondary crack extension is primarily dependent on ϕ (Cotterell 1972):

$$\sin\phi \sin\theta - \cos\phi(3\cos\theta - 1) = 0 . \tag{1.25}$$

The tangential stress at the crack tip is given by (Erdogan and Sih 1963):

$$\sigma_{\theta\theta} = \cos(\theta/2)/(2r)^{1/2}[K_I \cos^2(\theta/2) - 3/2 K_{II} \sin\theta] . \tag{1.26}$$

Hence, maximum $\sigma_{\theta\theta}$ follows θ at the initiation of the secondary curved branch under various ratios of K_{II}/K_I (as a function of ϕ) as well as at a later stage when the fracture straightens at a principal stress direction. Cotterell demonstrates that θ may change considerably also when the shape of the initial crack changes from a slit to an ellipse.

Fig. 1.11 a–d. Paths of secondary cracks due to stresses inclined to parent fracture. **a** Uniform tension. Resolved components of loading into modes I and II. Stress is inclined to the direction of the parent crack (*heavy line*) and θ is the angle produced between that line and the first increment of the secondary crack. (Lawn and Wilshaw 1975 a). **b** Plot of θ versus ϕ comparing curves obtained by various authors for tensile loadings. (After Erdogan and Sih 1963, Williams and Ewing 1972 and Finnie and Saith 1973). **c** Compressional stresses inclined to a parent crack in PMMA produce two secondary cracks which appear as curved branches *cb* at an angle θ of 90°. (After Barquins et al. 1989). **d** The tip zone of an inclined parent crack under inclined uniaxial compression σ_1 showing secondary curved branch fracture *cb*, echelon microcracks *e* and unloading fracture *ub*. (Petit and Barquins 1987)

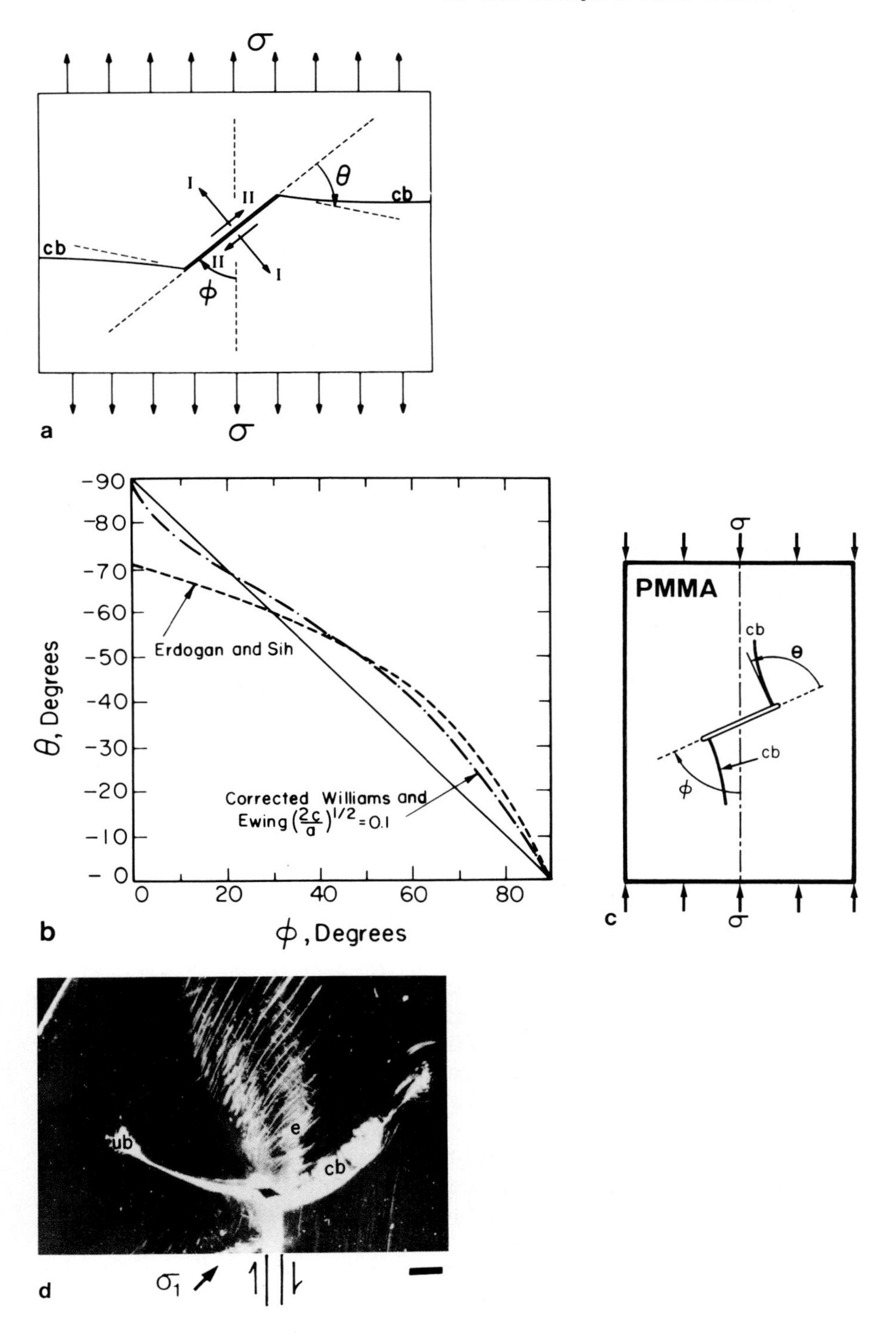
σ
I
II
θ
cb
cb
II
I
φ
σ
a
-90
-80
-70
-60
-50
-40
-30
-20
-10
-0
0
20
40
60
80
θ, Degrees
φ, Degrees
Erdogan and Sih
Corrected Williams and
Ewing $\left(\frac{2c}{a}\right)^{1/2}=0.1$
b
σ
PMMA
cb
θ
cb
φ
σ
c
ub
e
cb
σ1
d

Williams and Ewing (1972) observe that in fact, the curve by Erdogan and Sih (Fig. 1.11 b) predicts that if $\phi < 30°$ the growth of the secondary crack should be below the horizontal line that would develop normal to the applied tension, whereas for $\phi > 30°$ above this line. Williams and Ewing repeated with some modifications Erdogan and Sih's tensile experiment on PMMA. Their results from $\phi = 0°$ to $\phi = 90°$ (not shown here) mostly seem to fall between Erdogan and Sih's curve and the horizontal line. Finnie and Saith (1973) modified Williams and Ewing's curve (Fig. 1.11 b).

1.1.6 Fracture in Compression

Griffith (1924) observed that redistribution of remote compressional stresses around microcracks results in localized tensile stress concentration (see Sect. 1.1.1). The stress concentration increases from a circular hole through an elliptical hole further to a narrow crack. Glucklich (1963) calculated that, for example, if a narrow elliptical crack that has a length ratio of major to minor axes of 1 : 128 is exposed to a compression σ at an angle of 35° to the major axis, it would suffer from an aproximate tension of $36\,\sigma$.

It has generally been observed that when a centre cracked (or notched) plate is exposed to remote uniaxial compression (Brace and Bombolakis 1963; Glucklich 1963 and McClintock in Erdogan and Sih 1963) at an angle to the length of the initial crack, the extension of the initial crack under this loading has several characteristics (Fig. 1.11 c): (a) the stress concentration zone shifts slightly aside from the tip of the initial crack, (b) crack extension is not along the initial crack but it kinks away as a "curved branch", cb, possibly following the maximum $\sigma_{\theta\theta}$ theory, and (c) the crack straightens then and takes the direction perpendicular to the greatest tension or least compression. After the shift in (a) and deflection in (b) caused by the shear, the opening mode takes over continuously in (c).

The maximum tensile $\sigma_{\theta\theta}$ around the boundary of a very slender elliptical crack due to the asymmetric uniaxial compressive stress σ is given by (Cotterell 1972):

$$\sigma_{\theta\theta}/\sigma = (c/\varrho)^{1/2} \sin\phi(1 + \sin\phi) \ , \tag{1.27}$$

when c is half length of the crack and ϱ is the radius of curvature of its tip.

Ingraffea (1977) performed experiments on rocks under compressional conditions showing the path direction of the curved-branch from the tip of the initial crack. The angle produced between the initial crack and the initial increment of the curved branch was $-70 \pm 2°$ as expected by the $\sigma_{\theta\theta}$ maximum theory.

Petit and Barquins (1987) applied biaxial compression on a slit in PMMA (polymethylmetacrylate) small plates at an angle of 30° to the direction of σ_1. A week solvent was added to the slit to lower the surface energy and thus encourage crack initiation and propagation. When the ratio of principal stresses σ_3/σ_1 was low, slow development of microcracks at the slit tip was followed by the rapid formation of the curved branch and the slower development of an axial zone of en echelon cracks e at the tip (Fig. 1.11 d). When the ratio σ_3/σ_1 increased, the echelon zone was still present, whereas the formation of the curved branch was progressively inhibited. The unloading induced inversions in the principal stresses so that an "unloading branch", ub, perpendicular to σ_1 was developed.

Barquins et al. (1989) tested the behavior of a slit in PMMA plates ($50 \times 32 \times 5$ mm) under uniaxial constant compressive conditions. They found that for different ϕ angles that were produced between the load and slit directions the branching angle θ for the curved branch was at $90 \pm 5°$ to the slit (Fig. 1.11 c). This was more clearly visible for angles greater than 40°. In the latter the distance on the slit from the curved branch to the slit tip increased with ϕ and tended to be greater than half the slit width, apparently differing from the maximum $\sigma_{\theta\theta}$ theory which relates to branching at slit tip.

An essential difference in behavior under tension and compression is that in the former secondary growth is unstable, and in the latter fracture propagation is stable. That is, in tension, σ and the rate of U_e release and the crack extension force G increase with crack growth, whereas in compression σ is constant and the rate of U_e release is steady, and a maximum value of G is reached with further crack growth causing a decrease in its value (Glucklich 1963; Cotterell 1972).

In grainy heterogeneous materials like concrete (Glucklich 1963; Soroka 1977), stable crack propagation under tension stops when the crack comes in contact with the more rigid phase and a further increase in σ is required to maintain crack growth (Fig. 1.10). Under compression the required increase in σ is substantially greater if the crack is to cut the more rigid phase, but initiation of a new crack is possible under more moderate stress conditions. However, the growth of the new crack stops when U_s increases beyond a certain limit, which gives way to a growth of a third crack, and so on. The growth of many cracks absorbs energy and avoids early failure. This contributes to the difference between tensile and compressive strength. Under compression the material withstands greater stresses than in tension, which explains Griffith's observation (1924) that "in such tests of stone and like materials as are available the crushing strength is from 7 to 11 times the tensile strength" ...

McClintock and Walsh (1962) suggest a modified Griffith concept accounting for crack closure under compression. They introduce a coefficient of internal friction between crack faces.

1.1.7 Experimental

1.1.7.1 Fracture Tests

Different experimental techniques enable the determination of various fracture mechanics parameters using a wide variety of specimen shapes, crack geometries and loading configurations (see Fig. 1.12 and methods in Corten and Gallagher 1971).

The constant force "dead-weight" loading (Fig. 1.12c) is such that the applied force remains constant as the crack extends.

Under constant displacement "fixed grips" loading the ends of the stressed plate are fixed. When the crack extends, the stiffness of the plate decreases, the load relaxes and the elastic energy drops. In these experiments equilibrium conditions create stable crack growth where the crack extension force G is a decreasing function of the fracture length in contrast to unstable growth under

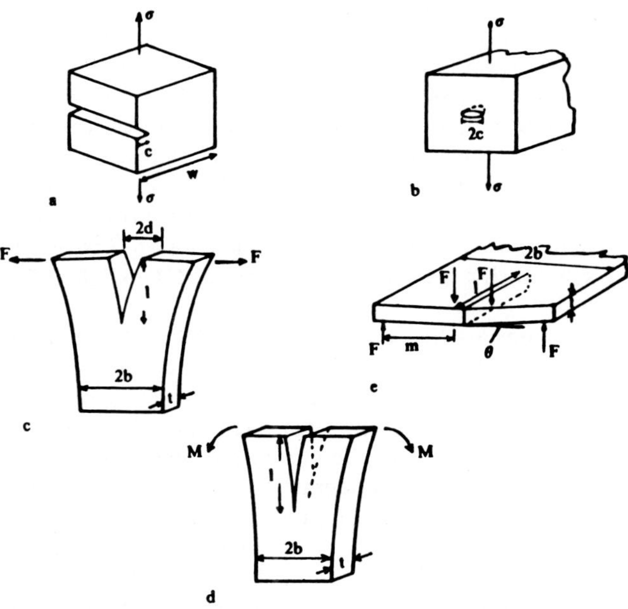

Fig. 1.12 a – e. Fracture toughness parameters for various configurations. In all the equations given below: E is Young's modulus, μ is the shear modulus, υ is Poisson's ratio, and $K^2 = GE$ for plane stress and $GE/(1-\upsilon^2)$ for plane strain. **a** Through-the-thickness edge crack in a finite specimen in uniaxial tension

$$G = [(1-\upsilon^2)/E]\,\sigma^2 \pi c\, f(c/w) \ ,$$

where $f(c/w)$ is the correction factor, $[1.12-0.23(c/w)+10.6(c/w)^2-21.7(c/w)^3+30.4(c/w)^4]^2$.
b Half of a penny-shaped crack in the edge of a semi-infinite specimen in uniaxial tension

$$G = [(1-\upsilon^2)/E]\,(4/\pi)\,\sigma^2 c$$

c Double cantilever specimen

constant displacement, 2d $\quad G = 3Ed^2b^3/4l^4$

constant force, F $\quad G = 12F^2l^2/Et^2b^3$

(both for $l \gg b$)

d Double cantilever specimen

constant moment, M $\quad G = 12M^2/Et^2b^3$

e Double torsion specimen

constant force, F $\quad G = 3F^2m^2/\mu t^4 bs$

(N. B. this is independent of crack length)

constant displacement, θ $\quad G = \mu\theta^2t^2bs/12l^2$,

where $s = 1 - 0.63\,t/b$ (Holloway 1986)

constant force that follows equilibrium according to the Griffith energy balance concept (both assume uniform tension).

1.1.7.2 Fracture Configurations Caused by Combined Modes

Figure 1.13 shows the results of two rotational modes about Cartesian axes of an advancing crack. The first (a) is a "permissible" type (Ryan and Sammis 1978). This type of rotation is typical of mode I and mode II coupling (Fig. 1.8). A rotation of the crack plane about the x axis ("tilt" configuration, Lawn and Wilshaw 1975a) is permitted and the crack plane remains continuous, hence a "permissible" type. This deformation often results in the development of rib markings (Chap. 2) (Fig. 1.13a). The second rotational mode (b) is "non-permissible", it occurs under superposed mode I and mode III loading. A rotation of the crack plane about the z axis ("twist" configuration, Lawn and Wilshaw 1975a) is induced. A continuous adjustment of the crack front is not possible, the instability results in a crack break up and segmentation into partial fronts, which form a piecewise adjustment (Sommer 1969). Fig. 1.13c demonstrates a series of en echelon segments separated by steps which result from the break up at Fig. 1.13b (en echelon segmentation and related aspects are being elaborated in Sect. 3.1.6). A third angular rotation, about y, leaves the crack plane unchanged. A different mixed mode loading of a plate is described in Section 1.1.6.

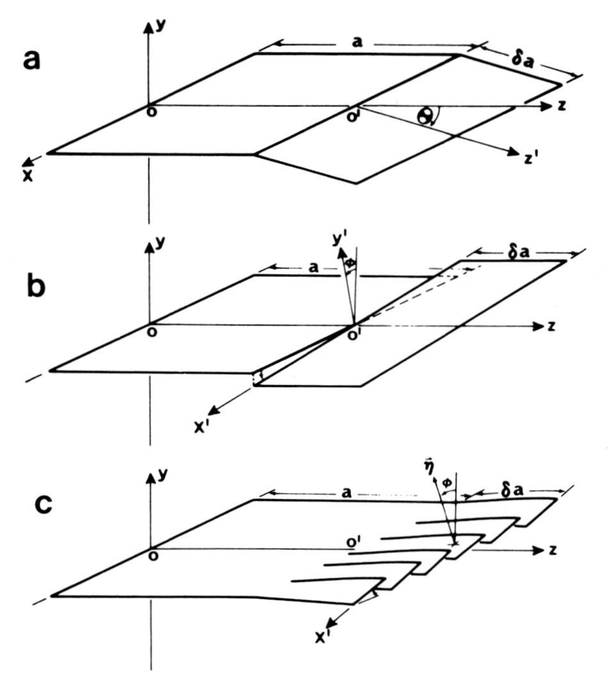

Fig. 1.13a–c. a "Permissible" rotation of the leading edge of an advancing crack (under combined mode-I and mode-II loading). **b** "Non-permissible" rotation (under combined mode-I and mode-III loading). **c** Compromise rotation (of case **b**) involves a piecewise twist of the crack plane and the formation of fracture lances. The direction of crack advance is *from left to right*; this corresponds to an upward-directed crack-advance increment in the lower colonnade. Therefore, the segment labelled *a* in **a**, **b** and **c** represents a prior increment of crack advance, whereas δa represents the current set of increment possibilities. (After Ryan and Sammis 1978)

1.1.7.3 Fracture by Indentation

The basis for treating indentation and deformation due to contact at a point is the elasticity analysis of Hertz (1881). This treatment is divided into static contact and dynamic (impact) loading (Fig. 1.14a–d). The subject of Hertzian indentation, its implications and some technical applications have been reviewed by Lawn and Wilshaw (1975b). This theory also has wide applications in the earth sciences (e.g. Anderson 1936; Orowan 1974).

A cone-shaped crack (Fig. 1.14e) is formed in a brittle solid when it is critically loaded with a spherical elastic indenter. Frank and Lawn (1967) described the Hertzian stress field in detail, showing that the shape of the Hertzian cone is governed by the nature of the strongly inhomogeneous, well-defined stress field beneath the indenter.

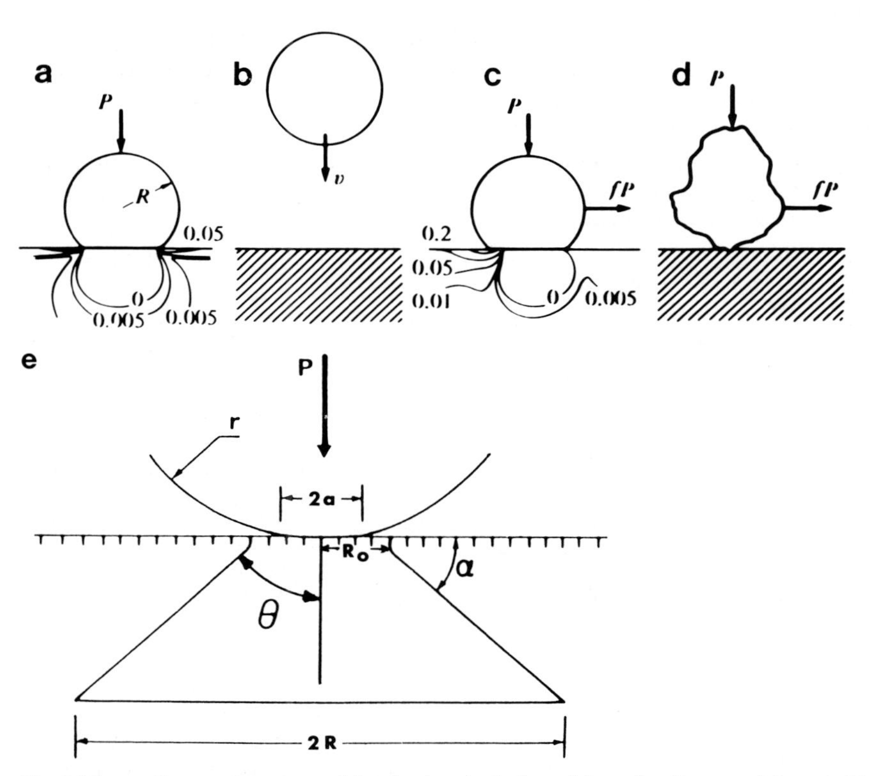

Fig. 1.14a–e. Contact damage model. **a** Static spherical particle, radius R, normal force P (Hertzian contact). **b** Impacting spherical particle, velocity v. **c** Sliding spherical particle, with tangential force fP (f = coefficient of friction) superposed onto normal force. **d** Irregular sliding particle. Equi-stress contours (units of mean normal stress) indicated in cases **a** and **c**. (Lawn and Wilshaw 1975a).**e** Hertzian cone fracture parameters, where P is indenter load, a is contact radius, r is indenter radius, α is cone angle,θ is crack angle and R_o and R are initial and final radii of the cone respectively. (After Lawn et al. 1975)

A cone is initiated in a ring crack which is formed close to the contact with an indenter at the surface of the indented material. Repeated applications of Hertzian stresses at the same locality may produce a series of additional cones close to the initial cone (Wilshaw 1971). Johnson et al. (1973) found that secondary ring cracks of smaller radii have frequently been observed to form during unloading of a steel indenter at its contact with glass. These authors have shown that when the indenter and indented material differ in their elastic properties, frictional forces are brought into play at the interface which modify the Hertz distribution of contact stress. An indenter which is more rigid than the indented surface is shown to lead to an apparent increase in fracture strength of the latter, and a less rigid indenter has the opposite effect.

The Hertzian fracture theory deals with stresses which develop due to pressure exerted by an elastic spherical indenter on a brittle solid. In magmatic diapiric processes the spherical indenter is replaced by a viscous liquid. The question then is, whether the Hertzian theory is still relevant for such modified systems. It has been shown (Bahat 1977a) that Hertzian cones were reproducibly formed in glass by thermally induced stresses without the use of an elastic indenter. Thus, it may be said that an elastic indenter is not a prerequisite for producing Hertzian fracture. A plastic "indenter" may produce similar results provided the pressure is maintained for a required time span, and the indenter is confined in a suitable system. This has been demonstrated experimentally. Perfect cone fractures were produced in the brittle material by liquid "indenters" (Bahat 1979c). Air indentation in Perspex has also been used (Bahat and Sharpe 1982).

In special geological environments like carbonatite extrusions (Bahat 1978), where intense pressures (of liquid and vapour) may equal the lithostatic load, the critical conditions of fracture can be determined from the Hertzian equations (e.g. Frank and Lawn 1967).

Deformation by Static Contact. The damage caused by a blunt indenter (Fig. 1.14a) is quite different from the damage inflicted by a sharp one (Fig. 1.14d). Some fractographic aspects of deformation possibly correlated with indentation are presented in Section 2.1.4.4. One notes that the force exerted by the indenter on the examined article generally has a normal as well as a tangential component.

Deformation by Dynamic Contact. Damage by impact can be caused by both blunt and sharp bodies. It is generally assumed that the entire kinetic energy of the impacting particle is converted into the elastic energy (Wiederhorn and Lawn 1979). The physics of the dynamic contact is considerably more complicated than that of the static contact.

1.2 Plastic Zones Ahead of the Crack

1.2.1 Indroduction

Models of elastic crack tips have varied between two shapes the elliptical tip (Inglis 1913; Griffith 1920) and the sharp slit (Irwin 1958), Figs. 1.3 and 1.9, respectively (see also Sect. 1.4.1.4). In brittle materials like silicate glasses, a departure from linear elasticity at the crack tip is confined to interatomic distances (Wiederhorn 1969). In materials that deviate from elasticity there is a distinct crack tip zone that may take two forms: (a) in ductile materials, essentially metals, there is a "plastic zone" ahead of the crack tip which may be seven to eight orders of magnitude larger than the above-mentioned interatomic distances in silicate glasses (Hahn and Rosenfield 1965). Fuller et al. (1975) show that a considerable amount of heat evolves at the tip of a fast-moving crack in glassy polymers. This heat extends the plastic deformation at the tip and results in the increase in the fracture toughness of the material. (b) In granular brittle materials like ceramics, concretes and rocks there is a "damage zone" of multiple microcracks in front of the crack tip (Hoagland et al. 1973). The plastic and damage zones dissipate large amounts of energy at the crack tips in their respective materials.

In ductile materials, for a given metal fracture toughness (K_{Ic}) is high and the sensitivity of the material to small cracks is relatively low. The deformed region, however, cannot support high stresses and the yield stress σ_{ys} is low. Conversely, brittle materials have low K_{Ic}, their σ_{ys} is high and their stability is more sensitive to small cracks.

The J-integral concept (Rice 1968) is beyond the scope of this chapter.

1.2.2 Various Manifestations of the Plastic Zone

Irwin (1948) and Orowan (1948) extend Griffith's theory to metals. They maintain the energy balance foundation and modify the crack-tip characterization. The popular term for their modification is the Griffith-Irwin-Orowan theory. This modification involves an "inner" non-linear small volume that immediately surrounds the crack tip and separates the crack tip from the "outer" linear elastic material.

A concept suggested by Irwin et al. (1958) is that the plastic zone at the crack tip adds to the reduction in the material strength caused by the crack itself. A treatment of this crack by linear elasticity requires that the crack length should be considered as slightly larger than it is. Accordingly, the half crack length c [Eq. (1.10)] is replaced by $c+r_y$, where r_y is the radius of circular plastic zone at the crack tip (Fig. 1.15), thus

$$K_I = \sigma\sqrt{\pi(c+r_y)} \ , \qquad (1.28)$$

when $2r_y$ is taken as the point on the crack where $\sigma_y = \sigma_{ys}$. r_y is given by

$$r_y = K_I^2/2\pi\sigma_{ys}^2 \quad \text{(plane stress)} \qquad (1.29)$$

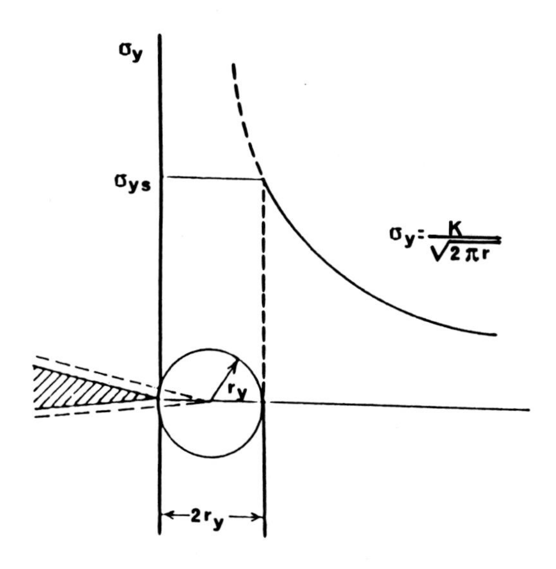

Fig. 1.15. Schematic Irwin's plastic zone at the crack tip. Effective crack length taken to be original crack length plus the plastic zone radius r_y. (Herzberg 1976)

$$r_y = K_I^2/6\pi\sigma_{ys}^2 \quad \text{(plane strain)} \ , \tag{1.30}$$

hence, r_y is three times smaller in plane strain than in plane stress (Sect. 1.2.4).

The Dugdale zone (Dugdale 1960) has been adopted by researchers in various forms including the Dugdale-Muskhelishvili DM model (Hahn and Rosenfield 1965) and the Dugdale-Barenblatt model (Lawn and Wilshaw 1975a). Barenblatt's (1962) idea was that the crack tip does not have an elliptic shape with a finite radius of curvature (Fig. 1.3). Rather the two crack faces close asymptotically at the tip where they exert intense cohesive forces between themselves. The Dugdale zone consists of a slit with an initial length 2c representing a crack in an infinite plate. A balance is maintained between the action of the remote opening stresses σ, transmitted through the linear material and the closing internal stress acting on the plastic zone. The plastic zones extend as long as $\sigma > \sigma_{ys}$ at the tips of the effective crack, which now consists of $2c + 2\bar{\varrho} = 2a$ (Fig. 1.16a, b). Hahn and Rosenfield (1965) show experimentally (Fig. 1.16c) the difference between the actual shape of the plastic zone and the calculated Dugdale zone in equilibrium, which is given by

$$\bar{\varrho}/a = 2\sin^2 A/2 \ , \quad \text{and} \tag{1.31}$$

$$\bar{\varrho}/c = \sec A - 1 \ , \tag{1.32}$$

where $A = \pi\sigma/2\sigma_{ys}$, and σ_{ys} is the yield strength of the material.

Hahn and Rosenfield (1965) demonstrate that the plastic zone in steel changes its shape, size and orientation with respect to the specimen geometry and crack orientation as a function of the stress level. At low stress levels the plane strain state prevails and a relatively small plastic zone extends normal to the plane of the crack, the plastic zone has the shape of slabs folded around

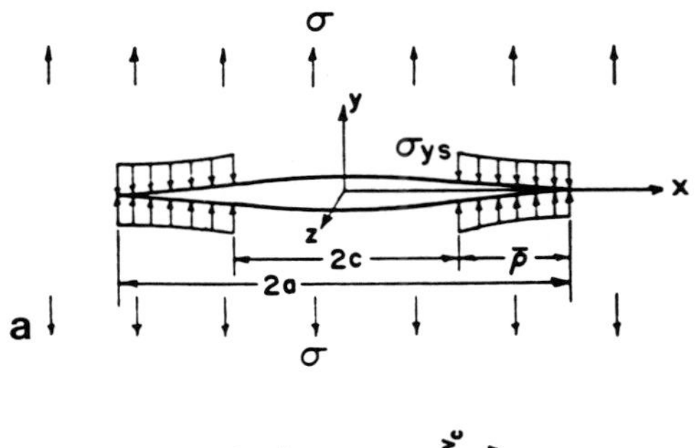

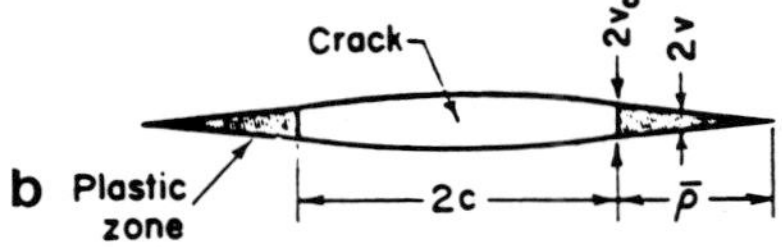

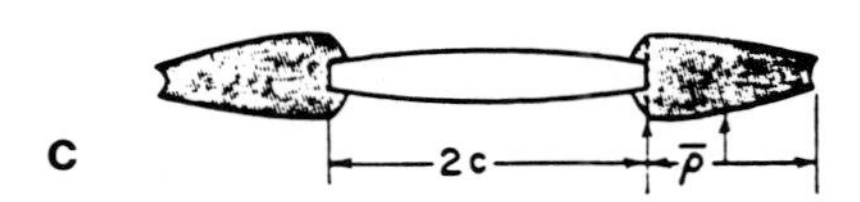

Fig. 1.16 a – c. Model of Dugdale-Muskhelishvili DM crack. **a** and **b** The DM model, where v is the half crack tip displacement in the y direction and $v_c \equiv v_{(x=c)}$, **c** diagram of the actual crack, see text for further explanation. (Hahn and Rosenfield 1965)

an axis which lies perpendicular to the tensile axis (Fig. 1.17 b). At high stress levels the plane stress state dominates and a large plastic zone develops into a size approximately equal to the specimen thickness. The shape and orientation of the zone is determined by the directions of the planes of maximum shear stress and forms two slabs inclined 45° to the tensile axis (Fig. 1.17 a, c). Transitions between the above two limiting conditions occur under intermediate stresses.

The fracture toughness K_c may be determined directly

$$K_c = (2 v_c^* \sigma_{ys} E)^{1/2} , \tag{1.33}$$

where v_c^* represents the half critical crack tip opening displacement at crack extension (Fig. 1.16 b).

The length of the plastic zone in metals (Fig. 1.15) depends on load and crack size. Ductile fracture is frequently classified as brittle when the failure stress is below the stress level for general yielding.

An attempt to treat the plastic zone within the framework of linear elasticity is that of the crack opening displacement COD (Wells 1963). This concept proposes that the COD, δ, will be directly proportional to the overall tensile strain e after general yield has been reached. Considering that the uniaxial tensile yield strain $e_{ys} = \sigma_{ys}/E$, and assuming that $c/r_y = e_{ys}/e$ Wells suggests the relationship:

$$2\pi e_{ys} c/\delta = e_{ys}/e , \quad e > e_{ys} \tag{1.34}$$

after incipient yield.

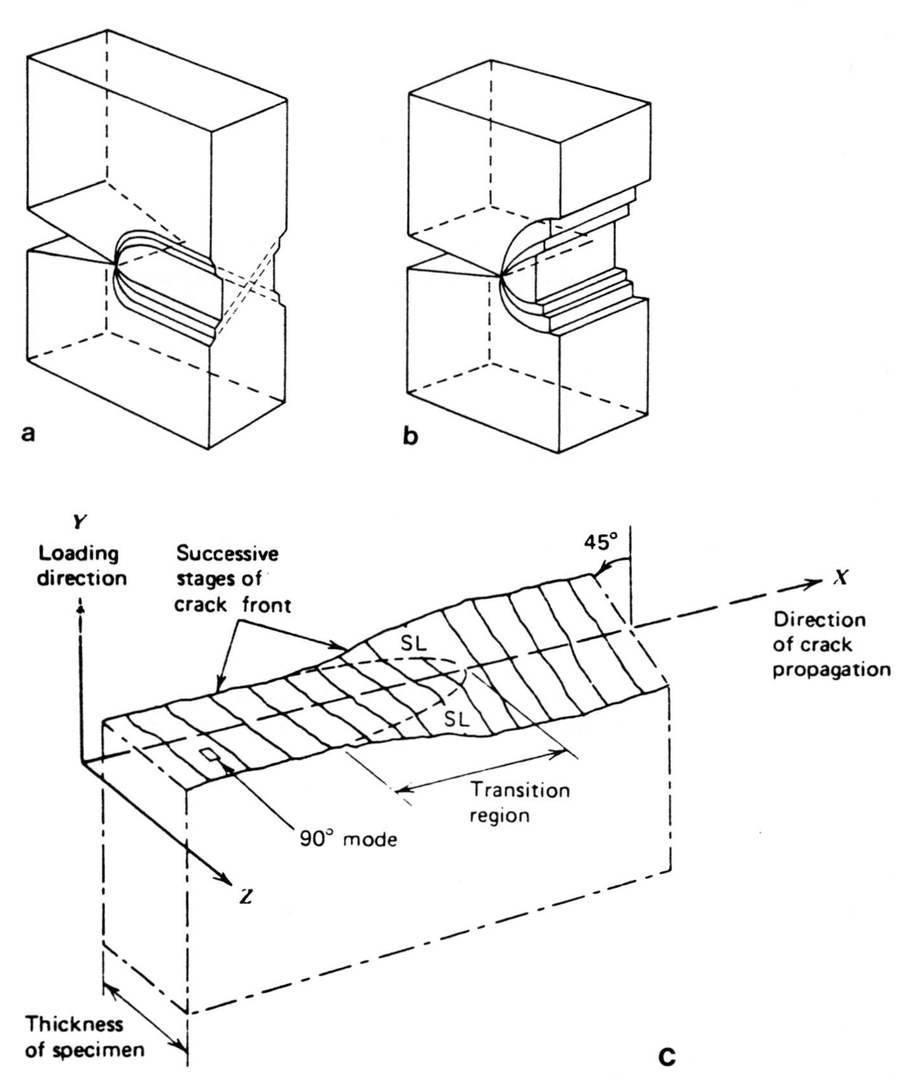

Fig. 1.17 a – c. Schematic crack tip plastic deformation in two idealized stress states. **a** Plane stress, 45° shear mode inclined to the tensile axis. **b** Plane strain, slip bands are normal to the surface of the plate on which the crack tip is seen. (Kanninen and Popelar 1985). **c** Diagram showing fracture mode transition from flat fracture under plane strain conditions (*left*) to slant fracture under plane stress conditions (*right*). (Hertzberg 1976)

1.2.3 The Size of the Plastic Zone in Silicate Glasses

Wiederhorn (1969) estimates the plastic zone size in silicate glasses from the Dugdale model of plastic flow at a crack tip. It is assumed that the solid is a non-work-hardening solid and that the plastic zone extends a length ϱ in front of the crack tip as depicted in Fig. 1.16. The material within the plastic zone is assumed to be at the yield stress of the solid. The length of the zone is

$$\bar{\varrho} = (\pi/8)(K_{Ic}/\sigma_{ys})^2 \tag{1.35}$$

and the half crack opening displacement at the crack tip

$$v_c = 4\sigma_{ys}\bar{\varrho}/\pi E \ . \tag{1.36}$$

Equations (1.35) and (1.36) are valid provided the breaking stress is much less than the general yield stress of the material, a condition that holds for glass, and Wiederhorn uses available experimental data and finds that $\bar{\varrho}$ for soda lime glass and silica glass are 2.6×10^{-9} m and 6.4×10^{-10} m, while the half crack tip opening displacements v_c are calculated to be 4.5×10^{-10} m and 2.2×10^{-10} m, respectively. He observes that in the absence of large amounts of plastic flow, energy absorption during crack motion in soda lime glass is 4.5 J/m^2, which is small compared to that obtained on metals, 10^4 J/m^2 and as a result, a crack once started tends to propagate catastrophically.

The new $\bar{\varrho}$ edges are non-linear zones, not quite cracks that can bear σ_{ys}, which tends to close them and balance the remote stress applied on the elastic part, which tends to open them.

1.2.4 Plane Strain and Plane Stress

The size change of a plastic zone in a plate is shown diagrammatically in Fig. 1.18. In the interior the state is that of plane strain where the stress is triax-

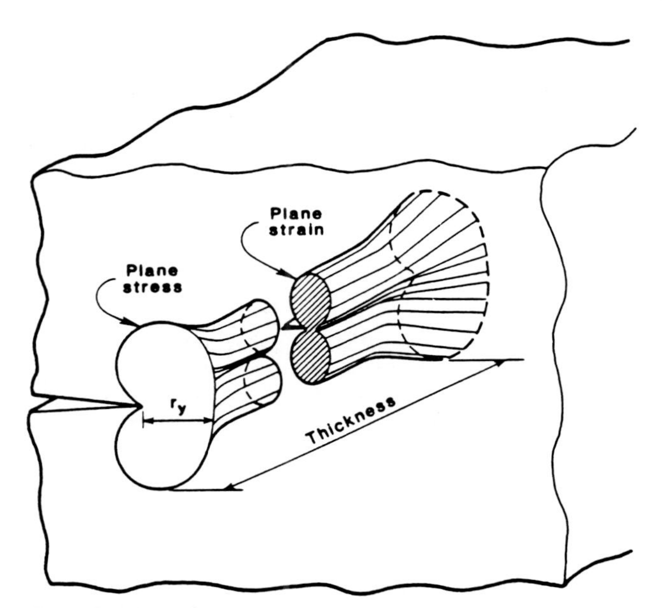

Fig. 1.18. Three-dimensional plastic zone, showing a small zone at the centre-plane strain, and a large zone at the edge-plane stress. (Kanninen and Popelar 1985)

ial, but the strain in one direction (along the thickness) is zero, and therefore the plastic zone is small. In plane stress, on the other hand, yielding is unconstrained in any direction, so that the plastic zone can equal the specimen thickness. The stress σ_3 is zero normal to the plane stress surface, but it gradually increases towards the interior.

There is always a state of plane stress at the surface of a plate where the plastic zone is relatively large. Here, yielding is controlled by a shear stress whose maximum is much larger for plane stress than for plane strain [$\tau_{max} = \sigma_1/2$ and $\tau_{max} = (\sigma_1 - \sigma_3)/2$, respectively] since it incorporates the full value of the local tensile stress (Knott 1973). Hence, the state of stress influences the size of the plastic zone, and shear stresses promote the enlargement of this zone (Hahn and Rosenfield 1965 and Tschegg 1983). A reverse relationship also exists, the size of the plastic zone influences the state of stress and large plastic zones promote the development of plane stress.

If the ratio of plastic zone diameter r_p to plate thickness B is 1, plane stress can develop. Plane strain occurs when $r_p/B \ll 1$. High K_I and low σ_{ys} result in a large plastic zone, since the latter is proportional to K_I^2/σ_{ys}^2. Consequently, fracture toughness tests require thicker samples for materials with high K_{Ic} (which also have low σ_{ys}). Generally, high strength materials have low K_{Ic}. The critical K_I value depends upon the thickness of the tested sample, as is shown diagrammatically in Fig. 1.19. There is some uncertainty in very thin plates. A K_{Ic} maximum occurs in thin plates at some Bo. This maximum is thought to be the real plane stress K_{1c}. There is a transition zone between Bo and Bs where the critical K_I has intermediate values. Beyond the transition zone at thicker plates the plane strain K_{Ic} prevails.

It has been observed for a number of metal alloys that in thin pieces where plane stress is dominant the fracture becomes slant under tension (Fig. 1.17c). According to the Tresca criterion, yielding occurs when the maximum shear stress τ_{max} attains $\sigma_1/2$ at surfaces which intersect in a line parallel to the intermediate principal stress axis σ_2 (axis x in Fig. 1.20a), and which form angles of 45° with the plane x−z that includes the largest and smallest prin-

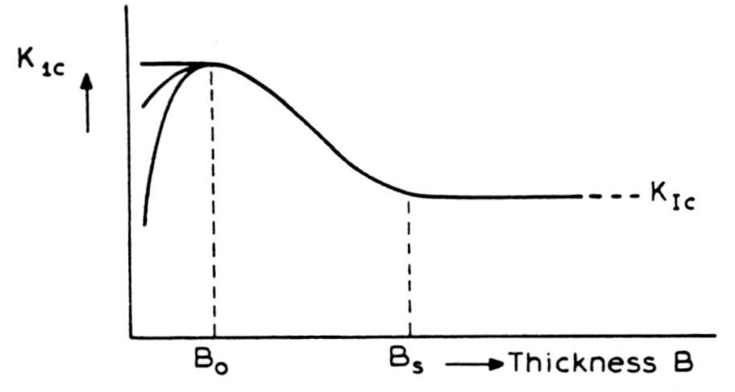

Fig. 1.19. The dependence of the critical Mode I stress intensity factor on specimen thickness. Maximum fracture toughness under plane stress conditions at B_o, a transitional region between B_o and B_s and plane strain beyond B_s. (Broek 1982)

cipal stresses (σ_1 along y and σ_3 along z, respectively). Deformation occurs by slip along these surfaces in metals. Extension of the specimen mostly shifts from mode I to mode III because lateral displacement occurs along the slanted surface. In practice, some crack opening is also associated with this shear (Knott 1973, Fig. 5.3) which will involve the movement of both screw and edge dislocations.

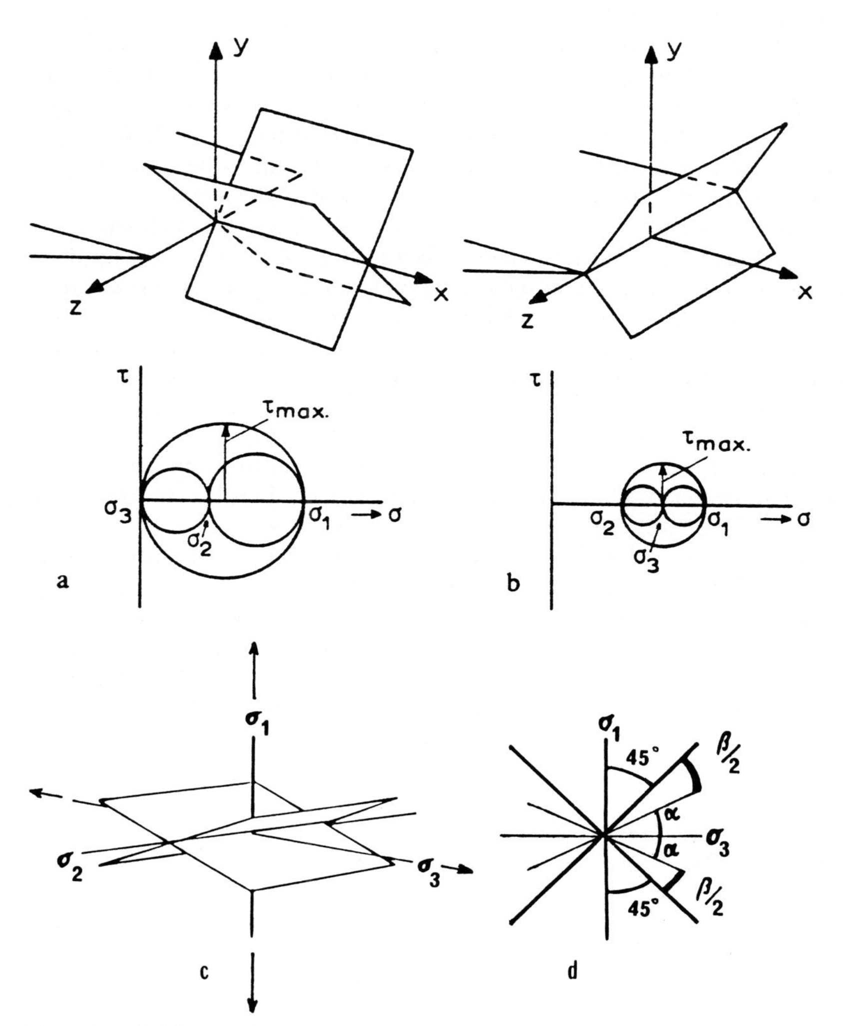

Fig. 1.20 a–d. Planes of maximum shear stress at the tip of a sharp crack for θ close to zero (Fig. 1.9). **a** Plane stress. **b** Plane strain. (Broek 1982). **c** The plane stress relationship between shear fractures and tensile principal stress axes in granular materials showing shear fractures that intersect in σ_2 and make an acute angle with σ_3. **d** Plane perpendicular to σ_2 in plane stress. Shear fractures (*thin lines*) make angle α with σ_3 and $\beta/2$ with the planes of maximum shear stress (*thick lines*) following the relationship $2\alpha + \beta = 90°$ (**c** and **d** after Park 1983)

Under plane strain conditions the strain e_{33} is zero. Yielding is constrained along the specimen thickness, which results in a rather small stress difference (Mohr circles in Fig. 1.20b). This promotes flat fracture normal to the tensile stress (along the thickness) and suppresses a slant fracture.

The planes of maximum shear stress in plane strain intersect in a line parallel to the intermediate principal stress (along z in Fig. 1.20b). Broek (1984) argues that $\sigma_3 = \upsilon(\sigma_1 + \sigma_2)$ and for plastic deformation which requires constant volume $\upsilon = 1/2$. The stress along z becomes $\sigma_3 = (\sigma_1 + \sigma_2)/2$, which is larger than σ_2 along x and smaller than σ_1 along y. Therefore these planes form angles of 45° with the xz plane but in a different orientation compared with that of plane stress (Fig. 1.20a). The planes of maximum shear stress are the planes of maximum void concentration and cause the most intense plastic deformation in ductile metals. According to Hertzberg (1976 p. 474), "it is important to recognize that both unstable, fast-moving cracks and stable, slow-moving fatigue cracks may assume flat, slant or mixed macromorphologies."

The Tresca criterion is adapted for metals that deform without volume changes and the angle of internal friction β is of no concern. Granular materials (rocks, ceramics and concrete) which suffer from dilatancy failure are characterized by the Coulomb criterion, which takes β into consideration, and in these materials slanting will depart from 45° by $\beta/2$, producing an angle α with the minimum principle stress (Figs. 1.20c, d).

1.2.5 Secondary Cracks in the Plastic Zone

Secondary cracks is the term used here for new cracks that initiate ahead of and adjacent to the parent fracture. Initial secondary cracks may develop into branching fractures or step cracks (Figs. 2.36 and 2.7 respectively), which are discussed separately. The occurrence of microcracks ahead of a primary fracture has been postulated by Griffith (1920, p. 179), Schardin (1959), Congleton and Petch (1967) and others. Preston (1926) postulates microcracks which developed adjacent to the moving parent fracture.

Secondary cracking may develop both within an elastic field and in a plastic zone. Lawn and Wilshaw (1975a, p. 133) propose a preference criterion between the two: for $\sigma_L^* \leq \sigma_{ys}$ the condition for elastic microcracking is satisfield before a plastic zone can develop, whereas for $\sigma_L^* > 3\sigma_{ys}$ the plastic stress concentration is insufficient to elevate the local tension to the fracture level, and the rupture becomes fully ductile. In this relationship σ_L^* is the strength parameter for unstable microrupture and σ_{ys} is the macroscopic uniaxial yield stress. Plastic zones are particularly conductive to secondary fracture. A sharp crack tends to blunt into a rounded notch in a plastic field by the local shear stresses, and the maximum tension σ_{yy} is shifted from the free surface of the notch (Fig. 1.2) to the plastic zone, with the highest stress concentration at the remote end of the zone. The tendency for the primary fracture to propagate is then replaced by an extensive growth of secondary cracking.

1.2.6 The Damage Zone

Rock experimentalists note that secondary cracks escort the primary fracture in a zone of finite width. Hoagland et al. (1973) test a notch in wedge loaded double cantilever beams made of limestone and sandstone (Fig. 1.21 stages a–d). They observed that initially (a) microcracks develop linearly at the stressed region around the notch. Continued loading of these specimens (b) results in an intense non-linear microcracking and acoustic signals become detectable. These may become associated with microcrack coalescence and linkage to the notch tip. Gross crack extension begins at the point of maximum load (c). At this stage expansion of the damage zone of microcracks may still be more rapid than the growth of the main crack. Finally, (stage d) a steady-state situation is achieved in which the main crack extends stably, pushing the damage zone ahead of it at the same rate. Hoagland et al. (1973) consider that the damage zone acts like the plastic zone ahead of the crack in metals by dissipating large amounts of energy and diminishing the stress concentration at the tip of the main crack. This explains why the energy dissipated in propagating a crack in rocks is typically much greater than is consumed for the single crystals of its constituents (see also Sect. 1.5.4).

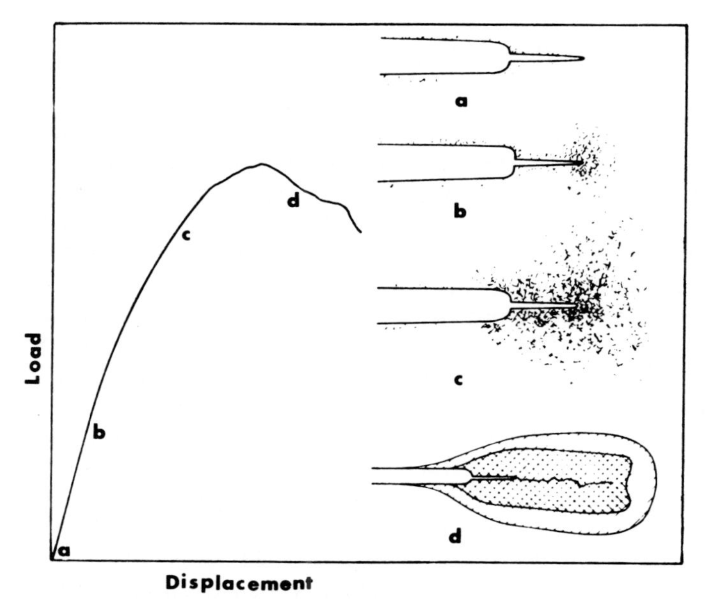

Fig. 1.21. Damage zone. A schematic representation of the stages of development of a damage zone around a machined slot wedge-loaded in rock. Initially only a few fresh microcracks are present, as in *a*. On loading, the rock first responds elastically and only a few microcracks are produced on weakest interfaces, *b*. Then in *c* the damage has become intense and spreads rapidly leading to inelastic behaviour. Finally, at *d* the main crack and the damage zone extend together. (Hoagland et al. 1973)

1.3 Atomistic Concepts of the Crack Tip

1.3.1 Thomson's Three Prototypes

Thomson (1986) divides cracks into three prototypes: (1) slit or brittle cracks which cleave, (2) wedge cracks which emit dislocations and (3) notches or plastically blunted cracks which merely activate external plastic flow in the surrounding medium. These different morphological classes also correspond to three different broad mechanisms of fracture to differing regions on the brittle-ductile spectrum of material response.

The slit corresponds to a crack which is atomistically sharp at the tip and can propagate rapidly. This crack may be purely brittle or it may absorb a finite number of dislocations on its cleavage surfaces, so that the crack surface behind the tip has a greater crack opening displacement (Fig. 1.22a). Brittle fracture is best known in silicate glasses, but, being a crystalline material, silicon presents a better characterization of a slit crack. In this material, as stresses at the tip are magnified to the tensile strength of the bonds, bonds are broken progressively. Brittle fracture in ceramic or in steel may be encouraged by low temperature, high strain rate and a particular chemical environment (like hydrogen in steel). No dislocation activity is observable at low temperature cracking in silicon or in fast fracture when the crack outruns its dislocation cloud. When the cracks grow above the ductile temperature, significant dislocation activity is present. A slit has an elastic enclave, which is a dislocation-free zone near the tip. Cleavage may occur without generation of dislocations from the crack tip itself. Thompson suggests that basically "cleavage cracks are a logically distinct class of crystal lattice defect".

According to Thomson (1986), "the wedge crack in its pure form is one where the crack is unstable against dislocation emission, and its characteristic shape is determined by the geometry and symmetry of the dislocation slip planes at the crack tip. In this case, the crack advance is not by opening of bonds, but by ledge formation as the dislocation is emitted. Thus one dislocation must be emitted for each Burgers vector increment of crack advance. The wedge is formed therefore by alternate 'sliding off' on the equivalent slip planes at the crack tip" (Fig. 1.22b). See further on crystal defects (e.g. Hobbs et al. 1976, p. 75). A wedge crack develops when dislocations are emitted on slip planes intersecting the crack tip. Thomson suggests that "the wedge is, fundamentally, a void (made by punching out dislocation loops or agglomeration of vacancies)". An important difference between a cleavage crack and a wedge is that the latter does not disappear when the external mode I stress is removed even though the stress singularity does, whereas in the former the atom bonds zip together again and the crack disappears.

Propagation of a notch involves several characteristic processes: plastic blunting increases the effective radius at the tip and inhomogeneities in the material promote microcracking ahead of the notch and hole (or void) growth (Fig. 1.22c). Blunting alternates with the absorption of voids to the advancing notch so that the plastic region around the tip is limited in size. Such a crack

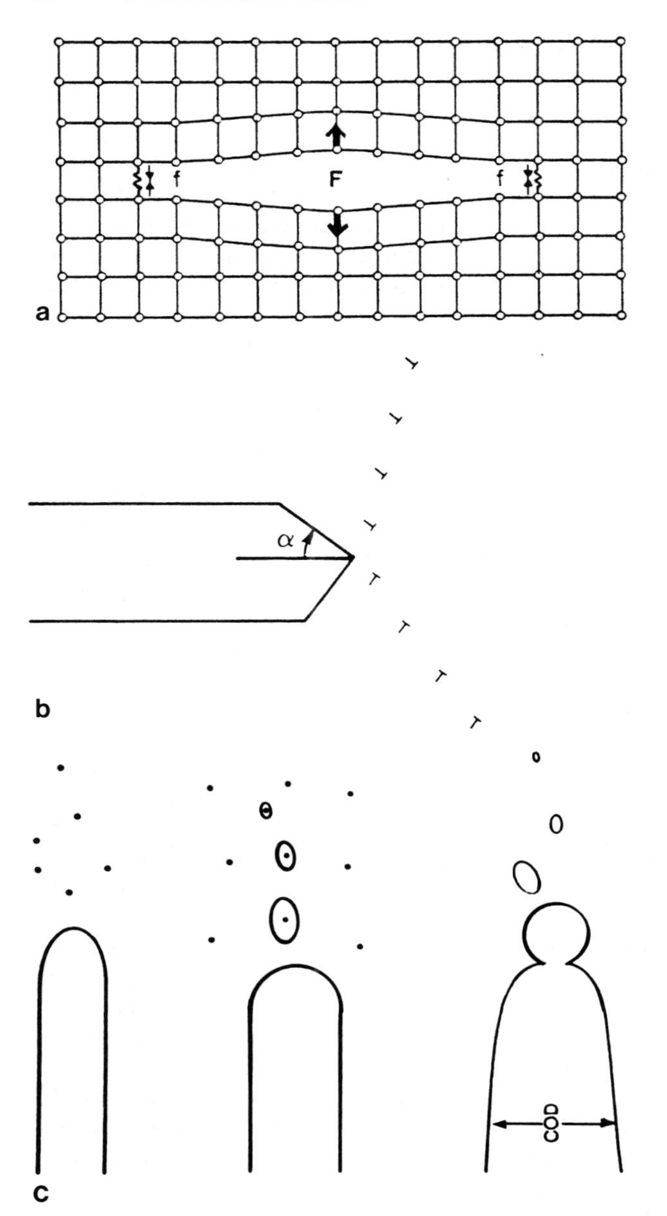

Fig. 1.22 a–c. Three crack prototypes by Thomson (1986) **a** An atomically sharp cleavage crack with external forces *F* exerted at the centre and non-linear attractive forces f at the tips. **b** As the stress is raised in an initially sharp slit crack, alternate emission from the tip on equivalent slip planes intersecting the crack create a wedge crack whose opening angle is complementary to the angle between the two slip planes. *T signs* designate edge dislocations. **c** Schematic drawing of the development of a ductile fracture from a blunt crack embedded in a material containing inclusions. Microcracks are nucleated at the inclusions which are blunted. Crack opening displacement is abbreviated as COD. (Thomson 1986)

can grow under stresses below the plastic yield. The dislocation is a lattice defect whose atomically sized core contains stresses equal to the theoretical shear strength of the constituent atomic bonds. Some dislocations are emitted from the crack tip, whereas others are generated in the material next to the tip. Small holes develop from the dislocations along crystallographic slip planes. As the holes grow in size, they partly coalesce with the crack and this is manifested by the necking on a larger scale. Some holes grow and coalesce with other holes and they are visible as dimples (Sect. 2.1.2.6). Often large voids are found to be nucleated around precipitated secondary crystals. These voids are connected by sheets composed of very small voids.

Dislocations are always produced from a source in pairs of opposite Burgers vectors. Dislocations with positive Burgers vectors are repelled and shield the crack from the externally applied stress field; however, antishielding dislocations with negative Burgers vectors are attracted toward the crack; many are absorbed on its surface and they enhance the effect of the external field.

Edge dislocations are generally associated with mode I and mode II and screw dislocations correlate with mode III. Edge dislocations are likely to be more strongly affected by K_{II} than by K_{I}. The material toughness is determined to a large extent by the strong crack-dislocation interaction which is promoted when stresses rise.

1.3.2 The LRT Atomistic Surface Force Model

Lawn, Roach and Thompson (1987) elaborate on the Barenblatt crack tip model (Sect. 1.2.2). They distinguish in the cohesive zone of a brittle crack the wide reactive "secondary zone" more remote from the tip where extrinsic interactions with intruding chemical species (such as water) are confined from the narrow protected "primary zone" adjacent to the tip. The latter is limited to about one atomic spacing where intrinsic binding forces operate without environmental chemical influences. These two zones zip monotonically along asymptotic cohesive cleavages that penetrate between atomic (oxygen) planes into the crystal lattice (Fig. 1.23). "Thermo-dynamically, energy expended in parting intrinsic bonds in the primary zone may be regained by adsorption of active species on to the now-active surfaces in the secondary zone". The boundary between the two zones shifts in cyclic load-unload-reload experiments in muscovite and tridymite (an essentially layered network crystalline polymorph of silica) towards the narrow zone. In these experiments unload may lead to healing. A shift of the threshold region is associated with a "reduction in applied loading needed to propagate cracks through healed as compared to virgin interfaces". Diffusion barriers control the fracture kinetics at low driving forces as they stand in the way of intruding molecules along the secondary zone and prevent them from penetrating to the primary zone.

Contrary to those models which presume a change in the actual tip structure (Sect. 1.4.1.4) the LRT model asserts that the criteria for crack extension remain uniquely expressible in terms of G. "While reaction products may be spatially isolated from the tip they may nevertheless contribute strongly to the fracture mechanics, by virtue of their influence on the surface forces behind the crack tip". Their model to some extent is a reminder of the liquid metal embrittlement concept (e.g. Kamdar 1977).

1.3.3 The Dissociative Chemisorption Model

On a atomic scale, the slow growth of cracks corresponds to the sequential rupturing of interatomic bonds at rates as low as one bond rupture per hour (Michalske and Bunker 1987). Michalske and his associates use silica glass for their atomistic model. The basic unit of silica glass is a close packed tetrahedron consisting of a central silicon atom surrounded by four oxygen atoms. Each oxygen atom is shared by the silicon atoms of two adjacent tetrahedrons, so that each tetrahedron is connected to four neighbours. The tetrahedrons form a random network of interconnected rings, each of which usually contains from five to seven tetrahedrons. Crack propagation involves the rupture of individual silicon-oxygen bonds. The smallest incremental distance the crack can move is the diameter of the silicate ring, which is from 0.4 to 0.5 nanometre (4 to 5 Å).

Michalske and Freiman (1982) and Michalske and Bunker (1984, 1987) postulate that water interaction with the advancing crack occurs in three steps (Fig. 1.24). First, a water molecule adsorbs to the crack tip. It then donates electrons from the oxygen atom to the formation of a bond with the silicon atom at the crack tip. Also, the water molecule donates a positively charged

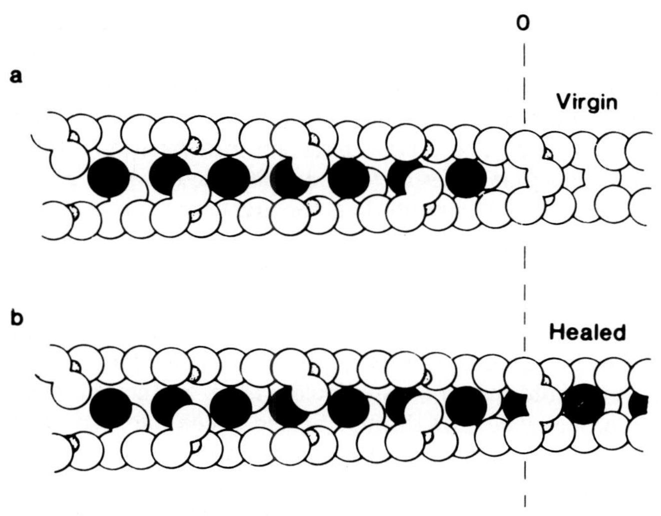

Fig. 1.23 a, b. Crack profiles for tridymite in water at G level of 200 mJ/m^2, for growth through **a** virgin material and **b** healed interface. *Open* and *dark large spheres* are oxygens and water molecules, respectively, *small spheres* are silicon atoms. 0– –0 designates the Irwin crack tip singularity. G is determined experimentally by a cyclic wedge loaded cantilever technique. Note a trapped water layer at the narrow spacing in the healed interface. (Lawn et al. 1987)

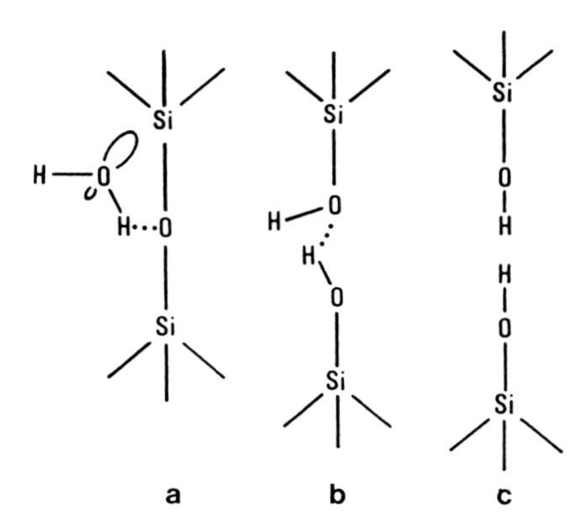

Fig. 1.24 a–c. The atomistic slow fracture model of Michalske and Freiman. The strained crack tip bond in three stages. **a** absorbs a water molecule; **b** is cleaved by dissociative reaction; **c** is converted to surface silanol groups. (Michalske and Bunker 1984)

hydrogen to form a hydrogen bond with the oxygen atom attached to the silicon atom. These result in some 20% strain and reaction enhancement (Freiman 1984). Then, the $S_i - O$ bond and the O–H bond in the adsorbed water cleave to form two hydrogen bonded silanol groups (hydroxyl groups attached to silicon). Finally, rupture of the hydrogen bond occurs. Hence the original bonds are replaced by two surface silanol groups. This model does not require prior dissociation of the water molecule [see below Eq. (1.37)]. The process is termed dissociative chemisorption. Evidently ammonia and methanol can also promote this process.

Michalske and his associates investigate how the chemisorption process could be enhanced by the applied stress. They use Fourier-transform infrared spectroscopy which enables them to distinguish and follow the various stages of the process. They observe that firstly, a silicon atom in a strained ring is more likely to adsorb electron-donating molecules such as water, ammonia and methanol. Secondly, all the chemical species that make cracks grow more rapidly in silica glass also dissociatively chemisorb on the strained ring. Finally, the rate of ring cleavage in a stressed surface of silicon is more than 10^5 times higher in their model when exposed to water vapour, compared with unstressed silica surface at that exposure.

The authors work with small rings consisting of two oxygen and two silicon atoms. Such so-called edge-sharing rings are practical for studying the effects of bond strain, because the bond angles and distances between atoms are greatly distorted from those of normal silica. In a reaction between an ammonia molecule and an edge-shared ring ammonia is adsorbed on a strained silicon site; and a dissociative chemisorption reaction breaks the Si–O–Si bonds. Such strained bonds react 10^5 times faster than unstrained ones.

It turned out that the size of the attacking molecule contributes greatly to the ability to promote crack growth. Methanol is more aggressive than water in rupturing strained silicon-oxygen bonds, but it is five orders of magnitude less effective in increasing the rate of crack growth than either water or ammonia. Water and ammonia have nearly identical molecular sizes (0.26 nanometre), but methanol is a much larger molecule (0.36 nanometres). Small molecules can readily enter a crack opening (which has a diameter from 0.4 to 0.5 nanometre) and cause reactions that rupture bonds, but larger molecules such as methanol have difficulty getting in. Methanol should diffuse into a crack tip at a rate four orders of magnitude less than the rate of water. Michalske and Bunker suggest that by taking into account the rate of molecular diffusion near the crack tip and the rate of dissociative chemical reactions on the strained Si–O bond, the relative rates of crack growth in silica glass exposed to different chemicals can be predicted. These authors also believe that the above general conclusions developed for silica can explain the strength behaviour of a wide range of brittle materials.

1.4 Kinetic Processes in Fracture

1.4.1 Subcritical Crack Growth

A pristine fibre of glass breaks instantaneously in extremely dry environments or at liquid nitrogen temperature (−196 °C) under stress conditions which approach the theoretical strength of the material (Sects. 1.1.1 and 1.1.2). In a moist environment at room temperature, on the other hand, delayed failure of subcritical crack growth (also termed fatigue) occurs at strengths 10 to 1000 times lower than the theoretical due to the presence of flaws in the solid (Griffith 1920). These flaws are thought of as very narrow (~0.05 μm width) and sharp cracks (Figs. 1.6a and 1.9) penetrating from the surface deep into the material which renders high stress concentrations at their tips [Eq. (1.2)]. The flaws may readily be introduced into the glass surface by normal handling. Static fatigue is the term used to describe the decrease in strength during the slow fracture growth under the relatively low tensile loading in the active environment (mostly water). The stress concentration at the crack tip which leads to bond rupture and fracture growth under this loading is known as stress corrosion.

Five aspects of subcritical crack growth are addressed in the following sections, these are: (1) manifestations of delayed failure, (2) chemical reaction at the crack tip, (3) aspects of fracture mechanics, (4) the fatigue limit and the shape of the crack tip, and (5) fatigue by alternating stresses in metals.

1.4.1.1 Manifestations of Delayed Failure

Basically, a crack in glass can start growing at extremely low velocities (10^{-10} m/s) and propagate through a wide range of rates for a long period of time (years) from a Griffith flaw size (2 μm long) to larger sizes until a catastrophic failure occurs. A person can hear a blast at night from the kitchen and discover in the morning that a glass vessel has been shattered spontaneously without an apparent cause. Residual stresses in the vessel had promoted a prolonged slow crack propagation, which is often termed incubation, that led to the sudden rupture. During a delayed failure the stress needed to crack glass decreases (Grenet 1899).

An exposure to water accelerates the crack growth, and static fatigue is a stress-dependent chemical reation of water with the glass surface at the crack tip. This phenomenon is similar to stress corrosion cracking in metals (Charles 1961; Wiederhorn and Bolz 1970).

Ancient flint workers noticed the influence of moisture and temperature on the "workability" of their materials. Historic Indians steamed their rocks before knapping (Patterson and Sollberger 1979).

Mould and Southwick (1959) investigated static fatigue curves (strength versus load duration) for specimens immersed in room temperature distilled water and in liquid nitrogen after the specimens had been subjected to various abrasion treatments that produced standardized damage in the glass surface. The low-temperature strength was independent of load duration, and for surface damage of simple geometry it was inversely proportional to the square root of the initial crack depth, consistent with the Griffith theory. Abrasions of different geometries produced differing static fatigue curves at room temperature. If, however, the strength values σ for each abrasion were divided by the low-temperature strength σ_N for that abrasion and plotted versus a reduced time coordinate $\log(t/t_{0.5})$ as the independent variable, where t is the time to failure at stress σ and $t_{0.5}$ the breaking time for stresses equal to half the low temperature strength for each individual static fatigue curve, all the data could be fitted to a single "universal fatigue curve" of the form $\sigma/\sigma_N = f(t/t_{0.5})$ (Fig. 1.25). Hence, $t_{0.5}$ depends strongly on the severity and geometry of the flaws in the surface. The rate of weakening under stress appears to depend ex-

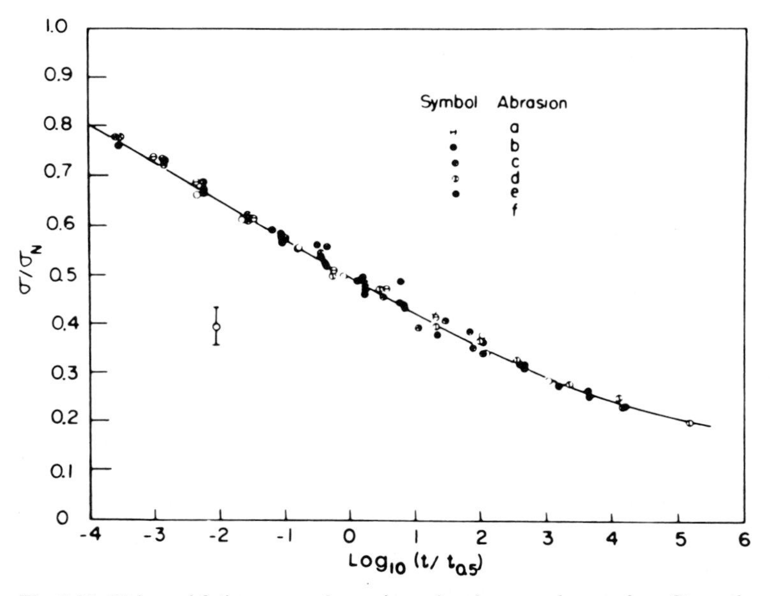

Fig. 1.25. Universal fatigue curve for various abrasions on glass surface. Strength σ divided by liquid-nitrogen strength σ_N versus logarithm of reduced load duration for each abrasion. *Vertical bar* indicates approximate uncertainty for individual points. (Mould and Southwick 1959)

ponentially on initial flaw size. It was also found that for two general types of abrasion, linear flaws weaken more rapidly under stress than do point flaws by a factor of approximately 50.

1.4.1.2 Chemical Reaction at the Crack Tip

Water reacts chemically with glass under stress destroying the Si–O bonds. Charles (1958a) considers that in silica glass the reaction is

$$\left[\begin{array}{c} \;\;|\qquad\;\;| \\ -\mathrm{Si}-\mathrm{O}-\mathrm{Si}- \\ \;\;|\qquad\;\;| \end{array}\right] + H_2O \rightarrow 2\left[\begin{array}{c} | \\ -\mathrm{Si}-\mathrm{OH} \\ | \end{array}\right] . \tag{1.37}$$

It is basically believed that hydroxyl ions attack the glass network, causing siloxane bond cleavage (Wiederhorn 1972).

The dominant theory of stress corrosion among many chemical mechanisms that have been proposed is the one attributed to Charles and Hillig. Ernsberger (1977) points out that these two authors produced in fact two theories. The earlier theory (Charles 1958b) proposed a model in which the microcrack gradually increases in depth without change of tip contour. The later theory (Charles and Hillig 1962) envisions a sharpening of the crack without propagation. The two theories relate the failure time t_f to the applied stress σ_L quite differently. According to Charles:

$$\log t_f = \bar{n} \log (1/\sigma_L) - k \, [\bar{n} \approx 16] \tag{1.38}$$

and according to Charles and Hillig:

$$\log t_f = -k\sigma_L - k'\sigma_L^2 + N(\sigma_L) , \tag{1.39}$$

where $\bar{n}$ is the stress corrosion susceptibility constant which actually varies with the condition of the surface, k and k′ are constants and N is a correction function.

In spite of the mathematically profound differences, both equations give adequate fits to the experimental data. Ernsberger suggests two reasons for this: the availability of adjustable parameters, and, more fundamental, that it is not practical to obtain experimental data at very short or very long times. He adds that "failure time is so extremely sensitive to stress that stress can only be varied by a factor of about two without getting out of the range of experimentally-measurable fracture times, and over such a narrow range almost any formulation can be made to fit the data".

Ernsberger mentions additional difficulties of the stress corrosion theory: that there is no satisfactory explanation for static fatigue in pristine glass, that no observations have been made on differences in stress-corrosion susceptibility among glasses of different compositions, and the unexplained appearance of fatigue effects in glass under high stress level in a vacuum environment (Wiederhorn et al. 1974a). Wiederhorn (1978) also addresses some of these difficulties.

Following Charles and Hillig (1962), Wiederhorn and Johnson (1973) propose that if crack growth in glass results from a stress-enhanced chemical reaction, the kinetics of crack motion can be described by reaction-rate theory. The velocity of crack motion is proportional to the rate of chemical reaction at the crack tip and can be described by

$$V = V_o \exp\left[-\Delta E^+ + (\sigma \Delta v^+/3) - (v_m \gamma/\varrho)\right]/RT \ , \tag{1.40}$$

where the pre-exponential V_o is the rate of the reaction at zero applied tensile stress determined by the chemical activity at the crack tip, ΔE^+ and Δv^+ are the activation energy and activation volume for the chemical reaction and σ is the principal crack-tip stress. The final term in the exponential represents a change in free energy of activation caused by surface curvature, where v_m is the molar volume of the glass, γ the interfacial surface tension at the glass-solution interface and ϱ the radius of curvature of the crack tip. R and T are the gas constant and absolute temperature, respectively. Equation (1.40) has been modified by various authors (i.e. see reviews by Wachtman 1974; Atkinson 1984; Freiman 1984).

Wiederhorn and Johnson (1973) find differences in the behaviour of different glasses, and conclude that in water the order of decreasing resistance to static fatigue should be silica, borosilicate, and soda lime silicate glass. They suggest that the pH at the crack tip influences the slope and shape of universal fatigue curves (Sect. 1.4.1.1).

An intriguing question is to what extent ionized water or, on the other hand, molecular water (Sect. 1.3.3), is the more important chemical variable in the subcritical fracture process.

1.4.1.3 Fracture Mechanics

Wiederhorn (1967, 1974) showed experimentally the change in crack velocity V as a function of the applied force and the tensile stress intensity factor K_I in the sub-critical range. He obtained curves that were characterized by three distinct regions (Figs. 1.26; 1.27). Results of slow crack propagation are usually plotted as log V vs K_I.

The slow crack growth in region I is manifested by a linear logarithmic plot. The behaviour of this region is attributed to stress corrosion. The rate of crack growth is controlled by the chemical reaction between the reactant and glass at the crack tip when the system is essentially at equilibrium. Region II is controlled by the rate of reactant transport to the crack tip. In this region V almost does not change and is independent of K_I. Freiman (1980) simplifies Wiederhorn's mathematics (1967) and describes the crack velocities in regions I and II by Eqs. (1.41) and (1.42) respectively:

$$V = (A X_p^n \exp b K_I)/n \tag{1.41}$$

$$V = C D_{H_2O} X_t/n \ , \tag{1.42}$$

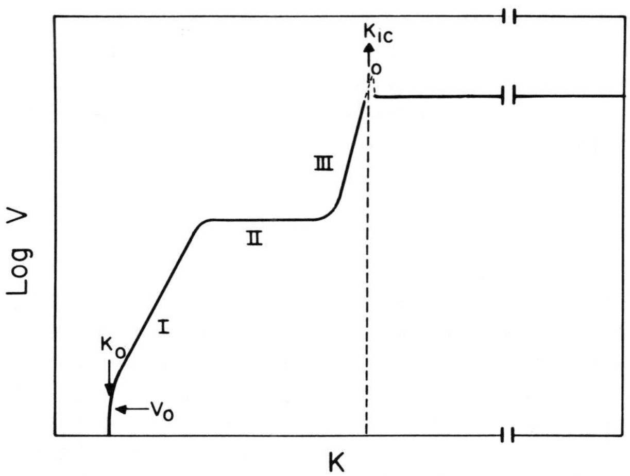

Fig. 1.26. Schematic variation of stress intensity factor K with crack velocity V. The subcritical curve left of K_{IC} is divided into three regions I, II and III. K_o and V_o are the stress corrosion limit and the crack velocity at this limit. (Wiederhorn and Bolz 1970). Supercritical growth occurs mostly along a terminal velocity plateau following a rapid increase in crack velocity at K_{IC}. (After Evans and Wiederhorn 1974). A velocity overshoot o immediately after K_{IC} before the plateau was predicted by Rabinovitch and Bahat (1979)

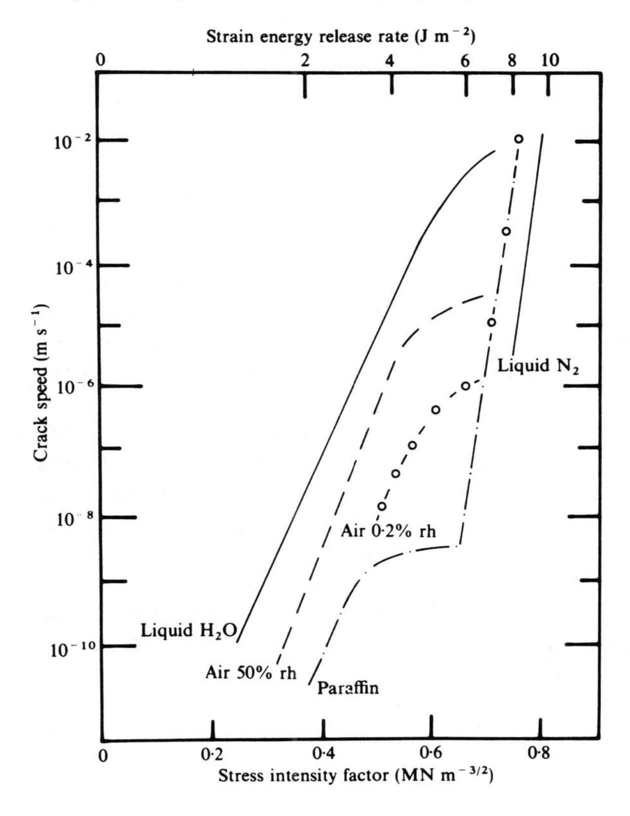

Fig. 1.27. Crack speed as a function of toughness (K and G) in the sub-critical regime for a soda lime glass. (After Wiederhorn 1967 from Holloway 1986)

where X_p is the partial pressure, n the order of the chemical reaction, D_{H_2O} is the diffusivity of water in the environment, X_t a boundary layer thickness and A, b and C are constants.

In region III crack velocity is quite high, it is insensitive to the reactant and to stress corrosion. The crack growth seems to be controlled by the stress in a mechanism which is not understood. Freiman (1980) notes that the reported activation energies for crack growth in region III range from 17 to 176 kCal/mol. However, experiments in vacuum have shown a dependence on temperature and the composition of the glass (Wiederhorn et al. 1974a).

Based on their data, Wiederhorn et al. divided glasses into two groups. Normal glasses, including soda lime silicate, aluminosilicate and lead glasses, exhibited sub-critical crack growth. On the other hand, silica and borosilicate glasses behaved "anomalously" in terms of their thermal and elastic properties and they fractured suddenly. For them subcritical growth did not occur (region III did not exist for borosilicate glass). Wiederhorn et al. concluded that the different behaviour of the glass corresponded to differences of crack tip structure. Following the "lattice trapping" model (Thomson et al. 1971; Hsieh and Thomson 1973) "in normal glasses a narrow cohesive region at the crack tip favors thermally activated crack motion, whereas in anomalous glasses, a wide cohesive region favors sudden fracture".

The K_o part of the line in Fig. 1.26 designates the location of the stress corrosion (endurance) limit below which fracture will never occur regardless of its duration (see next section). The behaviour of this threshold varies from material to material. Wiederhorn and Johnson (1973) confirmed the existence of this phenomenon in borosilicate glass, but they did not observe a definite fatigue limit in soda lime silicate glass. Wilkins (1980) and Atkinson (1984) found no such limit for crack velocities as low as 10^{-12}/ms or 10^{-9}/ms, respectively, in rocks, although Atkinson and Meredith (1987) reason that "the fact that the earth's crust is not rubble implies that a subcritical crack growth limit must exist for rocks" presumably caused by compression. The stress corrosion limit may be associated with crack blunting rounding off the crack tip in response to the reaction of water with the glass (Michalske 1983). Atkinson (1984) suggests that K_o is about 10–20% of K_{Ic}. V_o is the crack growth velocity at the stress-corrosion limit (Fig. 1.26). Atkinson (1984, 1987) adds that the threshold of crack growth K_o may be raised (to a value called K_{dc} which is the crack growth limit below which bulk diffusion creep dominates behaviour and crack growth is suppressed) if the stress is high enough for diffusion creep to be significant. Whereas the parameters K_{dc}, the K versus V curves in the subcritical range, K_c to some extent, and V_o are all temperature-dependent, K_o is not (Wiederhorn and Bolz 1970; Atkinson and Meredith 1987).

No effect of hydrostatic pressure on fracture stress up to 20 kbar could be detected by Wiederhorn and Johnson (1971) and the influence of hydrostatic pressure on sub-critical crack growth is not known (Atkinson 1984). Atkinson and Meredith (1987) suggest that increasing effective pressure should retard stress corrosion and shift the region II plateau to a lower crack velocity.

The K_{Ic} line shows that at a critical stage in region III the crack propagation assumes a catastrophic growth with a sudden jump in V (Fig. 1.26). At $K_I > K_{Ic}$ post critical crack propagation occurs (Sect. 1.4.3).

Many polycrystalline ceramics undergo delayed failure in a manner similar to glasses. Slow crack growth is shown to take place in aluminas of varying purity and grain sizes, lead zirconate titanate known as PZT and barium titanate (Evans 1972; Freiman et al. 1974). Crack velocity is shown by the latter authors to be a strong function of the K_I at the crack tip, and grain size appears to be the primary factor in determining crack propagation rates in alumina. In PZT the distinct effect of water on crack velocity can easily be seen.

There are materials which deviate from the typical three region slow fracture propagation shown in Fig. 1.26. Graphite, for instance, seems to undergo failure at room temperature only at large fractions of K_{Ic} along the third zone (Hodkinson and Nadeau 1975).

A considerable number of minerals and rocks behave anomalously and exhibit only region I. Atkinson and Meredith (1987) consider that generally "the more complex the microstructure of a material, the lower is its susceptibility to subcritical crack growth".

1.4.1.4 The Static Fatigue Limit, Crack Tip Shape and Crack Deepening

When a flaw is kept in water while relieved from any tensile loading it goes through a process known as ageing, which results in strength increase because the active chemical environment rounds the crack tip (Fig. 1.28). On the other hand, in contact with water while being stressed, the crack sharpens and lengthens, both weakening the glass and leading to static fatigue. Hence ageing and static fatigue are two competing processes. The counter-balancing between the two implies that for a given initial flaw size there exists an applied stress level below which crack growth does not occur (Davis et al. 1983). Thus is the static fatigue limit (Fig. 1.26) also known as the endurance limit (Shand 1954). Wiederhorn and Bolz (1970) determine the fatigue limit at $K_I = 0.25$ MPa $m^{1/2}$ for soda lime-silica glass in water at 23 °C for which $K_{Ic} = 0.8$ MPa $m^{1/2}$.

The ratio of time to failure to initial flaw size (t_f/a_i) at the fatigue limit is proposed as a possible failure criterion for a specific environment by Davis et al. (1983). If this endurance duration is exceeded under a given applied stress, survival is guaranteed for an indefinite time, an idea expanded from an earlier prediction by Holland (1936). Davis et al. find that (t_f/a_i) at 60 °C is shorter by 0.156 than in the 23 °C water environment. They also determine that the static fatigue limit is a constant fraction (0.27) of the inert strength σ_N of the pre-existing flaw. This means that for a particular soda lime-silica glass composition a failure criterion may be expressed in terms of the normalized strength ratio $\sigma/\sigma_N = 0.27$, which is equivalent to a K_o equal to 27% of K_{Ic}.

Doremus (1980) observes that in static fatigue the rate of tip sharpening is greater than the rate at which the crack size increases, and it is more significant in terms of failure (see Sect. 1.4.1.2). Doremus (1976) asserts that the Inglis

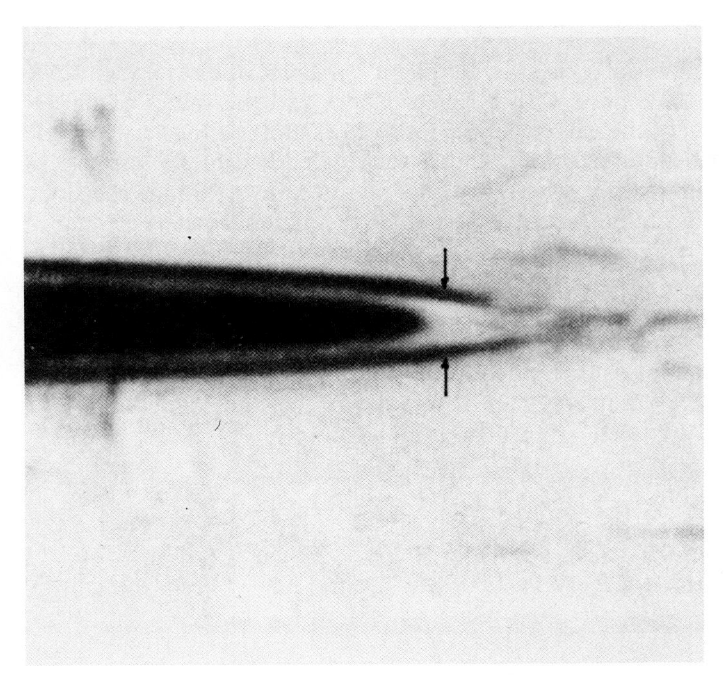

Fig. 1.28. A rounded (blunt) crack tip (having a finite radius) to a slowly propagating crack in polymethylmethacrylate. The *width at arrows* is approximately 8 μm. (Holloway 1986)

[Eqs. (1.1) and (1.2)] and Griffith Eq. (1.6)] criteria for fracture are inconsistent: it is possible that the applied stress is high enough to cause propagation by Eq. (1.2) but not by Eq. (1.6) and, in fact, the Griffith criterion provides a necessary but not sufficient condition for crack propagation essentially because the tip radius is not considered. Doremus equates Eqs. (1.2b) and (1.6) for the fracture stress, and with further substitution he obtains

$$\varrho_{min} = 8E\gamma/\pi\sigma_{th}^2 = 32b_o/\pi \ , \qquad (1.43)$$

where ϱ_{min} is the minimum tip radius for fracture and b_o is the interatomic spacing. This result suggests that ϱ_{min} is about ten times the interatomic spacing (about 16 Å) and is independent of other parameters (see also Lawn and Wilshaw 1975a, p. 143).

1.4.1.5 Fatigue Fracture in Metals by Alternating Stresses

Metallurgists study fatigue failure in materials where crack growth can occur at stress levels well below σ_{ys} by applying alternating stresses on the tested material in such a way that the crack growth rate da/dN, which is the increment of crack growth per cycle, can be related to the instantaneous value of ΔK, which is the difference between the maximum and minimum values of the stress intensity factor for each cycle ($\Delta K \equiv K_{max} - K_{min}$) according to:

$$da/dN = C(\Delta K)^n \ , \tag{1.44}$$

where C and n are material constants. In practice, the plot of da/dN versus ΔK values consists of three parts with different slopes and the values of da/dN vary as the tangents of the curve change (Fig. 1.29a). Both at the very low ΔK values (close to the endurance limit) and at very high ΔK values (close to K_{Ic}) the graph rises quite rapidly, whereas at the intermediate ΔK range the slope is moderate and n is about 2–4, irrespective of the material (see resemblance to the sub-critical regime in Fig. 1.26).

1.4.2 Failure Prediction

Crack velocities are measured directly or determined through changes in specimen compliance (Freiman 1980). It was shown that the data could be fit to an equation of the form

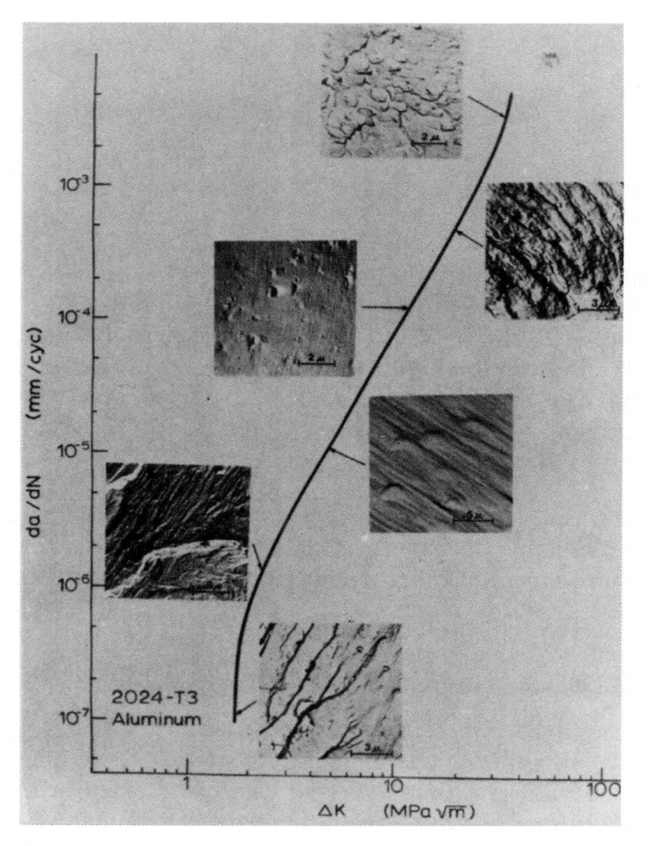

Fig. 1.29a. Electron fractographs showing fatigue fracture surface micro-morphology at various points on the *da/dN* versus *ΔKS* curve for 2024-T3 aluminium alloy. (Hertzberg and Mills 1976).

$$V = AK_I^n \ . \tag{1.45}$$

This empirical equation is often tested by ln V versus ln K_I plots. The time-to-failure, as a result of slow crack growth, can be determined from this relationship.

The time to failure t_f can be calculated for static fatigue directly from crack propagation data (Wiederhorn et al. 1974b). For a constant applied stress, σ_L

$$t_f = 2K_{Ii}^{2-n}/A\sigma_L^2 Y^2(n-2) \ , \tag{1.46}$$

where K_{Ii} is the initial stress intensity factor at the most serious flaw in the material, A and n are empirical constants, when it is assumed that the crack velocity is a power function of K_I [Eq. (1.45)], and Y is a geometric constant.

The most effective approach to failure prediction for glass, however, is proof testing (Wiederhorn et al. 1974b). The proof test is designed to impose a load on the component examined which is larger than the maximum expected service load. Thus, a limit is set for the maximum size of flaws that can be present in the component examined after the proof test has been completed. It follows that the initial stress intensity factor at the crack tip of the largest initial flaw K_{Ii} at the beginning of service is limited to a fraction of K_{Ic}, such that:

$$K_{Ii} \leq (\sigma_L/\sigma_p)K_{Ic} \ , \tag{1.47}$$

where σ_p is the proof stress. By a certain mathematical computation, Wiederhorn et al. (1974b) express the minimum time-to-failure, t_{min} in a functional form:

$$t_{min} = \sigma_L^{-2} f(\sigma_p/\sigma_L) \ , \tag{1.48}$$

which shows that t_{min} is inversely proportional to the square of σ_L and directly proportional to a function of the proof-test stress divided by the maximum in service stress. The authors however observe that to make long-term predictions of lifetime, crack velocity data must be extrapolated beyond the limits of measurement. For this, the V-K_I functional relationship must be known.

Gulati (1980) propose a representation of crack length versus time for the case of constant stress rate loading. The curve is divided into three regions analogous to the three regions of K-V diagram (Fig. 1.26). He considers that crack growth in a given material, and at a specified stress or stress rate, is extremely slow during most of the life time and increases exponentially as failure time or critical stress are approached. Gulati points out that at the end of the stable growth region crack growth is very rapid. Therefore, it is conceivable that the time variation of crack length in this region can help predict the onset of fast fracture.

Bahat and Rabinovitch (1988) derive t_f for a growth of an uplift joint in granite by using Eq. (1.45). Integrating Eq. (1.45) with respect to time for constant σ where $K_I = \sigma(1.2\pi c/Q)^{1/2}$ [see Eq. (1.18)], with c the crack radius, Q the modified geometrical factor of the flaw, and A and n temperature-dependent constants yields (Wiederhorn et al. 1974b):

$$c_i^{[(n/2)-1]} = 2/[A\sigma^n(n-2)t_f(1.2\pi/Q)^{n/2}] \ , \tag{1.49}$$

where $2c_i$ is the initial flaw length of the granite which approximates the maximum straight grain boundary in the rock. The initial stress intensity factor K_{Ii} can be calculated by

$$K_{Ii} = (1.2\pi/Q)^{1/2}\sigma c_i^{1/2} \ . \tag{1.50}$$

More details on the determination of fracture paleostress conditions in granite are presented in Section 4.2.1.5.

Dynamic fatigue is examined under two different sets of conditions. The applied stress is increased at a constant rate, or it may be oscillatory (Sect. 1.4.1.5). Das and Scholz (1981) propose a formula for predicting time-dependent rupture in the earth by earthquakes. The elastic rebound theory of earthquakes includes the assumption of a steady increase in the applied stress, at a very slow rate, until it would be released by the earthquake. Das and Scholz propose that the time-to-failure t_f (also called the nucleation time, which is the time from which the crack can be detected to start growing sub-critically to when it reaches instability) is given by:

$$t_f = l_o/V_o \cdot 2/\bar{n}-2 \ , \tag{1.51}$$

where l_o is the initial crack length (for two-dimensional cracks) and crack radius (for circular shear crack), V_o is the crack growth velocity at the stress corrosion limit, and $\bar{n}$ is the stress-corrosion index.

For oscillating stress it can be shown that the failure time t_f is linearly related to the time to failure under a constant applied stress t_o, given by (Evans 1973; Wachtman 1974):

$$t_f = g^{-1}t_o \ , \tag{1.52}$$

where g is a factor, depending on the ratio of the oscillatory part to the constant part of the stress.

Oscillation fatigue seems to be an important process of systematic jointing (Secor 1969, also see Chaps. 3 and 4). The development of methods for determining the time to failure for joints is a challenging topic for future investigation.

1.4.3 Supercritical Crack Velocities

Region III is a steeply inclined curve that ends up in a vertical line in the V versus K curve (Fig. 1.26), which indicates a sudden crack acceleration at K_{Ic}. At this K_I value the stable crack growth becomes unstable according to the Griffith energy balance concept and propagation occurs spontaneously without any additional stress. The crack, which up to that point advanced subcritically, starts to advance supercritically, following a transition which is often termed catastrophic. At this stage crack velocities reflect purely the mechanical elastic energy release (Schardin 1959) and effects of stress corrosion are either negligible or they may have a retarding effect (Kerkhof 1973). Beyond K_{Ic} the crack velocity attains a plateau (Fig. 1.26) limited by the terminal velocity (Schardin and Struth 1937) which is an upper limit to the crack velocity. Theories on crack propagation have been reviewed by Erdogan (1968). An established theory for this limit is still lacking, but it is generally thought that the information regarding the local stresses and strains that can be transmitted to the material immediately ahead of the crack tip depends on the supply of energy and is limited by the velocity of elastic waves.

It is not certain which form of elastic wave controls the terminal fracture velocity V_T, and V_T is generally considered to be between $\sim 0.38\ V_l$ (Roberts and Wells 1954) and $\sim 0.6\ V_l$ (Field 1971), where V_l is the dilatational stress wave speed, (typically in glass $V_l = 5 \times 10^3$ m/s). The value $V_T = 0.6\ V_l$ closely fits the theory that Rayleigh surface waves control crack propagation at the material surface. The velocity of these surface waves in a material having a Poisson's ratio 0.25 is $0.58\ V_l$. Although Schardin (1959) concludes that glasses exhibit a constant maximum fracture velocity independent of temperature and stress, Field (1971) predicts and Kerkhof (1957) observes that crack velocities much higher than V_T can be reached for special loadings. Pugh et al. (1952) observe shock velocity of 6150 m/s in soda lime plate glass which exceeds the velocity of sound by at least 10%, caused by the detonation of a high explosive charge, and a consequent crack velocity of 3500 m/s.

Schardin (1959) finds that in glass V_T correlates poorly with $(E/d)^{1/2}$, but this correlation improves considerably when corrections due to gradual changes in glass composition are introduced (where E is Young's modulus and d is density). A good fit is obtained on a V_T versus $(H/d)^{1/2}$ plot, where H is microhardness.

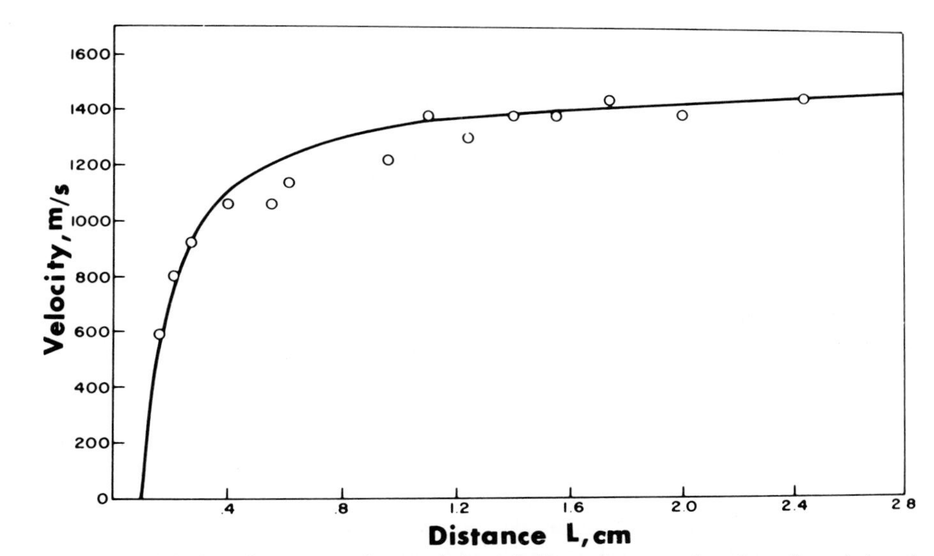

Fig. 1.29b. Velocity of a propagating crack in soda lime glass as a function of crack length. (Doremus 1973). The curve is based on Eq. (1.53). *Points* are experimental results from Field (1971, Fig. 11)

A plot of crack velocity V in a soda lime glass as a function of crack length l shows an approximate plateau of the terminal velocity V_T (Fig. 1.29b). Doremus (1973) compares this curve to a basic equation derived from previous investigations:

$$V = V_T(1 - l_o/l) \ , \tag{1.53}$$

where l_o is the initial length of the crack. The experimental results (by Field 1971) and the theoretical curve show a good fit, suggesting an asymptotical approach to the terminal velocity. Doremus, however, notes that previous similar comparisons by other investigators have found poorer agreements. See additional plots of terminal velocity by Berry (1960a, b) and Ramulu and Kobayashi (1985). Kerkhof (1973) suggests that V_{max}, the maximum measured crack velocity (essentially interchangeable with V_T), is given by the equation

$$V_{max} = 2(\gamma/dr_o)^{1/2} \ , \tag{1.54}$$

where γ is the specific surface energy, d is the density and r_o the mean ionic distance [approximately b_o in Eq. (1.8)]. This equation has been numerically satisfied within $\pm 10\%$ for most of the 42 different glasses tested having V_{max} between 700 and 2150 m/s.

Schardin (1959) observes a crack velocity decrease only when hackle marks (Fig. 2.30 a) appear. He considers that this is a result of the increase of surface energy. Field (1971) notes that the first serious roughening of the fracture surface gives a slight, but detectable slowing of the crack. He also considers that the new mirror-smooth surfaces that characterize crack branching indicate a slight reduction in crack speed.

Rabinovitch and Bahat (1979) suggest a single equation which describes all regions of crack growth from the subcritical to the postcritical stage:

$$(V^{-n} - 1)^3 - \xi K_I^{-1}(V^{-n} - 1) + U_2 = 0 \ , \tag{1.55}$$

where ξ, U_2 and n are material constants. Their prediction includes a possible instantaneous velocity overshoot at K_{Ic} followed by a decay to the final terminal velocity (Fig. 1.26). No such overshoot has been experimentally identified so far.

Dynamic fracture is characterized by the time dependent dynamic stress intensity factor or by the time dependent crack extension force (Irwin 1977). For a propagating crack, the fracture criterion is assumed to take the form in terms of K as:

$$K_I(t) = K_{ID}(V) \ , \tag{1.56}$$

where $K_I(t)$ is the time-dependent K, K_{ID} is the dynamic fracture toughness and the actual dependence of K_{ID} on V must be determined by experiment. V is the crack tip speed (Kobayashi and Ramulu 1985). Plots of experimental K_{ID} versus V (for steel and polymers) show that the crack generally propagates at a constant velocity when driven by a dynamic stress intensity factor which exceeds the static fracture toughness K_{Ic}. There are variations in the terminal velocity which are due to the interaction between the reflected stress waves in the specimen and the propagating crack tip (Kobayashi and Ramulu 1985).

1.4.4 Fracture Branching

1.4.4.1 Criteria for Branching

Research on crack branching seeks the required conditions which control the onset of crack branching. Lawn and Wilshaw (1975 a) consider three causes of crack branching: (1) the dynamic crack tip field distortion, also known as the critical velocity criterion, (2) initiation of secondary fractures, also known as the critical stress intensity criterion and (3) stress wave branching (Field 1971). Further references in the present context are made to: (4) the energy requirements for branching (Johnson and Holloway 1966, 1968, (5) the combined crack curving and crack branching conditions (Ramulu and Kobayashi 1985). Only the first two causes are elaborated here.

The first listed possibility of crack branching is based on Yoffe's theoretical study (1951), which predicts kinking of the crack prior to branching at about a third of the dilatational wave velocity. This is because at this critical velocity the maximum $\sigma_{\theta\theta}$ shifts angularly away from its original direction. The Yoffe crack velocity criterion has been questioned, since many investigators observed branching at considerably lower crack velocities (Congleton and Petch 1967), and the kinking angle 60° predicted by Yoffe has turned out to be large compared with experimental results (e.g. Kitagawa et al. 1975). Nevertheless, the mathematical rigour of the Yoffe model is its real advantage over other criteria for branching. Modifications along this line of research should perhaps be considered as unexhausted as yet.

Clark and Irwin (1966) suggest that the critical condition for braching is stress intensity rather than fracture velocity. Branching occurs when the K_I reaches the crack branching stress intensity factor K_b. Congleton and Petch (1967) arrive experimentally at the same criterion of branching and propose a model of nucleation of a Griffith crack ahead of the moving parent fracture at the location where the critical stress intensity is reached. Congleton (1973) suggests that there should be a certain ratio between K_b and the fracture toughness K_{Ic} of a given material.

The general relationship in tension for K_b is

$$K_b = Y \sigma_j r_b^{1/2} , \tag{1.57}$$

where Y is a combination of geometrical factors primarily related to the size of the crack relative to that of the specimen and crack shape, σ_j is the fracture stress and r_b is the radius for the initiation of macroscopic crack branching (Fig. 2.31). A more elaborate equation has been suggested by Smith et al. (1967) for rectangular beams subjected to bending. Reed and Bradt (1984) examine the breakage of complexly stressed large weathered and non-weathered sheet glass panels (36×91 cm) and find that even if the local stress state varies considerably, the glass fails in accordance with the relationship $\sigma\sqrt{r} = A$ [Eq. (2.7)] Reed and Bradt find that different loading rates or prior history had little, if any, effect on this relationship (see also Karper and Scuderi 1964). Kirchner and Conway J.R. (1987), on the other hand, examined the shapes of the mirro-mist boundaries in rectangular specimens fractured in tension and in flexure and found that the stress-intensity criterion K_b provided better predictions than Eq. (2.7).

Further elaboration on criteria for branching are beyond the scope of this section. However, three important questions on crack branching remain open: (1) does branching develop by the growth of initial microcracks inclined to the parent fracture, or from a kink in the direction of the propagating fracture? (2) does nucleation of branching occur ahead of the main fracture (Congleton and Petch 1967) or at the side of the main fracture? (Preston 1926, see Sect. 2.1.2.4), (3) is branching controlled by interaction of the new microcracks with the parent fracture, or by interaction among the microcracks themselves, or both? Direct detailed observations on mist (Fig. 2.31) are rare. Johnson and Holloway (1968) show microcracks in the mist zone (Fig. 2.24) which are located at the side of the parent fractures and are inclined to it. This observation seems to support Preston's prediction of branching that nucleates from the side of the parent fracture, apparently from microcracks inclined to it.

1.4.4.2 Duality in Criteria for Branching

Evidently duality in the criteria of branching is more the rule than the exception. Shand's early work (1954, 1959, 1965) reflects a certain duality regarding conditions of crack branching where he observes at the mirror boundary a critical stress (1954); a critical velocity of crack propagation and a critical or maximum value of local stress (1959), and a critical velocity (1965).

Later investigators who prefer the stress intensity criterion for branching (as suggested by Congleton and Petch 1967) over other criteria, generally are not quite satisfied with this criterion alone, but actually modify it by adding a second condition. Kirchner (1978) observes a strong correlation between the stress intensity at crack branching K_b and Young's modulus E. He finds that crack branching in 20 ceramics and glasses occurs at a strain intensity K_b/E close to $3.3 \times 10^{-5}\,m^{1/2}$, and suggests that K_b should be replaced by K_b/E as a criterion for branching. Rabinovitch (1979) considers that in order to explain fracture branching, an additional effect has to be added to the K_b criterion. Such an effect should divert energy from the kinetic part to the surface creating part. Hagan et al. (1979) add a second condition that $dK/dc > 0$. Namely, K must also be an increasing function of the crack length at the time of branching. This is particularly relevant when the crack propagates in a material under stress gradients or residual stress which can cause K decreasing as the crack length c increases.

Duality in criteria of branching is also suggested by Rice (1984) and Ramulu and Kobayashi (1985) as mentioned in the previous section and by Ravi-Chandar and Knauss (1984b).

1.4.4.3 Crack Arrest

Kobayashi and Ramulu (1985) distinguish between static and kinetic theories which explain crack arrest. The static arrest theory (Irwin and Wells 1965) see the arrest as an approximate reversal of propagation and suggests that a material property exists, known as the dynamic arrest stress intensity factor K_{Ia}, below which the crack ceases to propagate. This factor governs crack arrest in the same sense as K_{Ic} governs initiation. The kinetic approach, on the other hand, focuses on crack propagation process and considers arrest only as the termination of such a process. The theoretical aspect of this problem is largely still open today, but the kinetic theory appears to be the more accepted one today (see further, Kanninen and Popelar 1985, p. 213).

From the practical point of view, it is desirable to arrive at an arrest mechanism which will halt a crack from reaching catastrophe. Two major approaches are practiced. One relates to the exploitation of materials, the other to the integration of structural design techniques (Bluhm 1969). The former entails the development and use of materials which, by their toughness characteristics, help in arresting propagating cracks (Sects. 1.5.2 and 1.5.3). The utilization of a laminated structure which consists of alternating layers of elastic and plastic materials can be quite effective in achieving high toughness levels; arrest strips are also being used. Designers are developing structural configurations in which the stress levels near the crack tip are subdued.

1.5 Microstructural Aspects of Fracture in Polycrystalline (Grainy) Materials

1.5.1 General

The transition from an ideally brittle material like glass to a polycrystalline ceramic, concrete or rock involves several significant modifications. Instead of a simplified continuum approach to fracture propagation, the investigator has to take into account discrete microstructural elements, differences in chemical composition and mechanical properties of various grains. Accordingly, a wide range of fracture behavior influenced by grain resistance to cracking, grain boundaries, grain anisotropy, crystal cleavage, and crack interaction must be considered. Immiscible glasses and single crystals which contain inclusions also have their influences. The elastic properties of the polycrystalline material may significantly be affected by the microstructure. Very often straight cracks are replaced by zig-zag-like patterns on scales dependent on the grain size, with corresponding increase in the fracture surface energy γ (Wiederhorn 1969). The fracture toughness K_{Ic} can also be greatly modified by the microstructure (Sect. 1.5.2).

The treatment of microstructural fracture is conventionally divided into two, transgranular and intergranular. In a polycrystalline material with grains of low cleavage tendency, the crack path is influenced by the grain boundaries, but it traverses crystals without much orientation deviation. In a material which consists of crystals which are well cleaved, the crack orientation will be affected by both grain boundaries and cleavage planes. Lawn and Wilshaw (1975a, p. 112) estimate that G must immediately drop as the crack traverses the boundary between two misoriented grains. The drop is a function of the angle of misorientation. The tendency for intergranular fracture, that is, to propagate around grains along their boundaries rather than traversing them, increases as the angular misorientation between crystallites increases. The consequent tortuous crack path can account for somewhat greater variations in the fracture energies than evident in transgranular fractures, including reductions in the fracture energy of more than one half in certain crystals and in local G.

The intergranular failure mode is very frequently observed to increase with temperature, and the roughness of this fracture mode has been correlated with the fracture origin and the promotion of mist and hackle regions (Rice 1984). Transgranular section failure, on the other hand, is linked with river markings (Sect. 2.1.2.6) that help elucidate fracture origin, and it replaces intergranular fracture in compressive regions (Rice 1984). See further on intergranular fracture by McMahon et al. (1981).

1.5.2 Dependence of Mechanical Properties on Microstructure

There is a variety of ways by which ceramists can increase the fracture toughness K_{Ic} in brittle materials, by controlling the microstructure. These include an introduction of a ductile grain boundary phase, crack deflection, transformation toughening stress-induced microcracking and fibre composite structure (Beall et al. 1986). Examples which represent the first three relationships are given below.

An example of the interrelationship of microstructures, elastic properties and experimental techniques is shown by Beall et al. (1986). The K_{Ic} of monoclinic chain-silicate called canasite glass-ceramic is measured by five methods. Glass-ceramic is a ceramic body which is initially formed as glass and then transformed into a polycrystalline article by a careful heat treatment. The canasite crystals are highly acicular due to their chain-silicate atomic arrangement and form an interlocked grain structure.

The results for K_{Ic} increase at ambient room temperature of the canasite glass-ceramic by about two to four times the corresponding values typical for non-chain-silicate glass-ceramics. It is revealed that K_{Ic} decreased rather rapidly by a factor of almost five over the range of ambient temperature from 25 to 600 °C. This negative temperature dependence of K_{Ic} corresponds to a decrease in fracture surface roughness with increasing temperature which also correlates with a decrease in the degree of non-planar crack propagation. From 600 to 800 °C there is an increase in K_{Ic}, which is thought to reflect intergranular viscous flow of a glassy phase. The authors consider on the basis of these correlations and some additional observations that the high K_{Ic} values result from two primary mechanisms: (1) crack deflection, which arises from the combined effects of an acicular microstructure (crack propagation around the crystallites) and the preferred cleavage fracture (fracture through the crystallities), and (2) stress-induced microcrack toughening, due to the internal stresses caused by the cooling of the ceramic body after the heat treatment of the initial glass. The anisotropy in thermal expansion of the individual crystals leads to both intragranular and intergranular microcracking within the highly stressed zone ahead of the crack tip. The magnitude of the internal stresses is proportional to the temperature range over which the glass ceramic is in the elastic state on cooling from the crystallization temperatures. This explains why the stresses are maximum at room temperature and will decrease in magnitude with increasing temperature with a corresponding decrease in K_{Ic} (Fig. 1.30). It also becomes clear why this trend correlates with a decrease in the roughness of the fracture surface.

Gulati et al. (1986a, b) investigate the mechanical properties of K-richterite (potassium multiple chain silicate) glass-ceramics and find that the interlocking structure of these materials contributes to high magnitudes of strength, fracture toughness, fracture surface energy and threshold stress intensity afforded by the tortuous fracture path in the polycrystalline material. Improvements are also observed in the fatigue behaviour (high stress corrosion index), thermal shock resistance and impact resistance.

1.5.3 Crack Shielding

Crack (tip) shielding is the approximate inverse of stress concentration. It involves the reduction of the "effective crack driving force" at the tip compared to that of the remote field. Crack shielding has been successfully achieved by

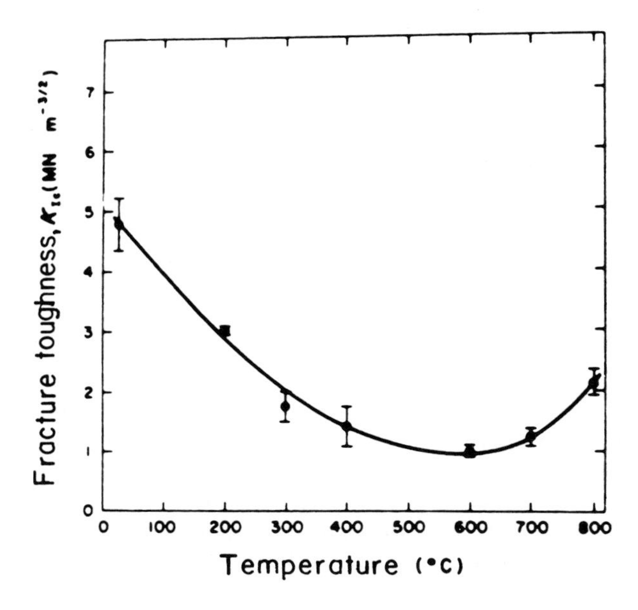

Fig. 1.30. Temperature dependence of the fracture toughness of canasite glass ceramic crystallized for 40 h, measured by the chevron-notch method. (Beall et al. 1986)

material toughening (raising K_{Ic}) via a series of extrinsic mechanisms (Evans and Faber 1984; Ritchie and Thompson 1985). An example of material toughening by stress-induced microcracks is presented in Section 1.5.2. Intrinsic mechanisms attempt to enhance the inherent resistance of the microstructure. "The common phenomena here are related primarily to processes which occur in the wake of the crack tip, whether associated with an enclave of previously deformed (e.g. plastic or transformed) zones, the formation of fracture surface asperities to yield an interlocking zone behind the tip, or the presence of fluid or debris within the crack. Such processes effectively shield the crack from the full (far-field) applied loading, yet since they act on the wake of the crack, their influence must diminish with decreasing crack length and increasing applied stresses" (Ritchie 1985). Crack shielding in a geological environment has been discussed by Nur (1982).

1.5.4 Fracture in Concrete

1.5.4.1 Strength Dependence on Microsructure

There is a distinction between concretes of common aggregates in which the coarse phase is stronger than the fine cement matrix and concretes of light aggregates in which the coarse phase is weaker than the matrix (Soroka 1977). In the former, fracture mainly occurs intergranular in the matrix, and the aggregates do not greatly affect the strength of the concrete. In the latter, on the other hand, fracture is both intergranular and transgranular, and the strength of the concretes of light aggregates depends both on the strengths of the course and fine phases (Bache and Nepper-Christensen 1965). Intergranular fracture occurs both through the matrix and along the interface aggregate-matrix (Alexander et al. 1965). In concretes of common aggregates fracture is mostly interfacial at least at low loads, and through matrix propagation occurs only at high loads.

Fractures commonly develop in the matrix between two close interfacial cracks preferably between two such cracks that occur at contact with the larger aggregates in the area (Hsu et al. 1963). This implies that the strength increases from the interface through the matrix and is maximum in the aggregate.

The fracture process is influenced by the size of the aggregate and its relative content in the concrete. When these two parameters increase, the strength decreases because the distance between adjacent interfacial cracks is shorter. However, the strength of the concrete may increase if only the relative content of aggregates rises (Soroka 1977).

Obviously, raising the interfacial cohesion between the aggregate and matrix is crucial to the concrete strength. The interfacial cohesion depends on the properties of both matrix and aggregate. Rough surfaces of the aggregate will improve the cohesion and smooth surfaces of the aggregate will reduce it. The chemical constitution of the interface is important. Alexander et al. (1965) find that the cohesion strength in volcanic rocks rises with the increase of silica in the aggregate because it improves the bonding with the calcium oxide in the cement.

There are conflicting observations regarding the relative importance of the matrix strength and the interfacial cohesion strength in determining the overall properties of the concrete. Alexander and Taplin (1962, 1964) find that changes in matrix strength are twice as important as changes in the strength of the interfacial-cohesion in affecting the concrete strength. On the other hand, Nepper-Christensen and Nielsen (1969) find that a concrete made of aggregate of glass balls is 2.5 times stronger than the same concrete in which the aggregate is smeared by oil which reduces the interfacial bonding.

The fracture toughness of a material increases with E [Eqs. (1.14) and (1.15)]. Kaplan (1959) finds that generally the strength of concretes increases in bending and in compression with the rise of E. This is possibly related to a stress redistribution associated with the increase in the aggregate rigidity. Less stress is locally applied on the matrix and more on the aggregate. Conflicting observations have been made as well (Mayer 1972).

The above correlations are qualitatively generally related to conditions of both compression and tension, but quantitatively they are different (Soroka 1977). For instance, the tensile strength of concretes is less sensitive than the compressive strength to the quality of the matrix. Therefore, the ratio of tensile strength to compressive strength is not constant, it decreases with the increase in the concrete strength. Generally, the ratio of bending strength to compressive strength varies between 0.10 to 0.20 in strong concretes and weak concretes, respectively, and the ratio between tensile strength determined by the Brazilian test (Jaeger and Cook 1979, p. 169) and compressive strength changes correspondingly between 0.07 to 0.11.

1.5.4.2 Crack Joining

Van Mier (1986) presents a physical model for concrete fracture loaded in uniaxial or multiaxial compression. His concept of initial growth of isolated microcracks that then join and form macrocracks somewhat resembles observations by Hallbauer et al. (1973) on fracture in argillaceous quartzite. Van Mier suggests that "the amount of energy released in a crack-joining event is considerably larger when compared to the extension of two isolated microcracks (under the assumption that the same amount of new cracks surface is formed)", and the surplus of energy released due to crack joining compared to the energy released from the extension of two single cracks is given by

$$\Delta U = [2l^2 + 4l\Delta l + 2(\Delta l)^2]\,\pi\sigma^2 Y/E \ , \tag{1.58}$$

where l, Δl, σ, Y and E are the microcrack length, microcrack extension, local tensile stress, geometrical factor and Young's modulus, respectively. When crack joining occurs, the energy release is governed by l instead of Δl as in the case of isolated crack propagation. The large increase of stress-relieved zones due to crack joining (Fig. 1.31) may lead to closure of neighbouring microcracks. Hence, whereas in homogenous materials (like glass) the first oc-

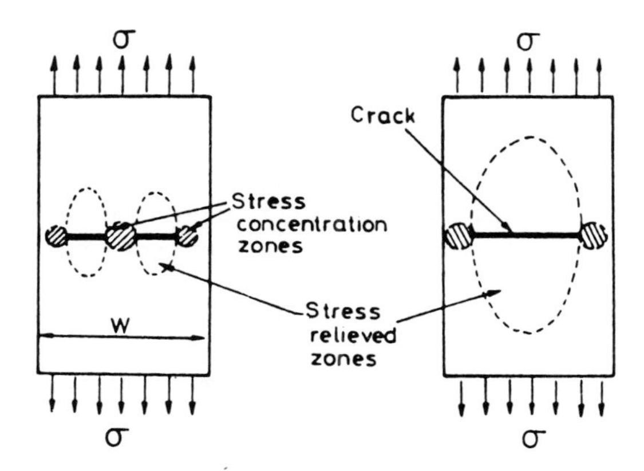

Fig. 1.31. Increase of stress-relieved zones due to joining of separate microcracks under tension. (Van Mier 1986)

currence of microcracking would immediately lead to unstable fracture, in heterogeneous materials (like concrete and rock) microcracking and crack-joining processes are the result of propagation and arrest mechanism. Van Mier considers that the same mechanism of microcrack-joining is observed in tension and compression, except that the process is considerably more unstable in the former.

1.5.5 Limitations and Deviations from Simple Microstructure-Strength Relationship

Although many correlations that were cited above appear to be straightforward, a more general evaluation of the dependence of mechanical properties of brittle materials on the microstructure is, in fact, both theoretically (Lawn and Wilshaw 1975 a) and experimentally (Rice 1984; Atkinson 1984) to a large extent intractable. Results for K_{Ic} are strongly dependent upon the test method employed. Commonly K_{Ic} increases with grain size and the crack length induced by the testing method, but at some threshold further increase in grain size involves a reduction in K_{Ic}.

According to Rice et al. (1981), the values for γ of isotropic materials are relatively independent of grain size. There is, on the other hand, a maximum in γ with grain size in noncubic materials (Fig. 1.32). Rice et al. explain that the major difference between cubic and non-cubic materials, to which the preceding γ behaviour can be attributed, is the presence of thermal expansion anisotropy in the latter. "Thermal expansion anisotropy leads to microstresses between grains of differing orientation. When a grain is larger than some critical size, the strain energy produced by the microstresses can produce microcracking, which will generally occur at a grain boundary. Microcracks nucleated at a crack tip can reduce the available strain energy at the tip of the primary crack, thereby making propagation of this crack more difficult. As the grain size increases from the threshold value, the extent of microcracking will increase, absorbing more energy, such that γ will also increase. However, when microcracking becomes too extensive, macrocrack growth can take place by

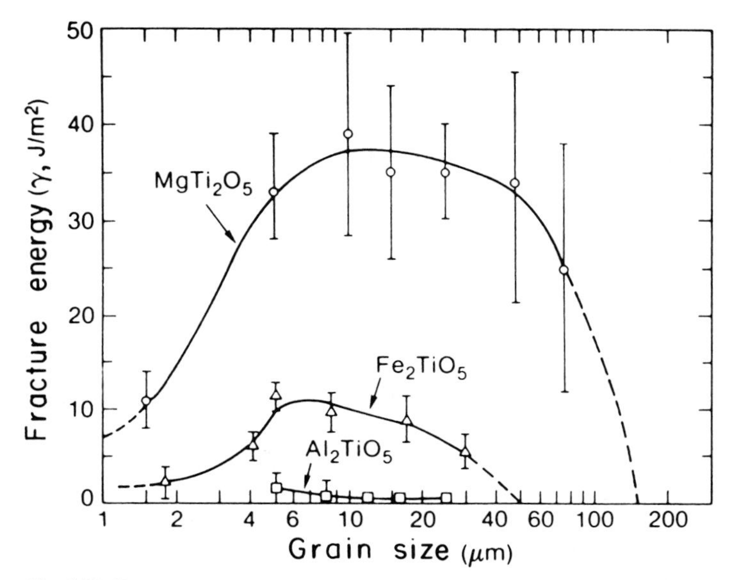

Fig. 1.32. Fracture energy (work-of-fracture) of $MgTi_2O_5$, Fe_2TiO_5 and Al_2TiO_5 versus grain size. (Rice et al. 1981)

linking of microcracks, thereby reducing the value of γ. These counter trends lead to a maximum in γ with grain size". At a large enough grain size, spontaneous failure of a piece will occur.

No maximum in γ with grain size is known so far in geological materials, and generally there is an increase in K_{Ic} with grain size, possibly because data on very coarse-grained rocks is not yet available (Atkinson 1984). In spite of complicating factors, generally, γ of grain boundaries is less than γ of cleavage due to many discontinuities along the latter. This may explain the common preference of intergranular growth along boundaries rather than transgranular cracking under subcritical conditions. Rapid cracking, on the other hand, is controlled more by inertia, which results in preference of less tortuous transgranular propagation.

The general transition from intergranular fracture to transgranular fracture associated with the increase in grain size may be linked with the maximum of γ with grain size (Fig. 1.32), which implies that the reduction in γ of cleavage is greater than the reduction of γ of grain boundaries when fracture propagates in a polycrystalline of coarse grains. Substantial transgranular failure found in some dense fine-grain Al_2O_3 (Rice 1984) is an example of a deviation from the above general trend.

1.6 Fracture in Rocks

The broad subject of fracture in rocks (Jaeger and Cook 1979) is limited here to short introduction to jointing, which is primarily a mode I phenomenon.

The treatment of jointing basically follows the general fracture principles which are outlined in the previous sections of this chapter. These include the assumptions that:
a) The rock to a first approximation is an elastic homogeneous, isotropic material, and therefore the linear elastic fracture theory is generally applicable.
b) However, deviations from linearity such as an enlarged process zone at the crack tip must be considered.
c) Geological fractures generally are larger or even much larger than cracks that are commonly studied by material scientists, hence the scale factor has to be taken into account.
d) It is legitimate and convenient to divide the fracture behaviour into several characteristic stages including initiation, propagation, interaction and arrest.

Two important parameters have to be added to the above considerations:
e) The rock is generally loaded by overburden pressure;
f) Internal fluid pressure (Sector 1965) probably has a major role in rock fracture.

Finally, the mechanisms that result in jointing are quite diversified (Price 1966, p.127), so that applications of the general fracture principles have to be pursued with caution, always constrained by field observations.

1.6.1 Joint Initiation Stage

Fracture in rocks like in glass or ceramics initiates at discontinuities such as flaws or small cracks as envisioned by Griffith. This is because "the maximum tensile stress in the corners of the crack is more than ten times as great as the tensile strength of the material, as measured in an ordinary test" (Griffith 1920). Hence, the Inglis "stress concentration factor" (Sect. 1.1.1) governs the initiation of jointing. Discontinuities in rocks can vary considerably from grain boundaries that can be considered as microcracks (Simmons and Richter 1976; Padovani et al. 1982) through inclusions (Figs. 2.22 and 3.58) to cavities (Figs. 2.25 and 2.27c). Joints often start at layer boundaries or previous fracture surfaces which constitute prominent stress concentrators. Joint incipience is a distinct nucleation (or incubation) process.

The treatment of joint initiation is idealized from a modified internal crack (Fig. 1.3). In this model, fluid pressure is activated on the crack surfaces and remote compression σ_3 normal to the crack length replaces the remote tension. Also the traction parallel to the crack has been cancelled. Finally, the shape of the crack has been thought to be an oblate (Sneddon 1951, p. 486) or prolate spheroid (Fig. 1.33) and initiation sites on some joint surfaces seem to support it (Fig. 3.22). Secor (1969) considers that the failure criteria for tension fracturing of rocks with internal pore pressure P in plane strain is:

$$\bar{\sigma}_3 = -S = -[\pi E\gamma/2c(1-\upsilon^2)]^{1/2} , \qquad (1.59)$$

where $\bar{\sigma}_3$ is the least effective principal stress and $-S$ is the tensile strength of the rock (tension considered negative) for the general pore pressure condition $\bar{\sigma}_j = \sigma_j - P$, (Secor 1969), where $\bar{\sigma}_j$ and σ_j are the effective principal stress and

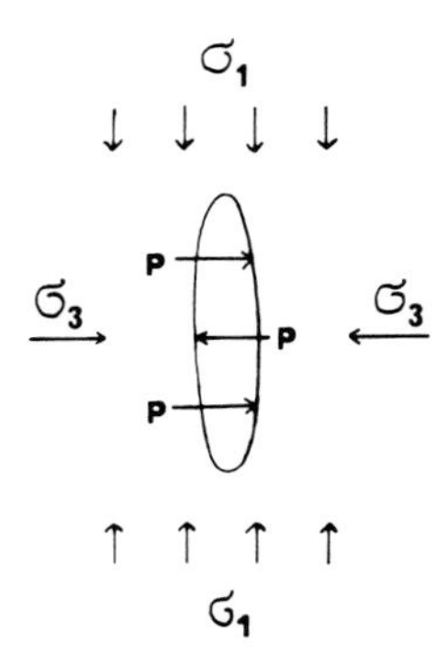

Fig. 1.33. An idealized prolate microcrack under remote maximum σ_1 and minimum σ_3 compressions and internal pressure P

principal stress respectively. Fracture initiates when a fraction of P exceeds σ_3 and becomes effective in wedging open the crack. This will occur when fluid in the crack reaches a critical volume (Secor 1969):

$$v_{crit} = 16/3\,[\pi\gamma(1-\upsilon^2)/2E]^{1/2}c^{5/2} \ . \qquad (1.60)$$

The cancellation of the traction parallel to the crack is done for convenience. It remains to be established that this is physically justified (Leslie Banks-Sills pers. commun.).

A requirement to have v_{crit} in the crack in order to initiate its growth may explain an observation made by Bahat (1988b). He investigated a population of joints in chalks whose lengths could be accurately determined by their fracture markings. No fractures shorter than 20 cm were identified in this population (see Sect. 4.2.1.1). Therefore, there has to be some mechanism that drives the crack to lengths greater than 20 cm. This questions the rather popular view that joint length distribution is governed by a power law (e.g. Segall and Pollard 1983). Segall (1984) considers that in the granodiorite that he investigates, the joints may have initiated from grain-scale cracks and each fracture had a finite length, with traces at the outcrop surface ranging from 1 cm to nearly 100 m, although he admits that one cannot observe the initiation process in the field. An alternative view is that a joint must have a minimum length commensurate with the requirement of a critical fluid volume before an actual wedging can occur.

Another intriguing aspect of joint initiation is the wiggly nature of some plumes as they curve from their initiation point at the layer boundary before attaining a steady propagation inside the bed parallel to bedding. The conditions that govern this transition are not known (see Sect. 4.2.1.1).

1.6.2 Joint Propagation

It is assumed that although the stress field during the incubation stage is heterogeneous (Sect. 2.1.3.4 and Fig. 2.30b), propagation occurs along a principal stress direction essentially normal to the minimum compression σ_3. It is often convenient to reverse signs and consider tension as positive. If the liquid inside the crack exerts pressure P which creates a tensional stress at the crack wall that exceeds the remote compression σ_3, the difference $P-\sigma_3$ may substitute the stress σ in Eq. (1.10) and the mode I stress intensity will be

$$K_I = (P-\sigma_3)[\pi c]^{1/2} \quad \text{(for a through crack)} \tag{1.61 a}$$
$$K_I = (P-\sigma_3)[\pi c]^{1/2} 2/\pi \quad \text{(for a penny-shaped crack)} . \tag{1.61 b}$$

Joints can propagate under fatigue conditions (Kendall and Briggs 1933) when $K_I < K_{Ic}$.

The crack extension force G_I is given for plane strain by (Irwin 1957):

$$G_I = K_I^2(1-\upsilon^2)/E . \tag{1.62}$$

Combining Eqs. (1.61 a) and (1.62) produces:

$$G_I = (P-\sigma_3)^2(1-\upsilon^2)\pi c/E . \tag{1.63}$$

Segall (1984) shows mathematically that G of a propagating joint in a set varies depending on five factors: the increase in crack length and corresponding increase in applied extensional strain both contribute to an increase in G. On the other hand, the effective stiffness of the system decreases as fractures propagate, causing G to decrease. The effects of elastic interaction with neighbouring cracks may increase or decrease G, depending on the fracture geometry. For parallel enchelon cracks, interaction leads to an increase in G if the crack tips do not overlap, and to a pronounced decrease in G if they do.

The mechanism proposed by Secor (1969) for fracture propagation at depth consists of numerous episodes of crack propagation interspersed with periods of quiescence during which the pore fluids from the surrounding rock percolate into the crack and assume conditions of $P > \sigma_3$ that enable the wedging process. The alternating rise and fall of fluid pressure that control joint propagation correspond to an increase and decrease in G, respectively.

The periodic increase and decrease of pore pressure and the consequent wedge opening of the crack can perhaps be pedagogically simulated by a simple experiment. A pot half full with boiling water is well closed by a lid. The pot is heated and vapour pressure gradually builds up in it. At a critical pressure the lid is slightly opened at one side and the vapor escapes. The lid then falls back to its closing position. The lid cyclic opening and closing takes place at regular intervals. Each opening represents a wedging act of a rhythmically propagating joint. Opening occurs when $P > \sigma_3$, where P is the vapor pressure and σ_3 is the weight of the lid, and closing reflects the $P < \sigma_3$ condition.

In the rhythmically propagating burial joint each cycle may be idealized as a fixed grips loading (Sect. 1.1.7.1) associated with a cyclic decrease of G following the wedging process, in contrast to a constant loading which characterizes certain stages of long uplift joints (Bahat and Rabinovitch 1988), where G increases (up to a certain extent) with the fracture length.

The discovery of rhythmic plumes along the joint surface (Fig. 3.2c) convinced Bahat and Engelder (1984) of the importance of the periodic crack propagation model proposed by Secor. Secor's model seems to fit quite well the development of burial joints (Chap. 5), but it is doubtful whether this mechanism can also explain the growth of uplift joints. The propagation styles during these two fracture stages are quite different. Whereas burial fracture is primarily a translation of remote compression into a local tensile wedging, uplift joints result from remote tension, quite likely caused by bending (Bahat and Rabinovitch 1988).

A joint propagates normal to the mode I loading and when shear stresses are also influential, mixed modes I + II and mixed modes I + III result in characteristic fracture patterns (Figs. 1.13 a and 1.13 b, respectively). The resulting fracture paths in rocks are fractographically quite well recorded (Chap. 3).

The stress fields of neighbouring joints cause them to interact with each other. Consequently, joints deviate from their original paths, they curve along characteristic trajectories and occasionally initiate secondary cracks (e.g. Hodgson 1961 b; Bahat 1987 b). Conditions for fracture interaction may differ quite considerably from one joint set to another (Sect. 5.9.2).

1.6.3 Joint Arrest

An unresolved problem is why a dynamic fracture arrests when it advances close to its maximum velocity (Schardin 1959, Sect. 1.4.3). For joints that are thought to develop under sub-critical conditions quite close to their K_0 value (Fig. 1.26) the problem is simpler. Four conditions for joint arrest may be considered:

1. The effect of decreasing the fluid pressure P [Eq. (1.61)]. The joint arrest is well manifested by the almost disappearance of the plumes after reaching their maximum intensity at the fan perimeter (Fig. 3.2 c) in burial joints.
2. The increase in the remote minimum principal compression σ_3 with depth

$$\sigma_3 = \upsilon/(1-\upsilon)\,dgh \ , \qquad (1.64)$$

where d is the rock density, g is the acceleration of gravity and h is depth; or the increase of σ_3 with lithology (e.g. Engelder 1985).

3. A contact of the joint with a free surface. Referring again to Engelder (1985): joints that originate within a thick siltstone do not cross the interface between that siltstone and the adjacent shales.
4. When the uplift process stops, tangential tension ceases and uplift joints arrest.

Chapter 2 Fractography in Technical Materials

2.1 Fracture Morphology – Basic Geometry

2.1.1 Introduction

Glass can fracture in different ways and from various causes. Figure 2.1 shows, in plan view, a series of possible fracture geometries. When molten glass is poured onto a cast iron table, rolled into a sheet, and then chilled before annealing, it develops internal stresses which cause fracture. A long rectangular sheet will first split crosswise into more or less square pieces (Fig. 2.1 a). When fracture occurs slowly due to a thermal gradient, ripple marks on the new surfaces can take the form as in (Fig. 2.1 b) according to Preston (1926), or as in (Fig. 2.1 c) according to Orr (1972). A fracture surface produced by bending is shown in Fig. 2.1 d. When a tempered glass, so developed under a controlled chilling procedure, is fractured, cracking is violent and involves a characteristic sequence of events resulting in an appearance as in Fig. 2.1 e. The failure starts at the centre of the plate and a herringbone texture develops. Rib markings start at the peripheries of the ragged herringbone zone and curve towards the two edges of the sheet. Initially these ribs are straight or circular and they then approach the edges of the sheet tangentially. It should be recalled that, after chilling, there are residual stresses in the glass. These are such that the total compressive stresses balance the total tensile stresses across the sheet. Therefore, after the completion of the fracture stage in Fig. 2.1 e, the compressive zones along the surfaces are changed into tensile zones which can lead to the development of echelon-like cracks normal to the edge of the sheet and along its periphery (Fig. 2.1 f).

These are just a few examples of fracture processes in glass with the characteristic fracture morphologies. Fractography is the branch of science which studies the morphology of fracture surface and deciphers, from fracture markings, the process of crack propagation and the mechanical factors involved. The detailed examination of the various basic geometries of fracture markings, with a reference to the evolution of fractography as manifested by new observations of various investigators, is the objective of this chapter.

2.1.1.1 Some Early Work on Fractography and Fracture Markings

Woodworth (1896) was perhaps the first to describe fracture markings (feather-fracture) in glass. De Freminville (1907, Figs. 6, 14 and in Appendix A) showed

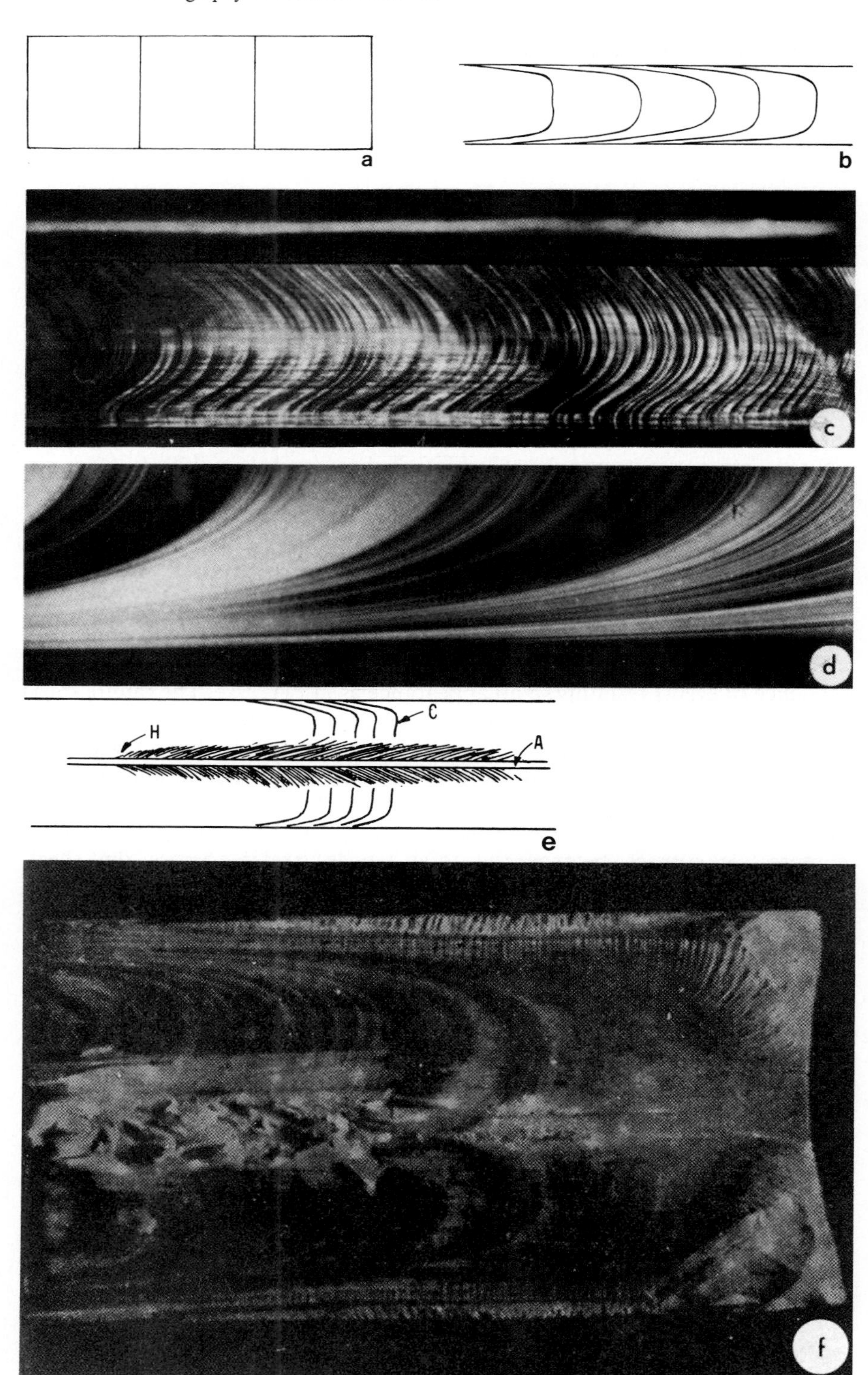
a
b
c
d
C
H
A
e
f

fracture markings that developed in glass due to impact. He distinguished between fracture morphologies that initiated directly at the point of impact (Fig. 2.2a) and indirectly from a point further across the tested specimen with respect to the impact location (Fig. 2.2b). The direct fracture surface may be either flat or conical and is basically divided into three distinct elements, the origin, the conchoidal zone of concentric undulations, and the tearing zone. The undulations that develop due to indirect fracture (Fig. 2.2b) are less regular than those obtained by direct fracture, but the distinction between the above basic three elements is maintained.

De Freminville (1914) expands his observations quite significantly by introducing five concepts that are well appreciated today: (1) the foyer declatement which is in fact the mirror plane (Figs. 2.2a–d and 2.3); (2) the petites ondulations conjugées (Fig. 2.2c) which are also known as Wallner lines; (3) the characteristic section of striae (Fig. 2.4); (4) he recognizes the reoccurrence of mirror planes on the fracture surface (his Fig. 36); and (5) De Freminville (1914, p. 985) notes the symmetry relations between the striae and the undulations on the fracture surface. Figures 2.2a and 2.3 are perhaps the earliest descriptions of the characteristic elements of the morphology of fracture surface as we know it today (Fig. 2.2d).

De Freminville (1914) does not distinguish between striae (his Fig. 33) and hackles (his Fig. 35). However, he discriminates, in his Fig. 7, between the large undulations that considerably affect the conchoidal fracture and the small undulations (the Wallner lines) that are hardly visible and have no influence on the morphology of the fracture surface.

Preston (1931) suggests that a single simple law governs the propagation of cracks in brittle and non-brittle materials. "This law is that the fissure as it advances endeavors to maintain its direction at right angles to the principal tension at its advancing head. The direction of this tension may change radically due to the advance of the fissure, but at all times the fissure endeavors to accommodate itself to changes in the direction of the tension." Hence the fracture surface, with the above markings, provided information about the orientation of the largest principal tensile stress.

◄

Fig. 2.1. **a** The initial splitting of a long rectangular flat plate of glass under its own stresses. **b** Ripple marks on the surface of a non-violent fracture showing progression *from left to right*. **c** Ripple marks on the surface of a non-violent fracture showing propagation *from left to right*. **d** Ripple marks on the surface of a fracture produced by bending. The leading edge of the fracture front is at the *upper part* of the picture (**c** and **d** from Orr 1972). **e** Ripple marks *C* on the surface of a violent fracture showing progression *from left to right* with a herringbone *H* texture at the centre. The herringbone texture is divided by a thin median line of bright polished fracture *A* (**a**, **b** and **e** from Preston 1926). **f** Violent fracture at the *centre*, ripple marks on both sides of the violent zone and echelon (shingle-like) cracks at the *peripheries* normal to the edges of a glass plate. (Poncelet 1958). The width of the section is thought to be about 1 cm

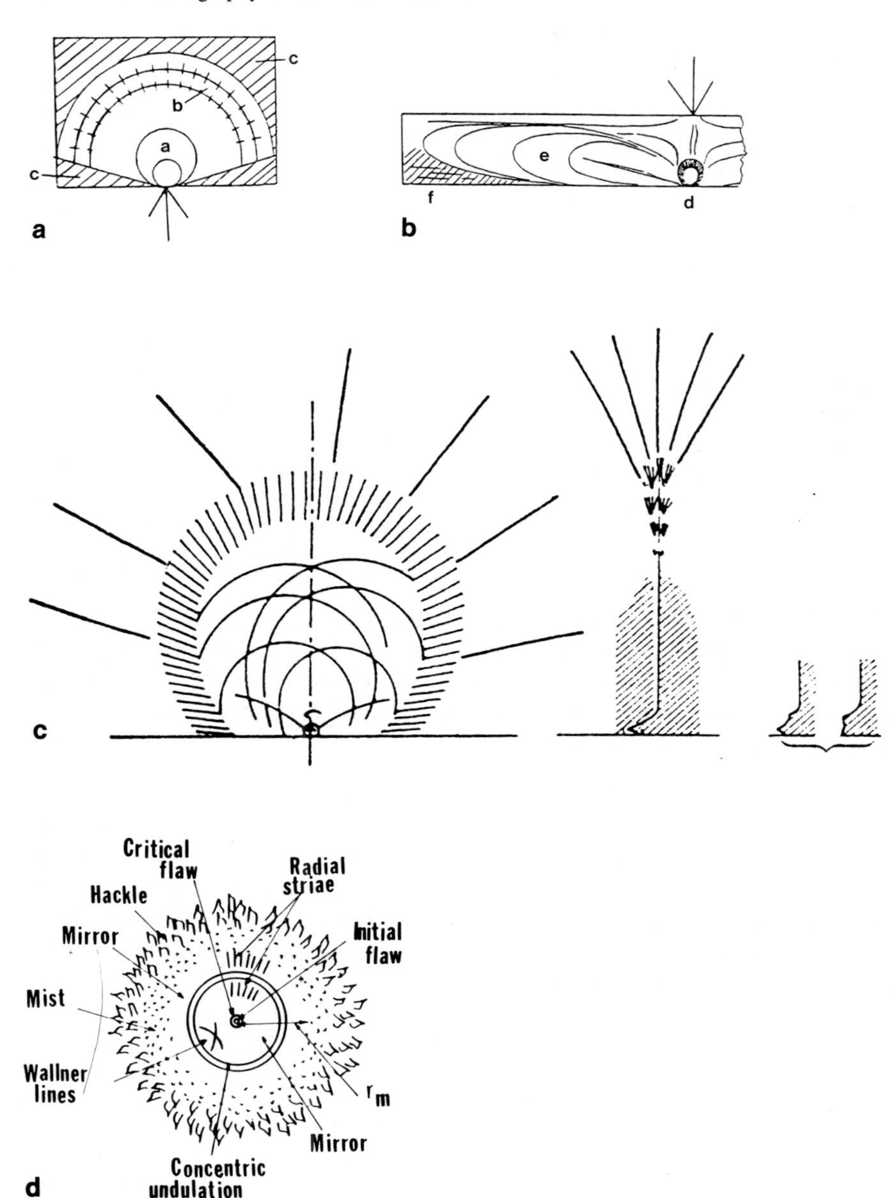

Fig. 2.2 a and **b**. Are "direct" and "indirect" fractures respectively, *a* and *e* on the figures are series of concentric ripple marks in the mirror plane, *b* shows radial striae (note also a radial crack between *d* and *e*), *d* is the zone of initial fracture, and *c* and *f* are sheared ares. (De Freminville 1907). **c** Plan view at *left* and side views at *right* of schematic mirror plane, mist (*short radial lines*) and hackle (*long radial lines*) zones. Several series of Wallner lines can be identified within the mirror. (De Freminville 1914). **d** Schematic representation of a fracture surface showing the present concept of fracture origin (pre-existing flaw) mirror plane, radial striae, concentric undulations (ripple marks), Wallner lines, mist, hackle, and r_m the mirror radius (see Fig. 2.31)

2.1.1.2 The Ubiquity of Fracture Markings in Various Materials

Surface morphologies of various shapes and styles are known to be associated with fractures in various synthetic brittle, quasi brittle and non-brittle materials, including silicate glasses (De Freminville 1907; Preston 1926; Terao 1953; Shand 1954), ceramics (Kirchner and Gruver 1973; Mecholsky et al. 1976), non-silicate glasses such as bitumene (De Freminville 1914), vitreous carbon (Nadeau 1974), polymethylmethacrylate (Wolock et al. 1959; Bansal and Duckworth 1977; Cotterell 1965; Clark and Irwin 1966; Doll 1975), polymerized transparent resin (Shinkai and Sakata 1978), metallurgical materials (many publications by Zapffe, including Zapffe et al. 1948), steel (Irwin 1962) and chromium (Bullen et al. 1970), germanium single crystals (Haneman and Pugh 1963), jellies (Preston 1931), cellulose acetate (Kies et al. 1950) and rubber (Bhowmick 1986). Fractography was also applied to minerals like quartz (Payne and Ball 1976; Hamil and Sriruang 1976) and sapphire (Congleton and Petch 1967).

Furthermore, modern technology provides us with fracture mirrors about 10 μm diameter, or even smaller, on surfaces of cracked fibres (Fig. 2.30a) which are more than a million times smaller than some fracture mirrors in sandstones (Fig. 5.19).

Also, the fracture processes and fracture morphologies associated with chipping or flaking of stone are of great interest to archaeologists in their studies of the manufacture and use of prehistoric stone implements (e.g. De Freminville 1914 in his Fig. 19, and Kerkhof and Muller-Beck 1969). But this is a two-way traffic. "Flint workers" experiment with obsidian and other lithic materials and enlighten the material scientists with their observations (e.g. Tsirk 1981).

2.1.2 Fracture Categories

2.1.2.1 Fracture Origin

Fractures commonly start at flaws on edges or surfaces and less frequently they initiate at internal inclusions. Glass defects that can be considered as flaws are bruises, scuffs, scratches or percussions but also stones, cords (Sect. 2.3.1), seeds or manufacturing defects (Frechette 1972). The location of the fracture origin can usually be identified as the locus where other markings, such as radial striae, meet. Ripples that define the front on the fracture surface are in various instances commonly convexed towards the direction of fracture propagation and direct their concave face toward the fracture origin so that the direction of fracture propagation can be readily established. There may be more than one fracture origin when failure conditions activate more than a single flaw.

Some metals, however, are excluded from this description. Occasionally, microcrystals, growing within the crystalline texture of a single crystal, may

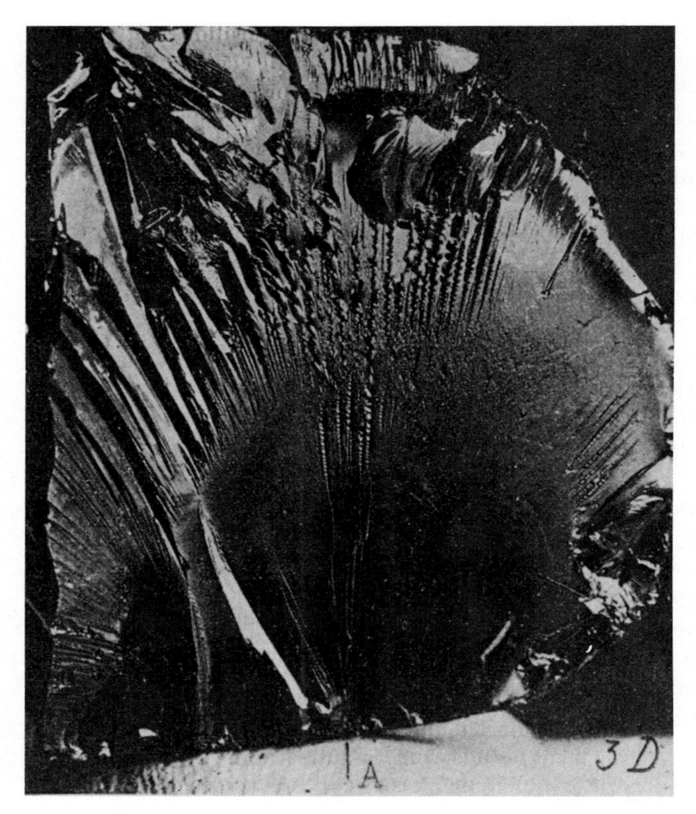

Fig. 2.3. A fracture surface in bitumene. (De Freminville 1914). Note a few radial striae within the mirror plane and a multitude of hackles beyond it. Many hackles appear as echelon surfaces, extensions of the striae

each show a separate fracture marking differently oriented, so that the identification of the fracture origin or tracing the direction of fracture origin becomes impossible (Zapffe et al. 1948, Fig. 2; Tipper and Hall 1953, Fig. 6a).

2.1.2.2 Striae, Lances, Cleavage Steps or River Patterns and Echelon Cracks

Striae, lances, cleavage steps, river patterns and feather features are approximately synonymous terms which are being used by material scientists to describe the fracture appearance which results from combined tension and shear stresses, and which is manifested by elongated lines which are parallel to the direction of crack propagation. These markings are associated with echelon cracks which have the characteristic overlapping arrangement of shingles on a roof (see Figs. 2.1f and 2.3).

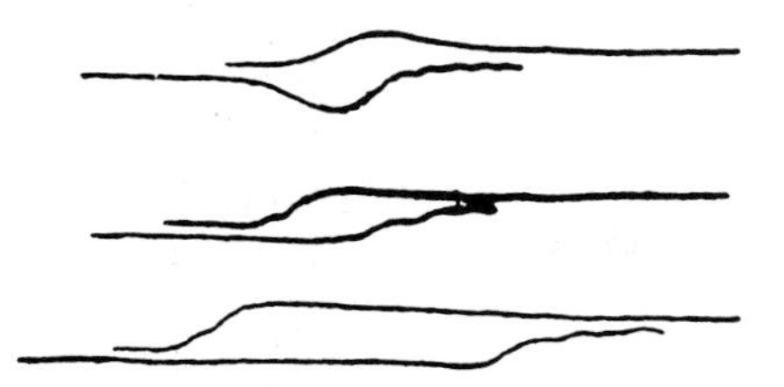

Fig. 2.4. Three cross-section types of striae. (De Freminville 1914)

Glass. De Freminville (1914) shows a fracture with surface striae in a bitumene glass (Fig. 2.3), and schematic sections of striae (Fig. 2.4). These somewhat resemble the sections observed by Woodworth (1896) along joint fringes of rocks (Sect. 3.1.1.1). Preston (1926) adapts De Freminville's French term stries to give the English term striations (Preston 1938). It is, however, preferred to use here the term striae in distinction from a somewhat similar term fatigue striations, which is used by metallurgists to characterise a different morphological feature (Sect. 2.1.2.6). Preston observes that the main direction of advance of a crack is lengthwise down the striae and that the striae and the wave front (rib markings) are orthogonal to one another (see also Fig. 3.19).

Preston's (1931) early explanation of the development of striae is cited below with slight modifications. In Fig. 2.5 a, taken from his work, we are supposed to be looking down through a rectangular block of glass; aa′ represents the trace of a crack which has begun at the lower face and is advancing up towards us. A front elevation of this crack is shown in Fig. 2.5 b. The principal tension at the crack front lies in the direction of the dotted lines in Fig. 2.5 a. Now let us suppose that in passing up above the neighbourhood of the plane pp′ (Fig. 2.5 b), the crack aa′ suddenly enters a region where the tension is inclined slightly to its former direction, as in Fig. 2.5 c, so that the trace of the crack as a whole is no longer at right angles to the tension. The crack as a whole cannot turn quickly enough and adjust itself to the new orientation of the stress since it is too extended. It can, however, break up into a number of adjacent echelon cracks (Fig. 2.5 d), each element of which can speedily adjust itself to be normal with the stress. Figure 2.5 e is an enlarged view of part of Fig. 2.5 d. It is obvious that the cracks cannot readily advance in such a state because, to extend the cracks, the sides TT′ must be pulled apart, and this cannot be done while the two sides are connected together by the solid glass at mm′. Preston then considers the nature of the stresses at m, Fig. 2.5 f. On the little square element shown, there is essentially a shearing action, the stresses $f_1 f_1$ acting in opposite directions on the two vertical faces. In accordance with the laws of mechanics, there must be a similar pair of stresses acting on the two horizontal faces as indicated by $f_2 f_2$. This system of stresses is equivalent to a tension in the NW-SE direction, with a compression in the NE-SW direction. That is, across the gap at m, the tension takes the inclined direction of Fig. 2.5 g. This is fairly obvious without any analysis of stress systems. As the cracks must advance their edges at right angles to this tension, their lips begin to curl over, as in Fig. 2.5 h. The element m is not much less solidly attached than before, and a certain flexibility is conferred to the glass, the cracks open-

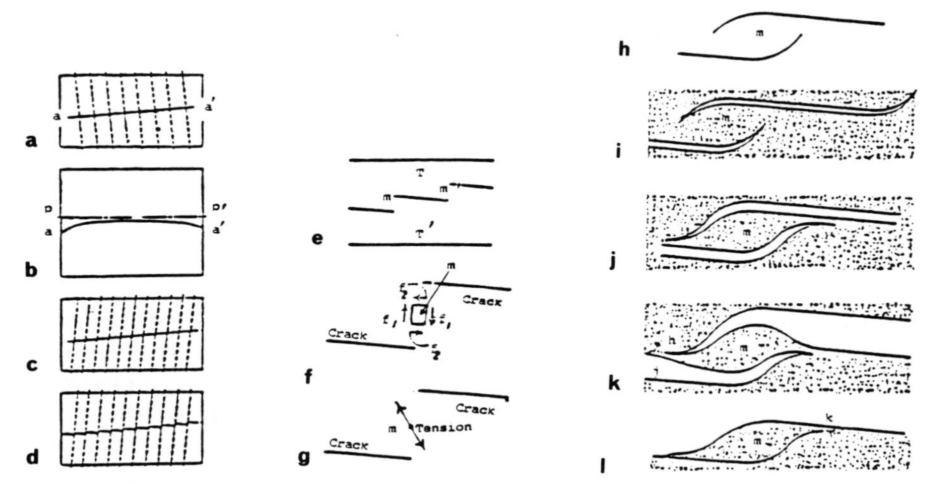

Fig. 2.5 a – l. The development of striae. (Preston 1931). **a** Crack aa′, formed perpendicular to the direction of the applied tension (*dotted lines*). **b** Side view of crack. **c** A change in the direction of the applied tension. **d** The crack splits into a number of elements. **e** An enlarged view of part of **d**. **f** Stresses acting on the connecting element *m*. **g** Direction of the principal tension at *m*. **h** Sideways extension of the lateral edges of the crack elements. **i** The crack elements can now open slightly and accordingly extend endwise. **j** Further lateral extension of the crack elements giving a tangential contour. **k** Two crack elements on the point of breaking through into one another at point *j*. **l** Rupture completed at *j*, showing "stria" *m* attached at *k* to the lower massive piece of glass

ing as shown in Fig. 2.5 i. As the points of the curved cracks approach the free surfaces of the existing cracks (Fig. 2.5 d), the tension once more becomes north and south, or perpendicular to the existing crack, and the advancing crack lip turns away tangentially to it, as has been seen in many previous examples (Fig. 2.5 j). Finally, the crack runs some distance along the line h in Fig. 2.5 k, and ultimately breaks out more or less raggedly at j, or k (Fig. 2.5 l).

It is noticeable that although De Freminville (1914) made a distinction between non-overlapping cracks (Fig. 2.4 a and 2.4 b) and overlapping cracks (Fig. 2.4 c), Preston overlooked this difference and concentrated on non-overlapping cracks (Fig. 2.5).

Murgatroyd (1942) simulated echelon cracks in a glass container wall subject to impact. He termed it a hackle structure. He observed an initial series

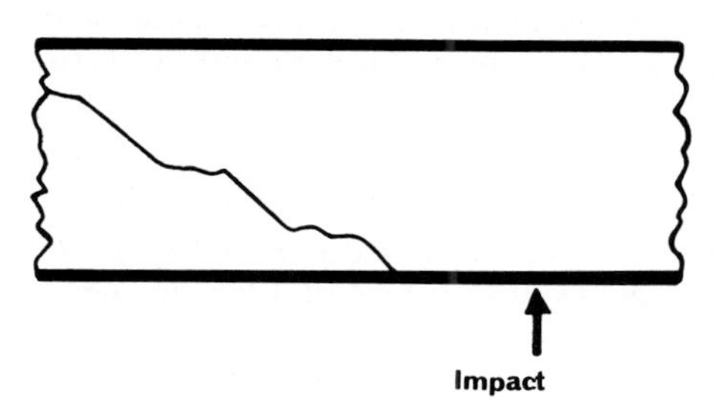

Fig. 2.6. Diagrammatic representation of the path followed by the fracture in a container wall subjected to impact. Straight echelon cracks develop first in a NW-SE orientation in the diagram and irregular bridges purposely enlarged develop later. (Murgatroyd 1942)

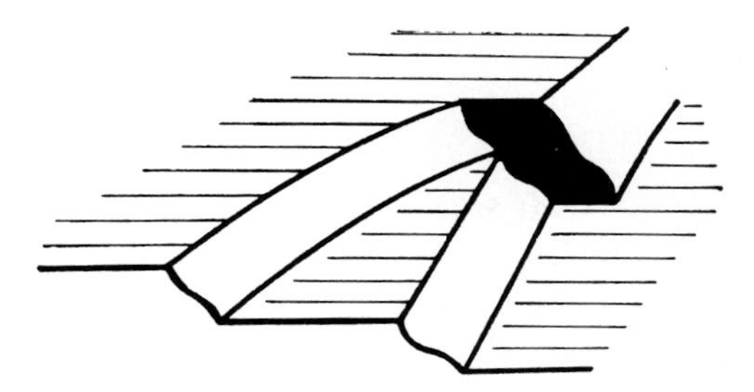

Fig. 2.7. Two steps merge to form a larger step. (Murgatroyd 1942)

of straight non-coplanar echelon cracks sub-parallel to each other at 45° to the glass surface which presumably developed under shear (Fig. 2.6). These cracks are later connected by irregular steps. Frequently, two steps may unite together forming a single step of larger dimension and of an asymmetric shape (Fig. 2.7).

Poncelet (1958) observed echelons at a periphery of a tempered glass when the initial compression along the periphery transformed into tension at an advanced stage of cracking (Fig. 2.1 f). One is reminded by the orthogonality of the echelons to the sample edge that the free surface adds additional constraints on the cracks.

Sommer (1969) applied fluid pressure to the lateral surface of a round glass rod whose ends were free from pressure and other constraints. Consequently, the fluid pressure was able partially to penetrate into randomly distributed notches or cracks at the surface of the rod, and fracturing occurred perpendicular to the axis of the rod. A superposition of a small amount of torsion which contributed some mode III loading in addition to the mode I caused by the side fluid pressure produced very symmetrical lances on the fracture surface (Fig. 2.8). The interpretation of this fracture process (Smekal 1953; Sommer 1969) is based on the assumption that the lances should be parallel to the principal direction of crack propagation. By the superposition of modes I and III the rotation of the principal stress axis is induced in the plane perpendicular to the direction of crack propagation. "Continuous adjustment of the crack plane along the entire crack front is not possible. Thus, the crack breaks into partial fronts which can adjust to the new stress direction. The 'lines' separating the partial fracture planes are the fracture lances". Hence, a step separates each pair of echelon cracks.

Sommer (1967, 1969) observed that lances can be produced on the fracture surfaces of circular glass rods loaded in tension by the superposition of a small amount of torsion. This fracture process is associated with a rapid increase in the tensile stress intensity factor, K_I, as well as with the prolongation of cracking, which encourages crack branching. Hence, a high mode I/mode III ratio encourages branching. On the other hand, echelon cracking would develop in a fracture experiencing increasing mode III/mode I ratio (see also Lawn and Wilshaw 1975 a, p. 72).

Basically, the mechanisms by which striae and echelon steps form are similar combinations of modes I and III, (see Sect. 1.1.7.2). Morphologically,

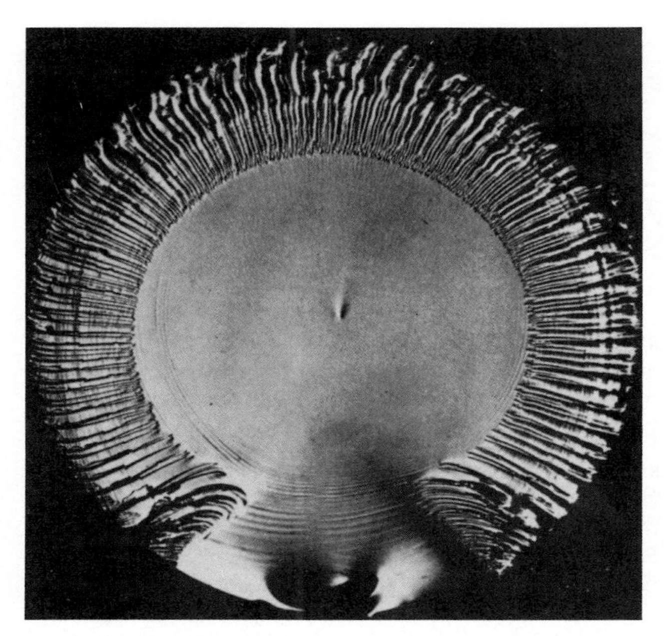

Fig. 2.8. The fracture surface of a cylindrical bar of glass 1.0 cm in diameter surrounded by lances. The fracture started at the *lower part* of the specimen in the photograph. (Sommer 1969)

however, they may be distinguished. Striae are formed within the mirror plane. In isotopic materials they are perpendicular to ribs and they may occur individually or in groups. Echelon steps are formed beyond the perimeter of the mirror plane and they may have no relationship to ribs. Their orientation adjusts itself normal to the boundaries of the sample (Fig. 2.1 f). Echelon steps characteristically occur in groups.

Ravi-Chandar and Knauss (1984a) investigated fracture markings in a polyester glass, Homalite-100, which resulted from loading simulated pressurization of a semi-infinite crack in a two-dimensional infinite medium. This simulation resulted in the known sequence of mirror, mist, hackle and branching zones on the fracture surface (Figs. 2.2c, 2.2d, 2.23a). On the mirror plane they observed straight lines that originated in voids and ran in the direction of crack propagation (Fig. 2.25). The lines propagated along different planes inclined to the main fracture surface and produced steps (striae).

Michalske and Frechette (1980) examined a crack that accelerated through the II/III transition range in a V vs. K diagram (Fig. 1.26). At low velocities the crack surface appeared mirror-like. As the velocity increased to approximately 10^{-2} m/s striae formed which they termed hackles. The striae gradually increased in number until they terminated sharply in a front termed cavitation scarp (Sect. 2.1.2.5).

Frechette (1972) observed that the development of striae and the tinkle of their splintering may occur long after the initial fracture of the glass.

Preston (1931) predicted that in the formation of secondary striae of comparatively small dimensions on the margins of the major striae, there was no reason why these striae should not have still smaller ones upon their surfaces, "and so ad infinitum". If one accepts that there are close similarities in the development mechanisms of striae in glass and in those of steps associated with echelon cracks at the joint fringes in rocks, then this prediction has turned out to be correct (Fig. 3.38).

Single Crystals. Perfect cleavage planes that lie parallel to crystallographic planes in single crystals, such as in mica, are well known. But quite often perfect cleavage does not develop and, instead of a single fracture plane, the fracture is being broken up into a set of parallel cracks that propagate at different levels, all being oblique to the cleavage plane (Fig. 2.29 a). Cleavage steps also known as river patterns develop between pairs of cleavage planes when the non-coplanar cracks are bridged by shear cracks or secondary cleavages. On a microscale, screw dislocations provide the most common causes of cleavage steps (Gilman 1958). According to Low (1959), "As a cleavage crack front passes a screw dislocation, the levels of the crack on either side of the dislocation are shifted with respect to each other. After the crack front has passed the dislocation, it moves on two different levels which are jointed by a unit cleavage step of the order of one Burgers vector in height". The screw dislocations are either initially present or may be introduced by plastic deformation which develops "at the tip of the advancing crack if the rate of crack propagation is low enought or if the temperature is high enough" (Low 1959). Low notes, however, that cleavage steps develop also in truly brittle crystals like silicon where dislocation images have not been revealed (Swain et al. 1974).

Ball and Payne (1976) examined the tensile fracture in quartz crystals and identified a mirror plane, concentric undulations and radial zig-zag steps. The cracks which propagated in a zig-zag fashion did so on two planes largely inclined to the stress axis, and the size of the steps increased systematically until fracture was complete or bifurcation occurred (Fig. 2.9).

Rice (1984) presented a series of photographs depicting various shapes of mirrors and associated step markings on the fracture surface of single crystals, the fractures having been obtained in flexure (Figs. 2.10, 2.11). A number of intriguing features which are distinct from fracture markings in glass or polycrystalline materials (see below) can be recognized on these surfaces:

1. Fractures frequently occur on crystallographic planes of low fracture energy and avoid planes of high fracture energy. Also, branching is relatively uncommon because branching cracks would deviate from planes favorably oriented for cracking.
2. Mirror planes are often decorated by two types of marking. These are radial and arched or cathedral-like, seemingly intersecting each other (Fig. 2.10). Both the radial and the arched markings occur in regular stepped geometries as distinct from hackles on glass fractures, which generally follow radial trends but do not maintain regular geometries. The radial and arched steps are in fact striae and possibly ripple markings, respectively, characteristic of glass frac-

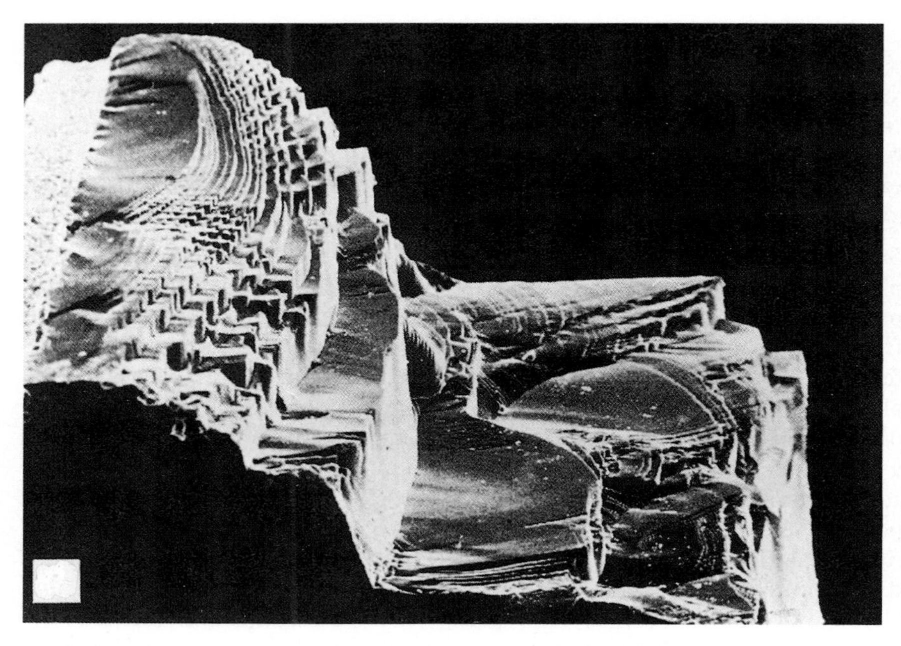

Fig. 2.9. A fracture surface in quartz showing mirror on the left and steps gradually becoming larger *towards the right.* The *white square* is about 0.06×0.06 mm. (Ball and Payne 1976)

tures. This becomes apparent in Fig. 2.11 where mist and hackle appear at the continuation of these striae. The angle of intersection between the radial and the arched markings is significantly different from that in glass.

3. Some of the steps in the striae overlap each other, whereas the arched steps seemingly do not (Fig. 2.10). If overlapping does occur in the arched steps, we have a case of mode III operating in the crystal in an orientation which would be "forbidden" in an isotropic material.

4. A mirror may be surrounded by echelon-like hackles on which secondary irregular hackles form (Fig. 2.12).

5. Narrow mirror tongues are common on single-crystal fractures and they tend to become narrower as the fracture stress increases (Rice 1984, Fig. 40).

6. Two fracture surfaces that intersect to form a common mirror may not be in a given crystallographic direction (Rice 1984, Fig. 30).

7. There is a tentative impression that has further to be verified. Radial striae seem to appear more commonly in mirrors that develop on single-crystal surfaces than in mirrors from glasses. I have examined many mirrors from glass bottles fractured by biaxial bending due to internal pressure. In these mirrors striae are relatively rare. In almost all photographs of fractured single crystals shown by Rice (1974, 1984) striae appear in abundance. Data from other authors (e.g. Low 1959) support this observation. These may possibly be correlated with the abundance of screw dislocations in the crystals, compared with

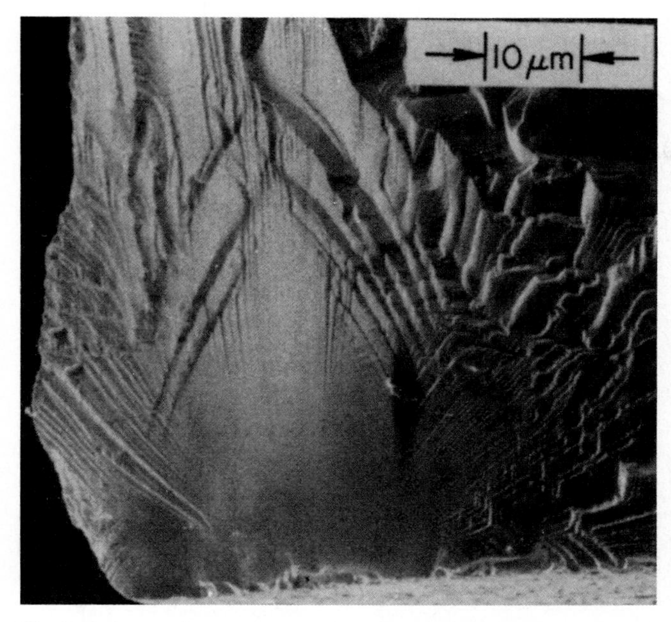

Fig. 2.10. A complex "cathedral" mirror in a $MgAl_2O_4$ crystal with ⟨111⟩ tensile axis and {110} tensile surface, showing radial straight striae and concentric but not circular arched steps crossing each other at angles less than 90°. (Rice 1974)

their limitation in glass. Perhaps also stress anisotropies in a crystal are more conducive to step development than in glass. It should be recalled that fracture processes are often delayed.

Polycrystalline Materials – Metallic and Ceramic. Cleavage steps may develop in randomly oriented polycrystalline materials because the change in orienta-

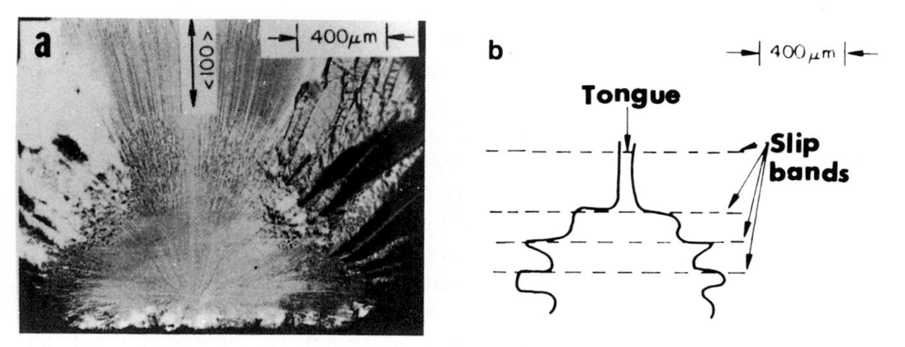

Fig. 2.11 a, b. A mirror on MgO crystals with radial striae that transform into mist and hackles as they propagate upwards (in the photograph **a**) with the exception of a mirror tongue at the centre where mist does not occur. This tongue, however, (in the sketch **b**) is a part of a second mirror beyond the belt of the mist. (Rice 1974)

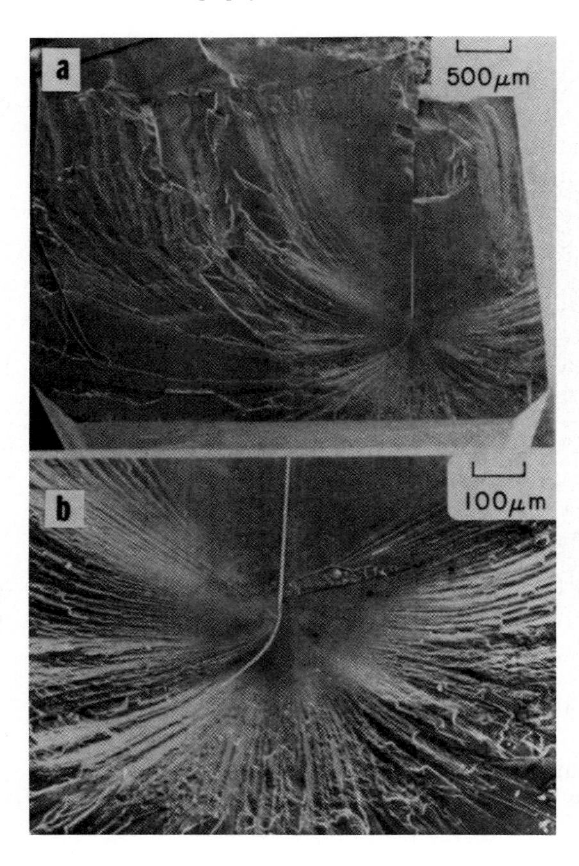

Fig. 2.12 a, b. A fracture originating from an internal crack (white vertical) in a CaF_2 crystal caused by flexure is shown in two magnifications: **a** The mirror at the centre is surrounded by radial echelon like hackles. **b** Secondary cracking in the hackles can be seen in the larger magnification. (Rice 1984)

tion at individual grain boundaries impedes the propagation of a single cleavage crack, causing the cleavage to be a combination of several microcracks. This would be associated with limited localized plastic deformation at grain boundaries. Crack propagation and growth of cleavage steps will be much less regular in polycrystalline bodies than in single crystal, and Low (1959) considers that "the finer the grain size, the more difficult would be the process of crack propagation".

The root zone from the parent to the step cracks in glass, single crystals, polycrystalline bodies and natural rocks (Sect. 3.1.6.1) characteristically consists of initial cleavage steps which are considerably smaller than the main cleavage steps. Quite often several small cleavage steps at the transition zone unite into a single larger cleavage step that propagates from the transition zone (Fig. 2.13). Kies et al. (1950) attribute the convergence of "tributaries" into "larger rivers" to differences in the rate of propagation of fracture. Such differences in local crack velocities are indeed reasonable in view of the fact that echelon cracking initiating from the parent fracture should require extra fracture energy and the reduced velocity of some initial echelons would compensate for this.

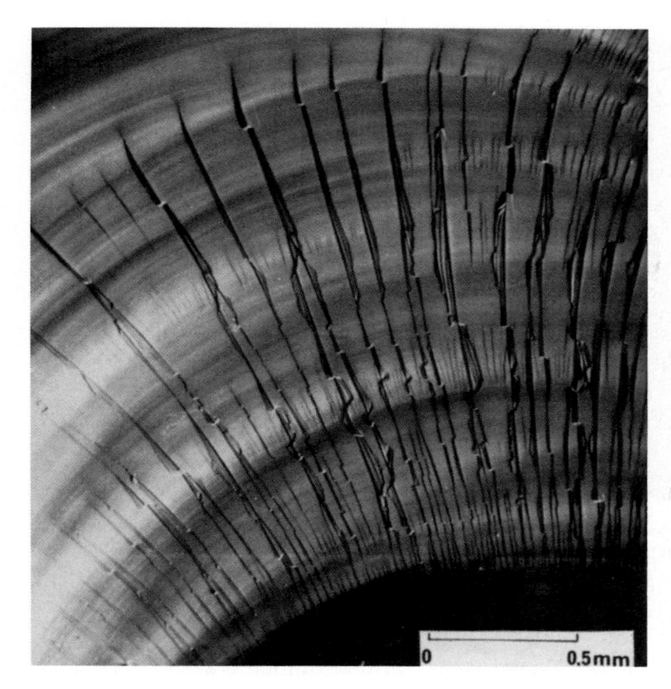

Fig. 2.13. Radial striae and concentric undulations orthogonal to each other on a fracture surface of soda lime glass. (Holloway 1986). Groups of several small radial steps unite at early stages of fracture propagation and form single larger steps (striae), resembling root zones in Section 3.1.6.1

Striae suborthogonal to ripple marks (undulations) are described in cast molybdenum (Zapffe et al. 1948). These markings are formed on transgranular cleavages due to thermal stresses that develop in the metal following rapid chilling from the melt. Periods of thermal pulsation can be recognized by crystallites that fan out along the curved boundaries of the ripple.

Tschegg (1983) conducted tests on the growth of fatigue cracks under conditions of cyclic torsion using specimens of AISI 4340 steel tempered at 650° with a gauge diameter of 12.7 mm and with circumferential notches. The tests were performed under stress intensity control, i.e. the nominal value of the alternating K_{III} at the crack tip was held constant. Tschegg observed the transition of a main mode III fatigue crack to a mode III+II step, which then developed into a mode I crack which was marked by plumes (Fig. 2.14) (see an elaboration on plumes in Chap. 3). These changes were observed from the outer surface toward the middle of the specimen. Tschegg considered that the change of mode of fracture was probably initiated by branch or step cracks which formed at non-metallic inclusions in the plastic zone ahead of the tip of the crack during stressing. At low values of K_{III} when the effective alternating K_{III} was < 18 MPa $m^{1/2}$), the mode I branch cracks which formed at approximately 45° to the specimen axis, grew faster than the Mode III crack and lead to a "factory roof" type of fracture marked by well-developed plume-

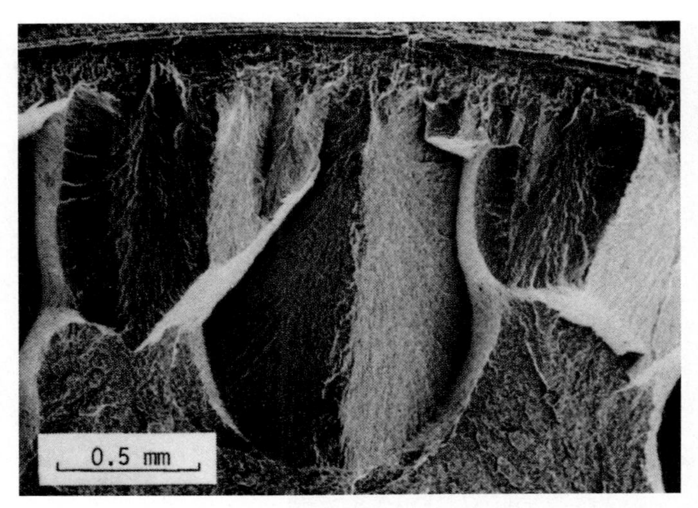

Fig. 2.14. SEM micrographs of AISI 4340 steel tested at an effective alternating K_{III} of 15 MPa m$^{1/2}$. Detail of "factory roof" fractures with plume-like markings on their surface, indicating predominant mode I crack propagation. (Tschegg 1983)

like micro hackles. For high values of K_{III} (effective alternating $K_{III} > 18$ MPa m$^{1/2}$), the Mode III + II steps, termed branch cracks by Tschegg "dominate the crack growth in the longitudinal and transverse planes of the specimen and a macroscopically flat mode III fracture occurs". This distinction somewhat resembles an analogous correlation by Sommer (1967).

Tschegg (1983) observed a "Mode III crack closure", an effect which depends on the applied torque, crack depth and specimen diameter. This effect involves the reduction of the effective range of K_{III} with crack length due to friction and interlocking interferences along the fracture.

In summary, striae, lances, cleavage steps and river patterns are synonymous terms which represent cracks that bridge echelon segments. Preston (1931) explains that echelons in glass are formed when the tensile stress changes its direction and the main fracture splits into parallel non-coplanar cracks. Later, pure shear which develops between each pair of such cracks becomes resolved into a tensile component operating at 45° to the echelons. Consequently, a further advance of the echelons results in steps bridging the pair. These steps are shown as irregular cracks that connect either overlapping or non-overlapping echelons (De Freminville 1914), and may be arced according to Preston. Murgatroyd (1942) shows that two steps may merge and form a single larger step. Sommer (1969) modernizes the early explanations of step formation in fracture-mechanic terms.

Sommer (1967, 1969) observed that echelon cracking would develop in a fracture experiencing an increasing ratio of mode III/mode I but a high ratio of mode I/mode III favours branching.

Cleavage steps are common on fracture surfaces of single crystals of ceramic materials and of certain metals at low temperatures. Striae occur as radial rectangular steps in the mirror plane of a fractured glass and they are distinguished from radial hackles that appear beyond the mirror. Striae sometimes are restricted to radial angular zones rather than being spread over the entire mirror.

A distinction can be made between striae that are radial with respect to the origin and echelon steps that occur at the specimen edge and are perpendicular to it. The free surface is an additional constraint on the latter (Fig. 2.1 f).

Branching occurs only as a continuation of hackles but not of striae. Occasionally, hackles appear in an overlapping texture like echelon cracks (Fig. 2.3) and a distinction between these two morphologies becomes difficult. Hackles (Sects. 2.1.2.4) and echelon segments (Sect. 1.1.7.2 and 3.1.6) are formed by different mechanisms.

2.1.2.3 Ripple Marks

Whereas striae are formed in stress fields which rotate the fracture front about the axis of the direction of crack propagation leading to fracture segmentation, ripple marks (rib marks or undulations) develop due to stress fields that progressively bend the crack front aroung the axis normal to the direction of propagation when both axes are on the fracture surface (Fig. 2.13). This bending does not require fracture segmentation (Sommer 1969; Lawn and Wilshaw 1975 a). Wallner lines (Fig. 2.15) form a distinct group of ripple marks.

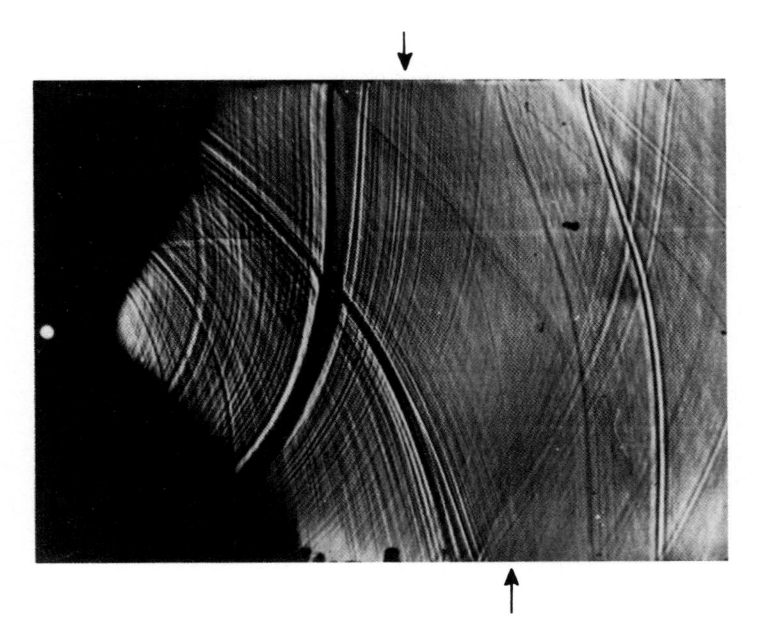

Fig. 2.15. Wallner lines on a glass fracture surface. The origin of the fracture is marked by a *white spot.* The fracture developed *from left to right.* The *arrows* show the first line marking formed by waves radiating from the crack tip which interact with the crack front. (Field 1971)

According to Preston (1926) ripple marks represent the "instantaneous position of a fissure advancing in an intermittent or oscillating fashion, and adapting itself to redistributed stresses". The two oscillating systems of the hesitating fracture front and the stress wave will interact and "will maintain and even increase in the oscillation for a time, thus producing the ripple marks" (Fig. 2.1 b). Preston, however, adds that when the fracture is proceeding explosively and the oscillations are far more rapid, if the rib texture traverses a hackled area, it sometimes appears to have the shape shown in Fig. 2.1 e.

Murgatroyd (1942) suggests "that the rib is actually a high point where a fracture which had been moving slightly upward before coming to rest resumed its course in a slightly downward direction when it recommenced". He observes that, in thermally stressed glasses, closely spaced rib marks develop when fracture occurs at relatively slow velocities (below 10^{-2} m/s) and disappear at higher velocities. He considers that the smooth-faces appear because fracture arrest becomes impossible at the high velocities. Murgatroyd simulates rib marks at still lower fracture velocities (3.3 mm/min to 1 mm/min) by an early version of the double-cantilever cleavage experiment and obtained smooth surfaces at higher fracture velocities. He suggests that "when the velocity of fracture reaches a certain value, it exceeds the velocity of the compression wave formed by molecular readjustment, and therefore no rest points occur in the forward movement of the fracture".

The fracture front oscillates when the stress at the crack tip remains substantially orthogonal to the crack front, although oblique to the crack plane (Poncelet 1958). Kerkhof (1973) calculated the direction of crack propagation as it rippled when the above obliquity was such that the mode I changed to a mixed mode I and II (Fig. 2.16). This occurred when the principal tensile stress σ_0 slightly rotated to a new position giving σ_1, producing the

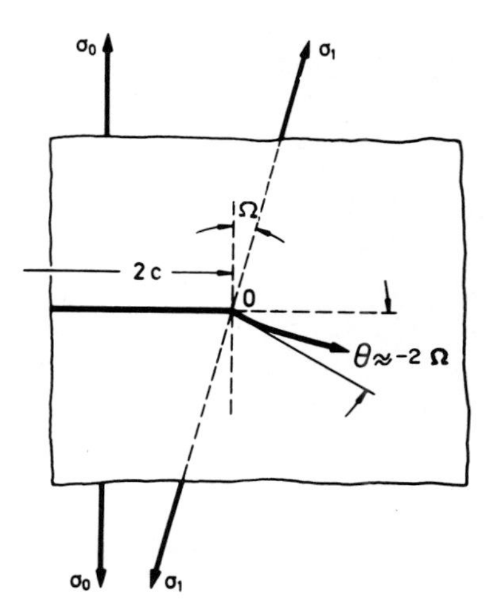

Fig. 2.16. Crack extension for oblique loading. (Kerkhoff 1973)

small angle Ω. The stress rotation at the crack tip resulted in a change in the direction of crack propagation of angle θ (Sect. 1.1.5) when:

$$\theta = -2K_{II}/K_I \ . \tag{2.1}$$

For a central crack of length 2c in a very wide plate the stress intensity factors are

$$K_I = \sigma_1 \cos^2 \Omega \sqrt{c\pi} \tag{2.2}$$

$$K_{II} = \sigma_1 \sin \Omega \cos \Omega \sqrt{c\pi} \tag{2.3}$$

and a combination of the above three equations yields

$$\theta = -2 \tan \Omega \cong -2\Omega \ . \tag{2.4}$$

Namely, the angle within which the running crack deviates is about twice the angle of rotation of the principal stress.

Arrest Lines. The rate of crack propagation is dependent on the stress intensity at the crack tip. If K_I is reduced while the crack propagates, the crack would reduce its velocity or even stop (Fig. 1.26). At low K_I values all other stresses, such as residual stresses, may have a significant influence on the crack and following Eqs. (2.1)–(2.4) slight changes in the direction of crack propagation may occur prior to crack arrest. This would result in a ripple mark which represents the shape of the crack front at the moment of arrest (Kerkhof 1975). Kerkhof, quoting works by H. Richter (1974), observed an arrest line that was induced in a plate glass disk at a crack velocity below 4×10^{-5} m/s. Kerkhof (1975), associates arrest lines, shown in his Fig. 5, with conditions close to $K_I \leq 23\ N/mm^{3/2}$, which is $\leq 0.73\ MPa \cdot m^{1/2}$.

Wallner Lines. Wallner lines, named after the scientist who first explained them, are delicate ripple marks on a fracture surface discovered by Freminville (1914) in the mirror planes (foyer d'éclatement) of glass (Fig. 2.2c). Their formation and some of their characteristics were demonstrated by Wallner (1939), Smekal (1950) and Shand (1954). A further example of Wallner lines on a glass fracture surface is shown in Fig. 2.15 for the region near the fracture origin. Figure 2.17a, b, and c shows schematic plan views of a fracture face and d is the side view of the specimen (after Field 1971). This figure depicts the positions of the fracture front at successive time intervals as it develops from the point O. Field's explanation (with adjusted figure numbers) is as follows: "If at P the fracture cuts through a defect, stress waves are produced. Since the distortional wave has tension associated with it, it disturbs the crack front along the locus of intersection, causing a line marking. In the figure the dotted arcs show the position of the stress wave at three successive time intervals: the line P, a, b, c is the Wallner line left on the surface (all other lines in this diagram are of course just construction lines). It is clear that if the distortional stress wave velocity for the solid is known, and also the points O and P, then the crack velocity can be evaluated. Figure 2.17 illustrates the appearance of Wallner lines for different crack velocities. Pa_1 is formed, if the crack velocity is low, Pa_2 if the crack velocity is high. If the crack is accelerating, the Wallner lines become steeper (Pa_2, Pa_2'). When the distortional wave forming the Wallner lines reaches a boundary it reflects: it is possible, therefore, to have a secondary Wallner line (QR), as depicted in Fig. 2.17c. The dotted line shows how the primary line would have continued if there had been no boundary: QR is the 'mirrored' part of the dotted curve".

The formation of a pair of Wallner lines W and W′ at a location of disturbance S is sketched in Fig. 2.18a (Kerkhof 1975), which represents an uncommon situation of a rectilinear crack front propagating with the velocity V_b. The elastic wave velocity starting at S is Vt (given by $\sqrt{\text{elastic constant}}/\sqrt{\text{density}}$). In this case $V_b/V_t \cong 1/3$; $V_b = 1150$ m/s and V_t in plate glass is 3460 m/s. The intersection points P_1 to P_4 and P_1' to P_4' of the crack front which is advancing upwards in the sketch, with the wave pulse (both propagating at constant velocities) produce the angle ψ with the simple relation

$$\text{Cos}\, \psi/2 = V_b/V_t \ , \tag{2.5}$$

which enables the determination of V_b.

If the propagating crack fronts are curved, curved Wallner lines are formed and V_b may be determined by one of the three schemes shown in Fig. 2.18b (after Kerkhof 1975). Kerkhof gives the ranges of velocities and stress intensity factors at which Wallner lines are produced as follows:

$$10\ \text{m/s} \leq V_b \leq 1500\ \text{m/s} \quad \text{and} \quad 27\ \text{N/mm}^{3/2} \leq K_I \leq 68\ \text{N/mm}^{3/2}$$

(between ≤ 0.85 to ≤ 2.15 MPa$\cdot$m$^{1/2}$ respectively).

The modes of curvature of Wallner lines change with crack velocity (Figs. 2.17, 2.19). In Fig. 2.19, divergent parabolic-like Wallner lines occur in an area of high crack velocity (50% of V_t) close to the fracture origin, and they become concentric circular with respect to the fracture front at low crack velocities (10% of V_t). A distinction between concentric circular arrest lines and parabolic-like Wallner lines on a surface of a rapid fracture is reasonably simple. Not always is the distinction between these two curved lines obvious, and wrong identifications of Wallner lines have been made in the past. Wallner lines are mostly limited to the mirror plane, but they occasionally occur in the mist zone as well.

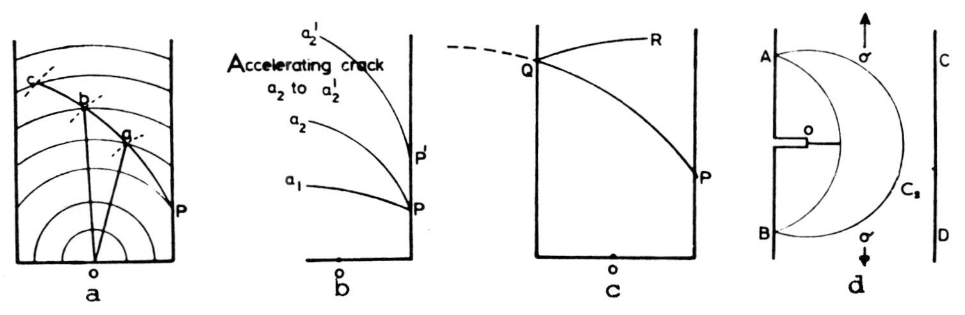

Fig. 2.17a–d. Schematic representation of Wallner line formation. (Field 1971)

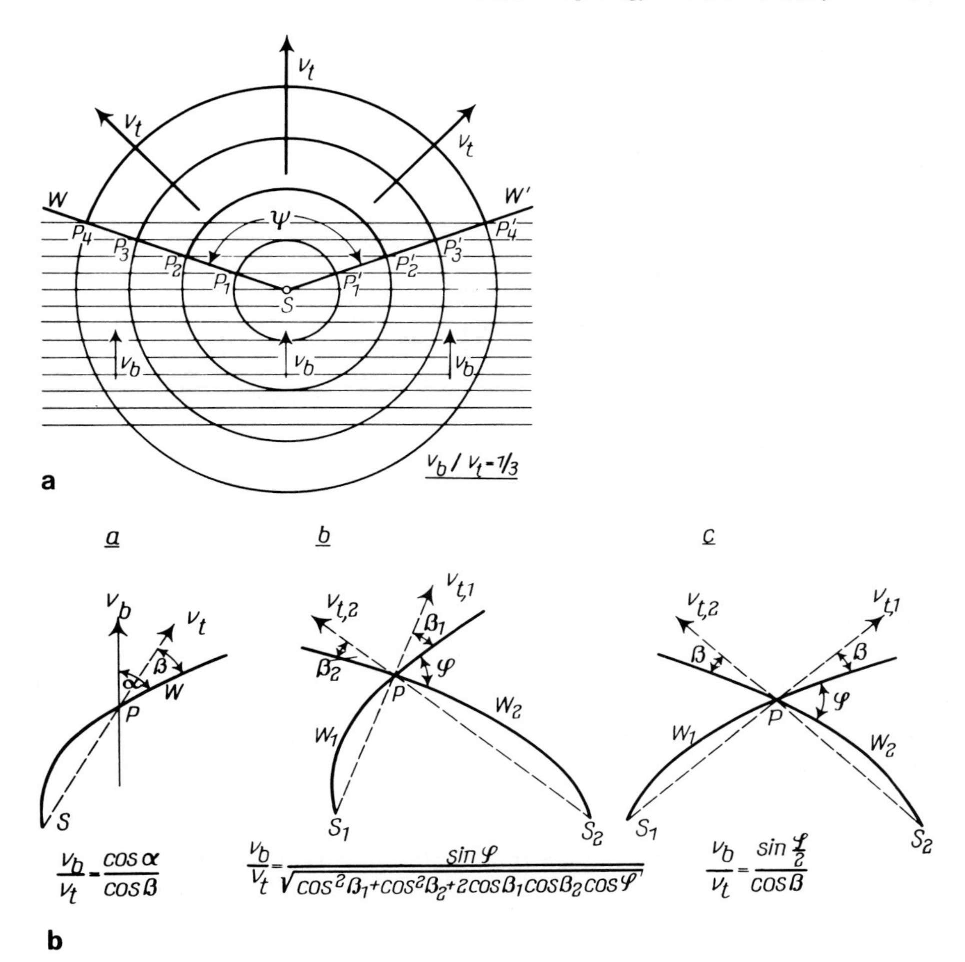

Fig. 2.18. a A schematic development of two straight Wallner lines, W and W', from a location of disturbance, S, when the velocities of crack propagation and wave pulse from S are V_b and V_t respectively, and intersection points are P_1, P_2, P_3, P_4 and P'_1, P'_2, P'_3, P'_4 at four equal time intervals. **b** Determination of crack velocity V_b from three possible types of Wallner lines, W, W_1, W_2 when V_t is the velocity of the transverse wave. The crack propagates upward in the sketch, and *arrows* marked V_t show the directions of wave propagation in the three cases. See corresponding equations for solutions (**a** and **b** from Kerkhof 1975)

The Wallner lines are generally very faint and their limited detectability in polycrystalline materials has made them of little value for the determination of crack velocity in studies of ceramics and rocks. However, they did turn out to be useful in investigations of quartz single crystals (Payne and Ball 1976). They also have been observed on the fracture surfaces of diamond and tungsten (Field 1971).

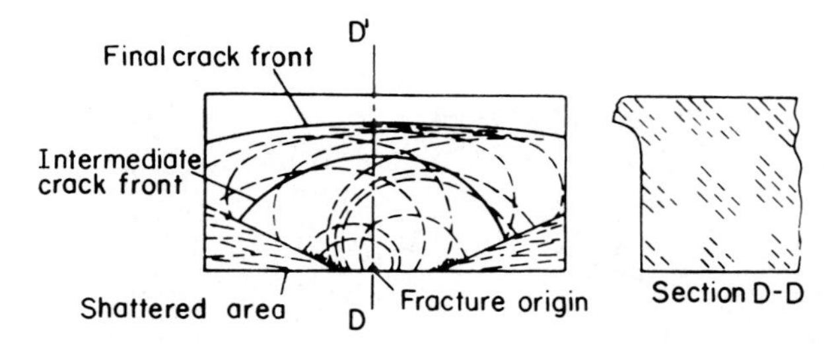

Fig. 2.19. Drawn fracture surface of rectangular bar of about 0.95×1.9 cm broken in bending, showing Wallner lines and crack boundaries. Note the *two triangles* of hackles (*shattered areas*) on the tension side. The crack velocity and nominal stress decreases with distance from D to D'. Due to greater stresses in the direction normal to D-D′, the crack spreads more rapidly in the lateral directions and the intermediate crack front becomes semi-elliptical rather than semi-circular. The closing together of the Wallner lines close to D' characterizes a slowing down of the crack velocity. (Shand 1954)

Cross-Section Through Ripple Marks. Sections taken parallel to the direction of crack propagation have revealed the profile of ripple marks. Geological investigations of ripple profiles in sediments have shown correlations between various dune profiles and water-stream or wind velocities. By analogy one would expect a multitude of ripple profiles on fracture surfaces. It is therefore surprising that so few studies have been carried out in this field. Murgatroyd (1942) observed a symmetric cusp-like profile (Fig. 2.20a) which suggested to him that the rib was actually a high point where a fracture, which had been moving slightly upward (in the photograph) before coming to rest, resumed its course in a slightly downward direction when it recommenced. An asymmetric cusp-like profile in glass has been observed by Bahat (1977a) (Fig. 2.20b). Cusp-like shapes are also fundamental figures in plan (Fig. 2.20c). Schematic sections of fatigue striations in metals show symmetric and asymmetric pro-

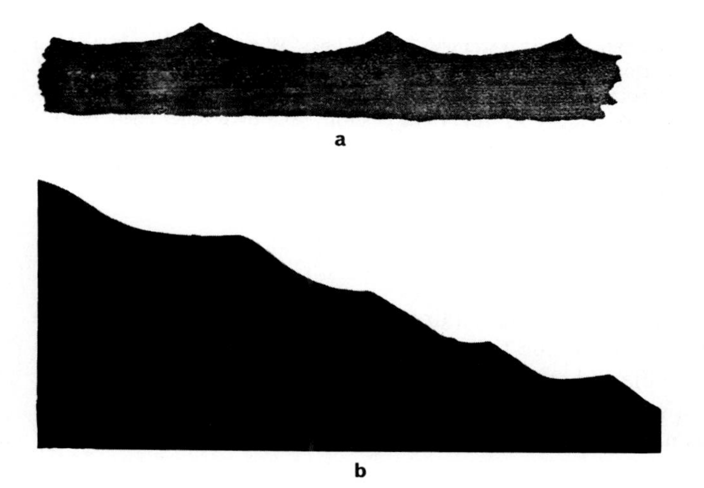

Fig. 2.20

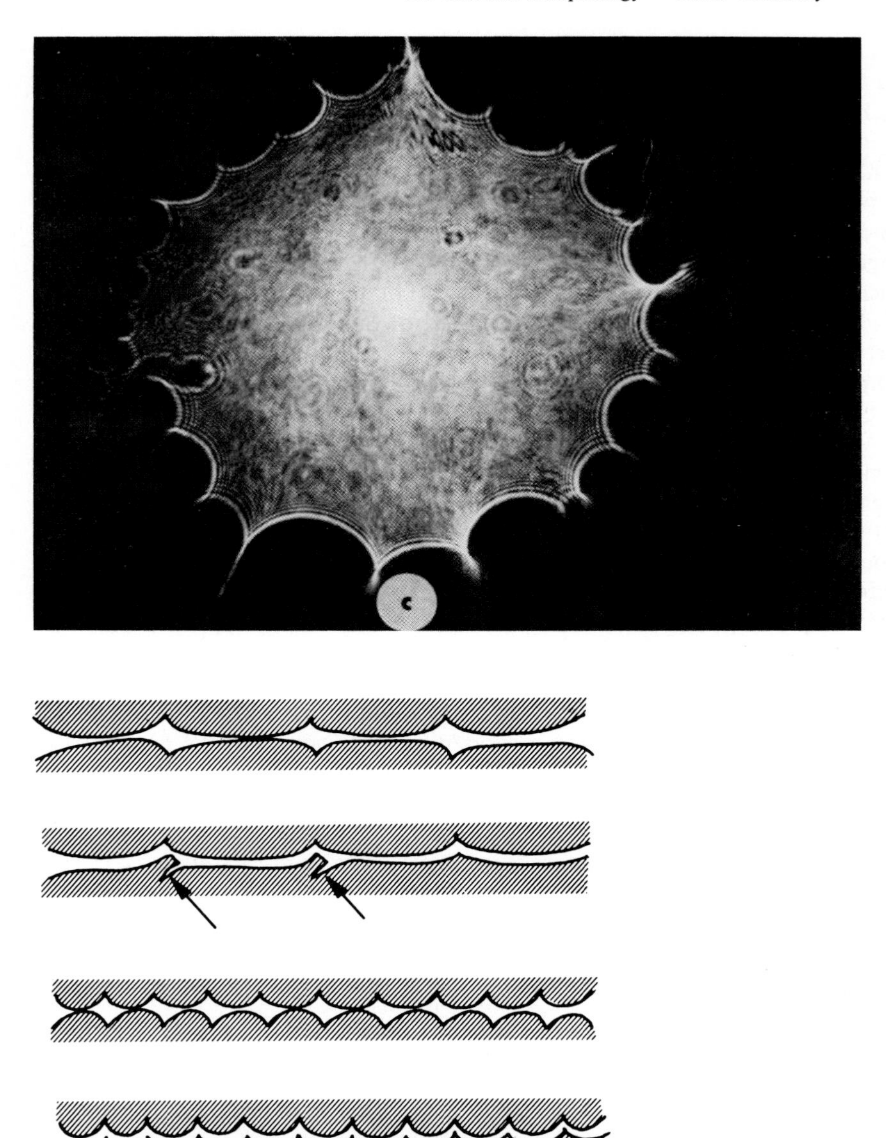

Fig. 2.20 a–d. Sections with sharp boundaries. **a** A symmetric cusp-like cross-section through rib marks in glass caused by a low energy impact. (Murgatroyd 1942). **b** The asymmetric cusp-like cross-section through rib marks in glass caused by thermally induced Hertzian fracture. The fracture origin is at the *extreme lower right side* of the profile. The width of the section is 3.3 mm. (Bahat 1977a). **c** A plan cusp-like pattern caused by far-field caustics from an irregular water droplet on a flat glass plate. (Berry 1976). **d** Four schematic profiles of ductile fatigue striations in a metallic specimen. *Arrows* show narrow kinks in profiles. The stress axis is vertical. (Laird 1967)

files (Fig. 2.20d). Hence, distinctions are expected between ripples from slow and rapid crack propagation, or between ripples that resulted from different crack processes. There is further elaboration of this topic in Section 3.1.3.2.

Ultrasonic Crack Modulations. Kerkhof (1973) attached a compound oscillator with a Y-cut quartz to the end of a glass plate in such a way that the continuous ultrasonic transverse waves emitted would cross the running crack orthogonally (Fig. 2.21). Consequently, the crack and wave interacted with each other on a parallel front at the same time. The resulting maximum principal tensile stress became a periodic function of distance and time and hence, the direction of the crack changed periodically due to a superimposed mode II component, and the profile assumed wave-like undulations which represented the crack front at regular intervals. A determination of crack velocity is given by: $V_b = \lambda_b f$, where V_b is crack velocity, λ_b is the wave length and f is the frequency. For instance, for a crack speed of 1000 m/s, the distance between two neighbouring fractographic traces of a wave with a frequency of $10^6\ H_z$ would be 1 mm. Surfaces of ultrasonically modulated cracks reveal that at low crack velocities distances between neighbouring ultrasonic lines are small and they gradually increase at higher velocities resembling previous observations by Murgatroyd. Experiments show that when the crack front approaches a zone of compressive stress, for example near an inclusion, there is a decrease in mode I and consequently the ultrasonic lines become very close to one another, indicating a decrease in crack velocity from about 0.7 to 0.1 m/s, and once the crack front crosses the inclusion a sudden increase occurs in the distance be-

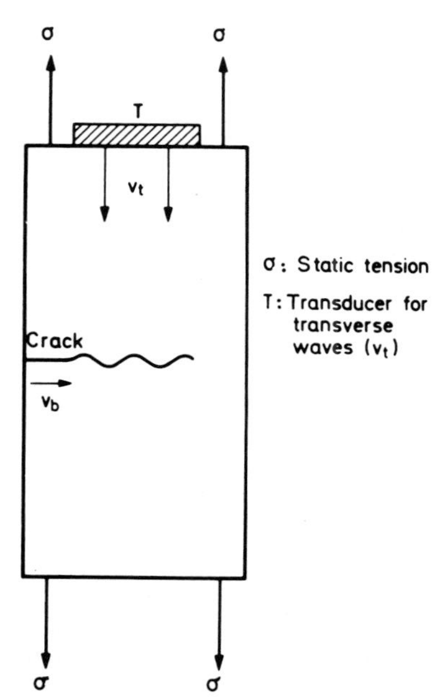

Fig. 2.21. Principle of crack modulation by transverse ultrasonic waves

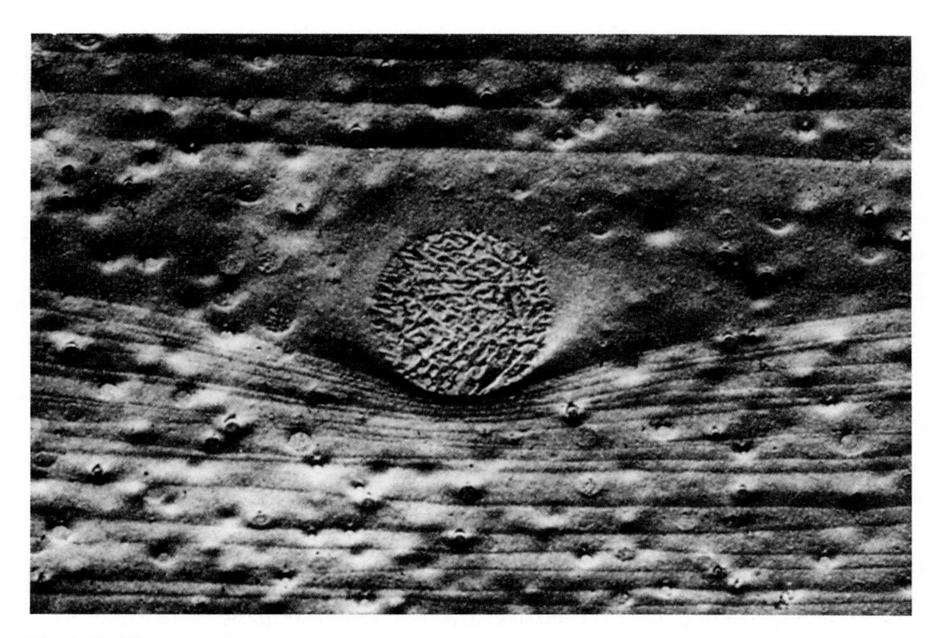

Fig. 2.22. Electron microscope photograph of an ultrasonically modulated crack surface in the neighbourhood of a spherical crystalline inclusion in opaque glass. Ultrasonic frequency, 1 MHz. Diameter of the large central inclusion, 38 mm. Direction of fracture propagation *from bottom to top* of the picture. Crack velocity decreases from about 70 to 10 cm/s at the rim of the inclusion

tween ultrasonic lines (Fig. 2.22). Note that generally in contour maps there is a reverse trend: an increase in distance between lines corresponds to a decrease in rate. The width of an ultrasonic line is about 10–40 μm and its height is ~0.7 μm (Green et al. 1977).

Kerkhof (1973) also used mechanical impact and pulse wave techniques in the application of wave fractography in his investigation of brittle fracture. With the aid of impact tensile loading, he managed to fracture glass at high energies. This provided him with the most characteristic morphological elements on a fractured surface, from the origin through concentric ultrasonic lines at increasing distances, to mist, hackles and bifurcation (Fig. 2.23a). The maximum distance between the ultrasonic undulations occurs in the mist zone, which he interpreted as an indication of maximum speed of propagation (about 1500 m/s, see Fig. 1.29b).

2.1.2.4 Fine and Coarse Hackles

De Freminville (1914, Fig. 56) and Preston (1926, 1931, Fig. 19b) distinguish between fine and coarse hackles at the periphery of the smooth fracture surface (the mirror plane) of a glass broken by a blow. Preston (1926) correlates the coarser hackle with the more "violent" parting of the glass. The term

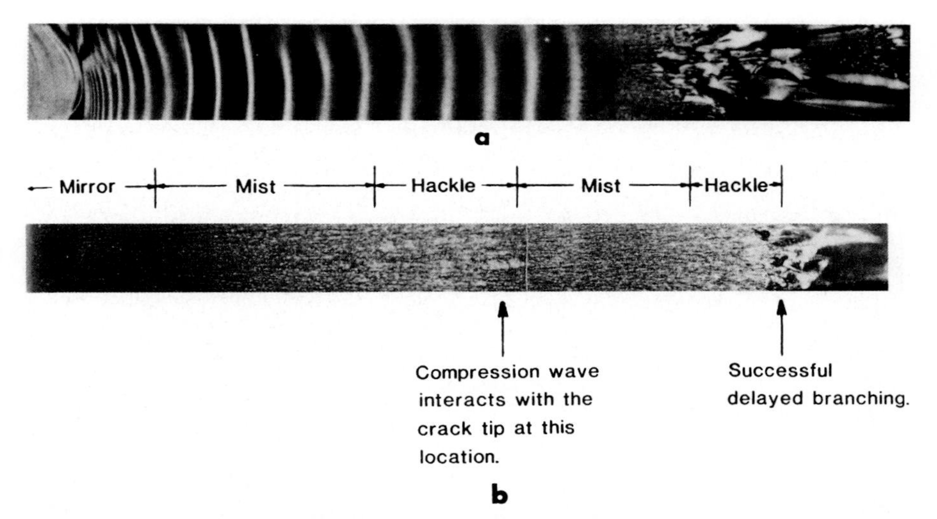

Fig. 2.23 a, b. Zones in the direction of fracture propagation from the origin through mirror, mist and hackle. **a** Ultrasonically modulated crack surface after impact tensile loading. Maximum crack velocity has already developed at the zone in the *middle* of the picture. At *right*, surface roughness, hackles and the commencement of branching are evident. Ultrasonic frequency, 1 MHz; thickness of plate, 4 mm, the height of the specimen in the photograph (Figs. 2.21, 2.22, 2.23 a from Kerkhoff 1973). **b** Repetition of mist and hackle in Homalite-100 due to the effect of a crack parallel compressive wave. (Ravi-Chandar and Knauss 1984 c)

violent seems to have been used by the early glass investigators, including Shand, in the absence of the more modern term stress intensity.

Preston (1926, Fig. 13) shows a photomicrograph (×44) of flaky morphology in glass with hackles and plumes on the fracture surfaces. This is accompanied by an interpretation that hackles represent incipient forking of the crack, the first stage in the development of radial branching. He suggests that the microscopic examination shows "that hackle consists of a fluting of the fracturing surface with the production of new fissures budding off from the primary one, and lying more or less parallel to it. If one of these fissures can get a sufficiently good start, so as to extend nearly from face to face of the glass, it can then extend lengthwise almost as well as the parent fissure. This can only happen at a peak of violence, when hackle has developed a great coarseness, and if one fissure has reached a development that can turn it into a fork the chances are good that a dozen others are equally ready". This indeed can be viewed as an "embryo" concept of microcrack nucleation parallel to the parent fracture, later adopted or modified by other scientists.

Johnson and Holloway (1968) identify two distinct types of microcrack in the mist zone on the fracture surfaces of soda lime silicate glass. One type predominates near the mirror boundary and the other is concentrated near the hackle boundary. Johnson and Holloway consider that both types of mist seem to occur where a small segment of the main crack divides leading to the development of surface and sub-surface cracks (Fig. 2.24).

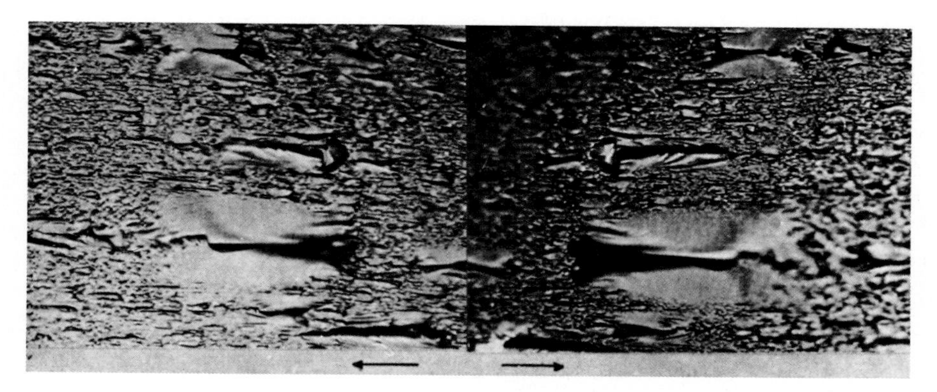

Fig. 2.24. Matching areas in mist zone on the two surfaces of a rod broken in bending. The *arrows* indicate the direction of fracture propagation, its length is 0.02 mm. A crack division into two depths in the material is revealed (and emphasized by *shadows*) along division lines of several large microcracks. (Johnson and Holloway 1968)

Ball et al. (1984) studied the mist region in a series of soda lime silica float glasses. They found the width of the mist region, W_{mist} to be inversely related to the square of the fracture stress σ_f which fits the mathematical prediction of Eq. (2.7) (see also Fig. 2.31):

$$W_{mist} = r - r_m = A^2 - A_m^2/\sigma_f^2 \; , \tag{2.6}$$

where r, r_m, A and A_m are the mist-hackle and mirror-mist boundaries and mist-hackle and mirror-mist constants, respectively. Rice (1984), on the other hand, observed that W_{mist} increases in fine-grain materials where hackle is multigrain in size, due to the reduced fracture energy associated with mist nucleation at grain boundaries. Ball et al. (1984) also analyzed the surface topography of the mist region by a computerized contour mapping technique and employed scanning electron microscopy to determine the surface roughness. They found that as the strength of the specimen increases, the surface roughness of the mist region also increases and the average root-mean-square roughness in the mist region is proportional to the square of the fracture stress.

Preston (1926) considered that the boundary between mirror and mist is quite indeterminate because "surfaces that appear polished to the naked eye are in fact hackly". This important problem has not yet been quite resolved. Various investigators (e.g. Speidel 1971; Carter 1971; Michalske 1979) suggest that hackles or microbranches can nucleate and develop when $K_I \ll K_{IC}$. It is, however, not always certain whether all authors who describe hackles within the mirror really refer to nucleation of hackles and not, in fact, to nucleation of striae. There is some confusion in the terminology (Sect. 2.2). For example, Ravi-Chandar and Knauss (1984a) examine fracture in Homalite-100. The mirror zone, which appears featureless under small magnifications, exhibits lines running in the direction of crack propagation (Fig. 2.25). These lines are not hackles, they are steps in the fracture plane that originated at voids.

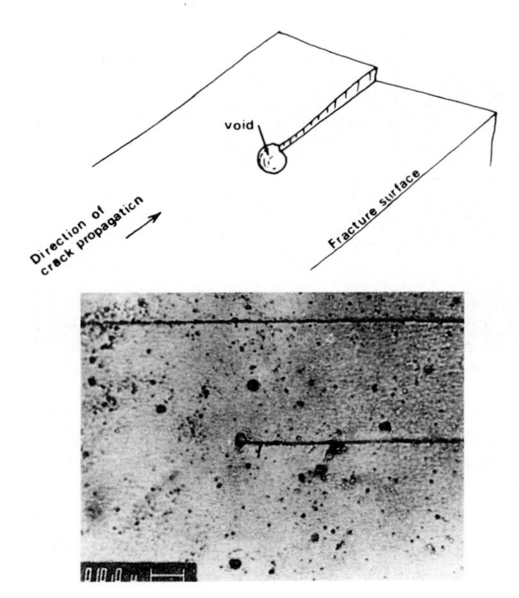

Fig. 2.25. A fracture surface in Homalite-100. A step develops from a void illustrated schematically above the micrograph. (Ravi-Chandar and Knauss 1984a)

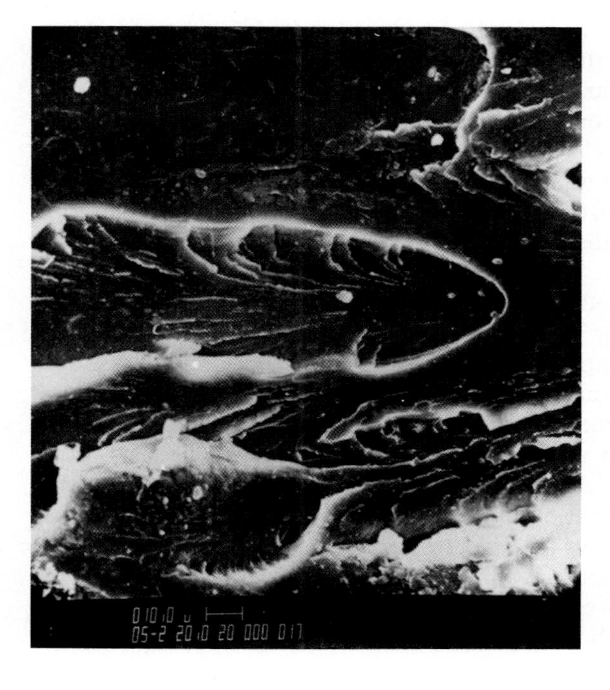

Fig. 2.26. Magnified view of the mist zone showing interacting plume-like hackles and parabolic markings. (Ravi-Chandar and Knauss 1984a)

In the mist and hackle zones Ravi-Chandar and Knauss (1984a) observed parabolic markings (Fig. 2.26) which resulted from small flaws that were activated by stresses operating ahead of the main propagating crack and which interacted with the crack front as it advanced. In the hackle zone these markings were generally larger and penetrated deeper into the material below the fracture surface than in the mist zone. It was seen that the maximum depth of the surface markings increased monotonically along the crack path. Since K_I also increases monotonically along the crack path, it is interpreted by the authors as the variation of fracture marking intensity with K_I.

Rice (1984) observed that in polycrystalline materials, the mist and hackle zones deviated from their typical morphologies in glass and became more flake-like. As grain size becomes substantially larger than the scale of the mist and hackle, these markings generally become less distinct. But he adds that "hackle can often still be discerned, at least at lower magnification, especially with transgranular fracture where mist appears to be replaced by fracture steps on fractured grains".

Rice (1984) points out that nucleation of mist on the scale of the grains, encourages initiation of mist at lower ratios of $r_m/2c_{cr}$ (Fig. 2.31), namely, under conditions of low fracture stress. The same possibly applies to hackles "as the grain diameter approaches the width of hackle ridges".

2.1.2.5 Additional Fracture Markings in Glasses and Polymers

There is a multitude of fracture markings that are known from technical (industrial) materials but they have only slightly or not at all been used in earth sciences. However, their potential application in tectonofractography should not be ruled out. Their brief characterization is given below.

Kies et al. (1950) expected to obtain various micromarking geometries on fracture surfaces resulting from interactions between neighbouring fracture elements. If the velocities of the straight fracture front of the primary crack and the circular fracture front of the secondary crack (which initiates at a point of discontinuity such as an inclusion) are the same, they propagate simultaneously out of plane at different depths in the material and their boundary is a parabola (Fig. 2.27a) (see Smekal 1950; Kerkhof 1975). A series of parabola-like fracture elements may appear as a succession of scallops, each scallop being the leading edge of a parabola. Kies et al. (1950) observed that the fracture front in cellulose acetate was being halted by the interacting parabola. Commensurately, a crack front would deviate from a continuous arc shape with lagging cusps at points where the fracture level changes and interaction with a parabola occurs (Fig. 2.27b). Kies et al. show parabola-shaped heads associated with a herringbone fracture cutting a cellulose acetate plate. A hyperbolic boundary develops when the secondary crack is faster than the primary one. An ellipse results when the secondary crack is slower than the primary, which runs ahead and engulfs it.

A fracture surface approaching a spherical inclusion or a bubble suffers from a perturbation due to the difference in the local stress conditions. The

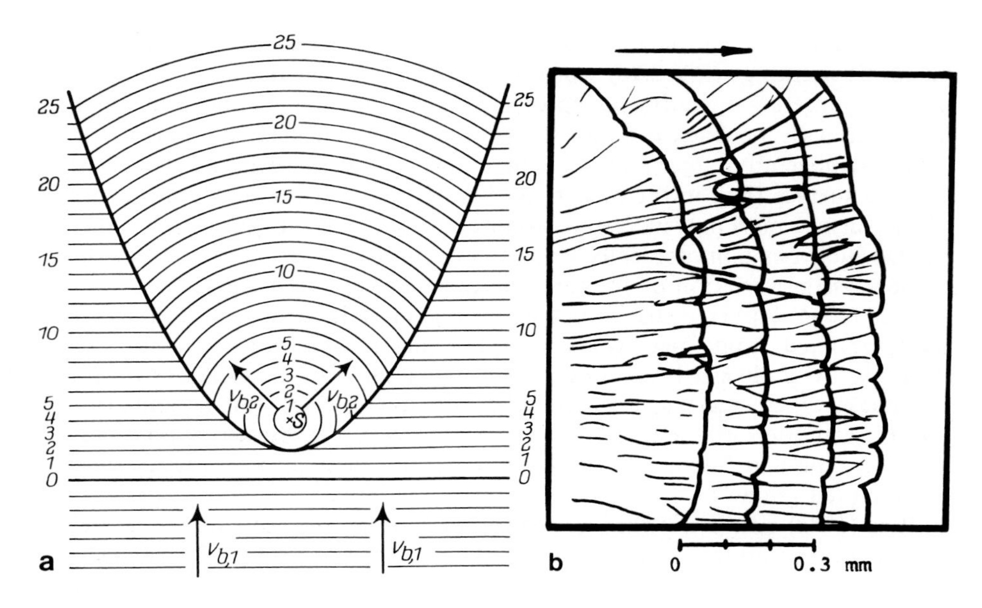

Fig. 2.27. a Conditions for the development of a parabola, as primary and secondary cracks intersect, when $V_{b,1}$ and $V_{b,2}$ are the crack velocities of the primary straight crack front moving upward in the figure and the secondary curved crack front that initiated at the disturbance *S*, maintaining the condition of $V_{b,1} = V_{b,2}$. *Numbers* designate time intervals. (Kerkhof 1975). **b** A fracture in cellulose acetate travelling to the right. Note the formed parabola-shaped heads with a focal point in one of them, and the lagging embayments along the crack front at locations where the microfracture level changes. (Kies et al. 1950). **c** Fracture surface intersecting a bubble and generating a texture of gull wings in its wake. The fracture propagates *from top to bottom* in the figure. The diameter of bubble is 0.25 mm. (Frechette 1972)

fracture front may reduce its velocity due to compression caused by the pressure of an inclusion (Fig. 2.22). Once past the discontinuity, the divided fracture front is once again restored into a single front. However, before doing so a cusp texture of "gull wings" develops in its wake (Fig. 2.27 c) because the wings tilt in opposite directions and they fail to meet at the wake (Frechette 1972).

Varner and Frechette (1971) and Quackenbush and Frechette (1978) revealed a wedge-shaped fracture marking symmetrically positioned about the mid plane of the sample (Fig. 2.28 a) in a sheet glass slowly broken at crack velocities between 2.5×10^{-7} and 5×10^{-6} m/s in organic liquids (alkanes). At the lower crack velocities in this range, the liquid, as it diffuses in from the nearest boundary of the specimen, can no longer maintain contact with the central portion of the advancing front of the fracture. The centre lags to allow additional time for the liquid to permeate the crack, resulting in negative curvature of crack front. As the negative curvature develops, there is a redistribution of the stress intensity and K increases at the centre relative to the edges. Gradually the stress builds up and fracture at the centre attains high velocities without the assistance of the liquid. Commensurately, the curvature changes from negative to positive. The wedge, also termed intersection scarp, separates outside areas which fail exclusively by the additional effect of the liquid, from inside areas where the fracture processes are liquid-free. Quackenbush and Frechette (1978) compare their measured result to other data on a log V vs. K diagram (Fig. 2.28 a) and arrive at the conclusion that the negative curvature in the moist environment occurs along region II and becomes positive again at the transition to region III. Region II (Fig. 1.26), in fact, is composed of components from region I outside the wedge and region III inside the wedge (Michalske 1984). According to Kerkhof (1975, after H. Richter), wedge-shaped fracture markings or "transition lines" develop in the velocity range 10^{-5} to 10^{-4} m/s and stress intensity range $15 \leq K_I \leq 25$ N/mm$^{3/2}$, which is $0.47 \leq K_I \leq 0.79$ MPa/m$^{1/2}$.

Michalske et al. (1978) investigated an intersection scarp that was formed at the boundary between wet and dry portions of a crack front which was propagating in an environment with limited access of water. They found that when the crack front encountered water on the specimen surface (Position 2 in Fig. 2.28 b), the wetted portion of the crack front jogged slightly ahead of the dry section because the wet section required less stress to propagate.

Michalske and Frechette (1980) studied fracture propagation at various velocities in glass plates and investigated the influence of water environment on the fracture process and surface morphology. They observed a concentric marking on the fracture surface, which they termed cavitation scarp (Fig. 2.28 c), which was generated at crack velocities of 10^{-2} to 10^{-1} m/s and which was indicative of a sudden increase in velocity to catastrophic levels (>1 m/s). The increase in crack velocity occurs when an accelerating water-filled crack breaks away from its liquid environment where it is held back by viscous forces, and attains the velocity corresponding to a dry environment for the same level of K_I. It is reasonable to imagine that the two sides of the scarp

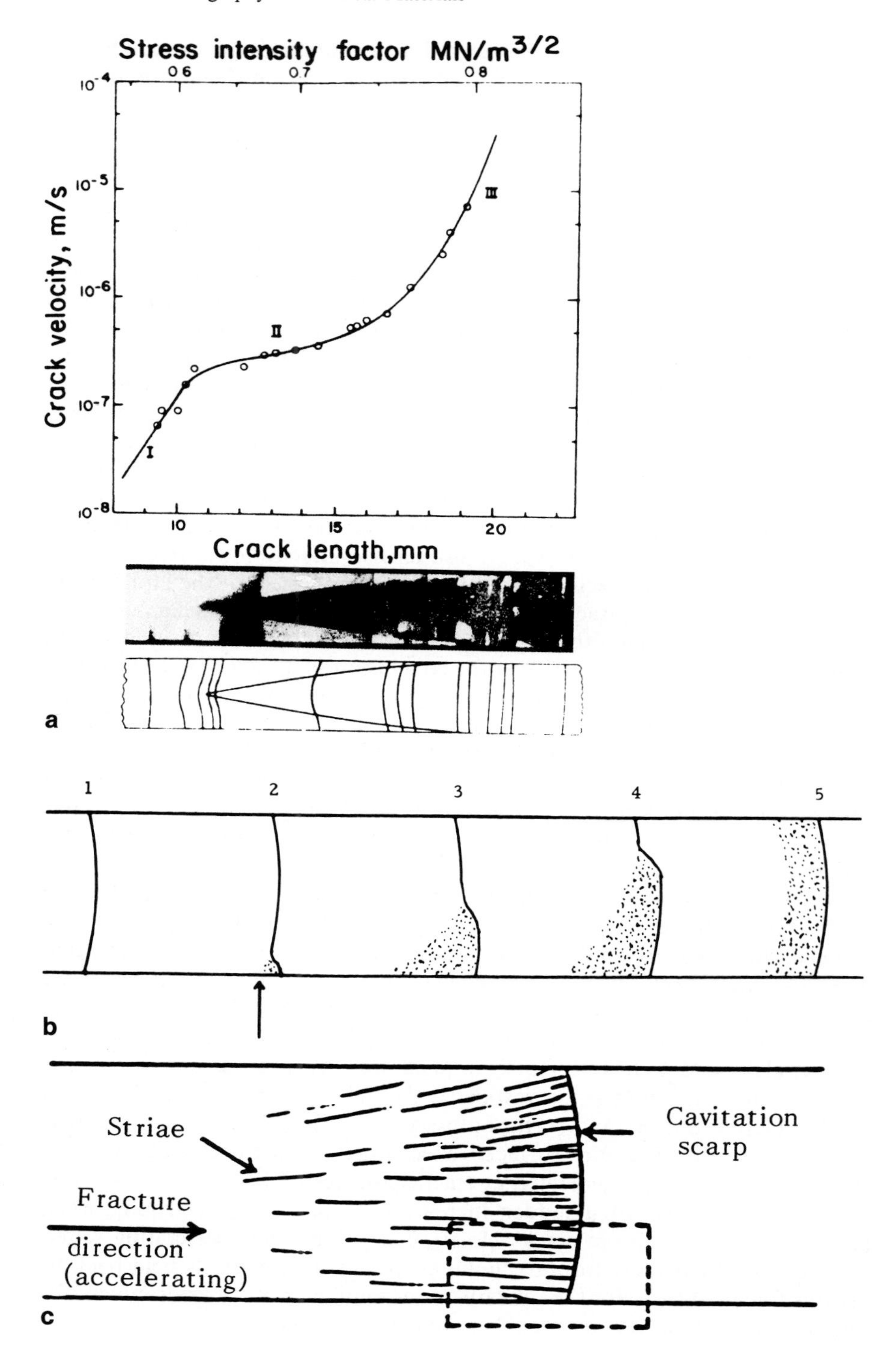
Stress intensity factor MN/m3/2
0.6
0.7
0.8
Crack velocity, m/s
10-4
10-5
10-6
10-7
10-8
I
II
III
10
15
20
Crack length,mm
a
1
2
3
4
5
b
Striae
Cavitation
scarp
Fracture
direction
(accelerating)
c

are somewhat out of plane with one another and that the scarp is the result of a convergence of these different planes that form morphological ridges on the fracture surfaces.

2.1.2.6 Markings on Fractures in Metals

In contrast to ceramic materials, most metals can exhibit considerable extents of plastic deformation under certain conditions of stress at ambient temperatures. Permanent elongation of tens of percent can often be obtained. However, under specific conditions of mode of stress, rate of application of stress, temperature of test, impurity content in the metal and chemical environment, the same metals can lose virtually all their ductility and fail in a brittle manner. Accordingly, metallurgists broadly classify fractures in metals as ductile or brittle. The morphology of the fractures formed in a ductile manner is quite different from that of brittle fracture. However, the ductile fractures can themselves vary, dependent on the metal or alloy and the conditions of the test. Likewise, considerable variations can be observed amongst brittle fractures. Whereas the critical dimension of a flaw required to cause catastrophic failure of a ceramic material or a rock is very small, often of the order of several tens of microns, that for metals is often several orders of magnitude greater. Flaws of critical dimensions are often intrinsic in glass, ceramics or rocks (Sects. 2.1.3.1 and 2.1.3.4) but are much less frequent in metals. The growth of flaws to critical size, which would result in ductile failure, expresses itself in a characteristic morphology. An increase in the load applied to a metallic specimen or component leads to the onset of plastic deformation, the formation of voids within the material, the linking of the voids and the final rupture of the item. It is essentially the lattice imperfections which allow the material to deform plastically. It is likewise the concentrations of these imperfections which, in turn, can prevent further deformation and in certain cases induce cracking (Honeycombe 1984, p. 33). Some metals of body-centred cubic B.C.C. or hexagonal close-packed H.C.P. structure, such as iron or titanium respectively, can fail under certain conditions, virtually without any plastic deformation, e.g. by brittle fracture. Since we have been considering the fracture morphologies of glasses and ceramics, perhaps it might be easier first to consider the morphologies of brittle fractures in metals.

◄

Fig. 2.28. a Wedge and fracture surface markings showing reversal of crack-front curvature (*lower figure*), fracture direction is to the *right*. The *black top and bottom boundaries* in the photograph (*central figure*) represent the surfaces of the specimen. Relation between K_I, and crack velocity (*upper figure*) of glass sheet broken in liquid decane. The scale of the crack lengths is common to both plot and photograph. (Quackenbush and Frechette 1978). **b** Stages (*1* to *5*) in fracture front development from a dry starter crack in a partly wetted specimen by double cantilever beam loading. *Stippled areas* indicate water trailing the crack front. *Arrow* shows initial area of water contact. **c** A schematic representation of the exposed fracture surface of glass accelerated through Region II/III transition in water. With increasing velocity, initially mirror smooth surface (*far left*) shows generation of striae, followed by a cavitation scarp and return to mirror smoothness at greater velocity (**b, c** from Michalske 1984)

Brittle Fracture. One form of brittle fracture in metals is termed cleavage. Cleavage fractures are bright to the naked eye and highly faceted or stepped when viewed under an optical or scanning electron mircoscope SEM. Figure 2.29a represents a view of the surface of a rapid fracture in a plate of S. A. E. 4130 steel heat treated to a tensile strength of 750 MPa. It is to be noted that the cleavage facets in all metals are flat, at least locally, due to their formation on specific crystallographic planes of the lattice (see also single crystals in Sect. 2.1.2.2).

Brittle fractures, at least on a macroscopic scale, are also obtained as a result of stressing under conditions of fatigue whether this be (1) by symmetrical reversed stressing, i.e. by the application of alternate cycles of tension and compression each of the same level of stress, or (2) by pulsating tensile stress, e.g. between two given levels of stress, or (3) by spectrum loading whereby the component is subjected to an often random mixture of tensile and compressive stresses of different and varied levels. The fractures so obtained can be quite complex. To the unaided eye, or at low magnifications, at least two and sometimes three zones can be observed on the fractured surface of a specimen which has failed due to fatigue, as in Fig. 2.29b, which shows a fatigue fracture in a spring made from a wire of stainless steel of type AISI 302. This photo-micrograph shows a "thumb-nail" zone which represents the progress of the fatigue crack on the surface of the fracture. Beyond this zone a radial structure can be observed representing the ductile failure of the specimen. Sometimes, although not in this specific case, the thumb-nail zone can itself be seen to consist of concentric bands known as beach marks representing stages of advance of the crack front. In the case of Fig. 2.29b, the point of origin of the fatigue crack is on the surface of the spring and virtually at the radial centre of the thumb-nail zone. The origin of the fatigue crack generally consists of some imperfection in the material, such as a particle of foreign non-metallic material, a gas pore, a corrosion pit or a surface indentation or irregularity, to name but a few possibilities. The formation and development of the fatigue crack, although considered to be a result of a complex movement of dislocations (Lynch 1979), is in fact accompanied by only a very small amount of macroplastic deformation, so that fatigue is considered as a brittle mode of fracture. However, once a critical length of crack is reached by stressing in fatigue, the component or specimen is no longer capable of sustaining the applied load and instantaneous failure. For initially ductile metals this final failure is also ductile (Fig. 2.29c) (see also Sect. 2.3.2.2).

Metallurgists distinguish between beach marks (Sect. 2.1.3.4), which are curved concentric arrest lines, and fatigue striations (Fig. 2.29d), which are generally straight and parallel but sometimes have slight curvatures, both being associated with fatigue fracture. Their main distinction is in their scale. Beach marks are of the order of millimetres to centimetres wide, whereas fatigue striations are in the range of 0.1 μm to several μm wide. Both types of marking represent the successive positions of the crack front due to cyclic stressing, therefore their direction is normal to that of the applied tensile stress.

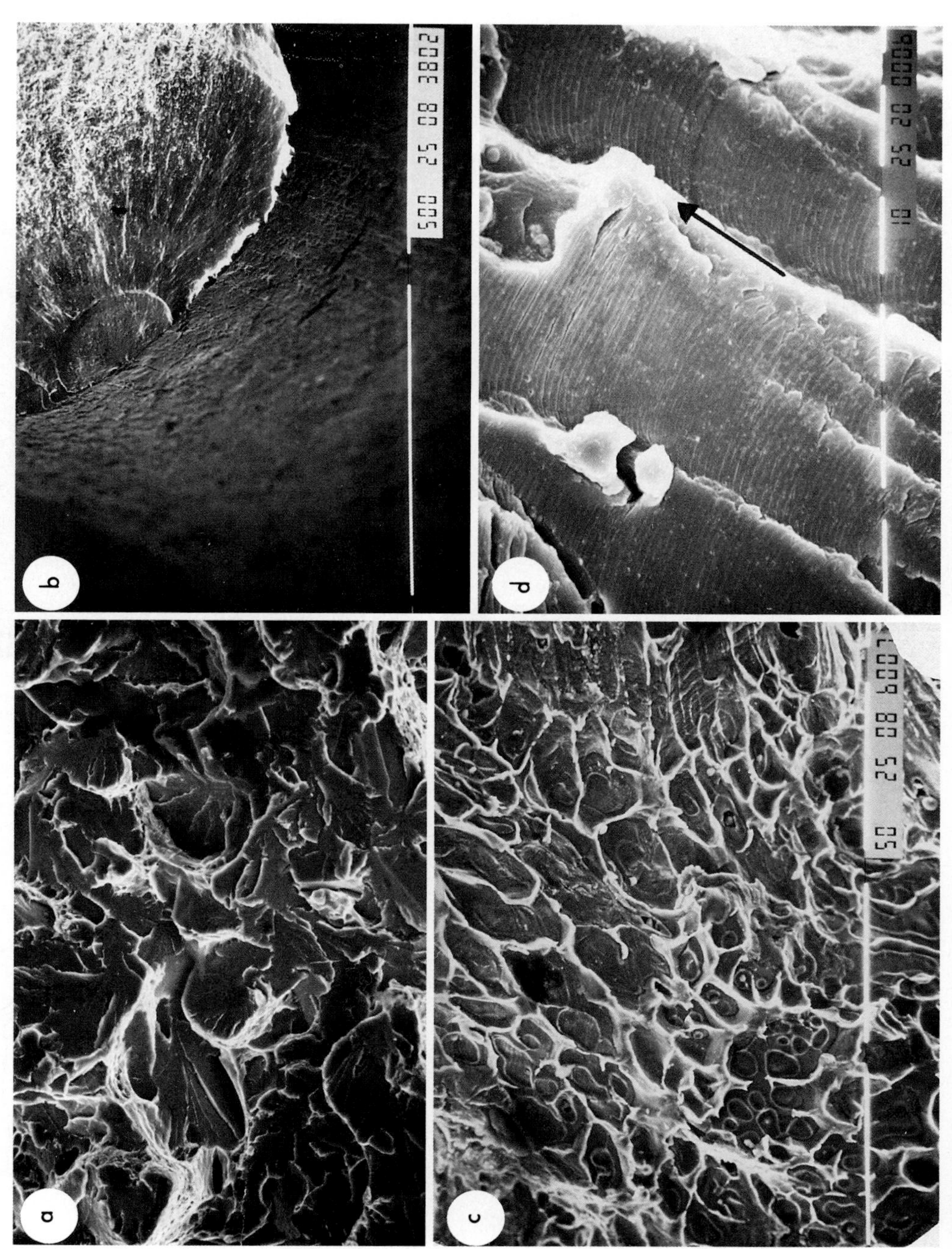

Fig. 2.29a–d

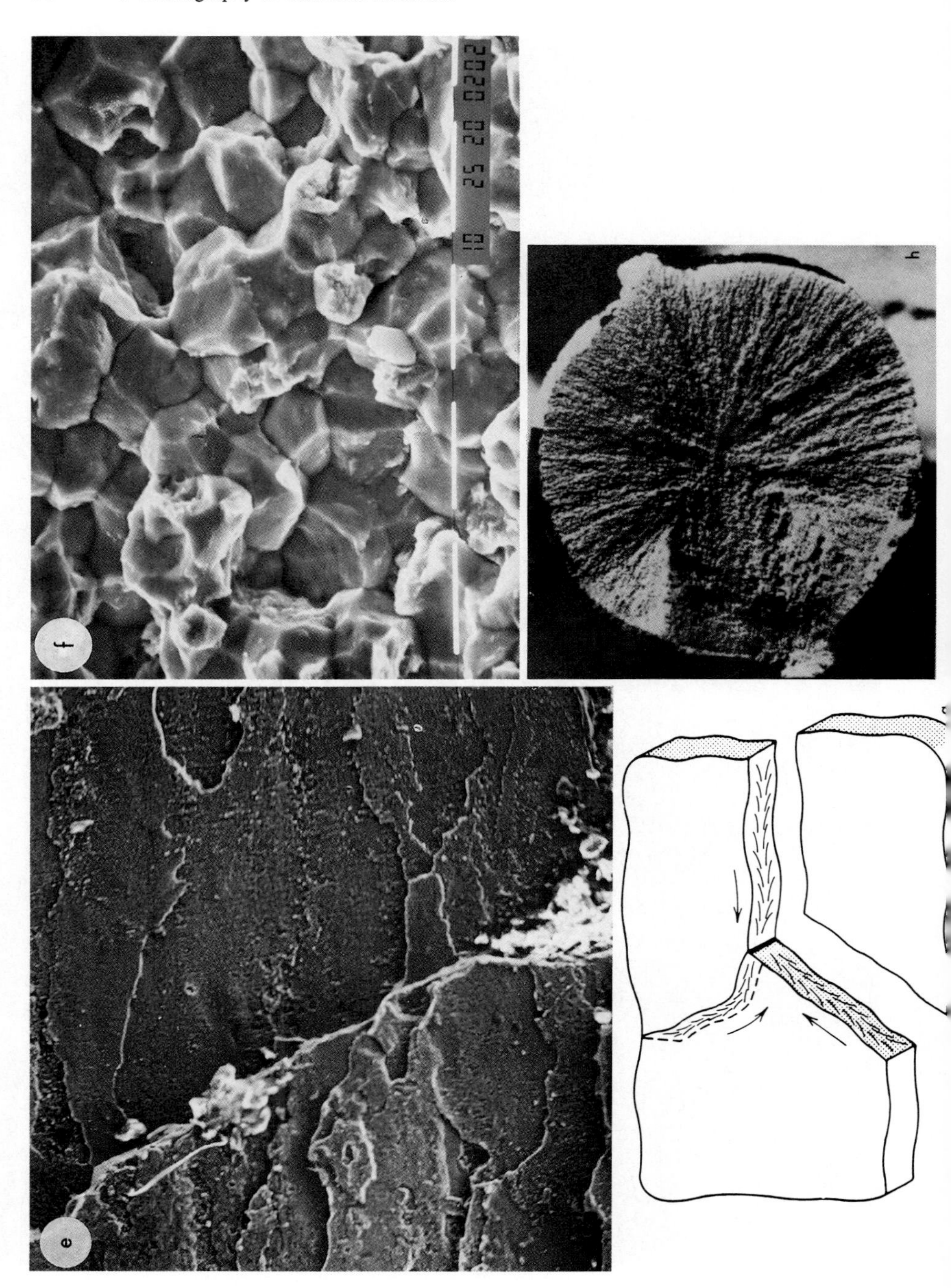

Each of the many fatigue striations (Fig. 2.29d) generally represents one stress cycle. Fracture propagation is incremental; load-rise results in crack growth, and load relaxation results in crack closure. The distance between striations can, on occasion, be used to determine crack velocity and it has also been suggested that stress magnitude can be deduced from their width (Luyckx 1980).

Two further examples of brittle fracture in metals which will be given here, and there are many more, are those due to stress corrosion and hydrogen embrittlement. The former is encountered in virtually all metals and alloys, while the latter is encountered primarily, although not exclusively, in low alloy steels of high tensile strength. Stress corrosion occurs due to the combined effects of an applied tensile stress and a corrosive environment where either factor alone would be insufficient to result in any significant damage to the material. In certain aluminium alloys of the Al-Cu-Mg (2000 type) and Al-Zn-Mg-Cu (7000 type) compositions, when heat-treated to a condition of maximum tensile strength, an applied tensile stress of the order of only about one tenth of the ultimate tensile strength of the material, and in the presence of moisture from the atmosphere, is sufficient to result in severe intergranular cracking due to stress corrosion as shown in Fig. 2.29e. The latter is a fracture of a flap actuator from an aircraft. The actuator, in the form of a cylinder, was manufactured from a 7075 type aluminium alloy heat treated to the T6 temper, which is that of maximum tensile stress but of minimum resistance to stress corrosion. Failure resulted from the combination of a hoop tensile stress of about 90 MPa, from hydraulic pressure experienced at the outside surface of the cylinder, and the unintended exposure of this surface to the atmosphere due to local failure of the protective paint system (Cina and Kaatz 1976). This is a brittle fracture. In austenitic stainless steels of the 18% Cr 8% Ni variety, cracking due to stress corrosion is transgranular but likewise brittle.

Hydrogen embrittlement is a cause of intergranular cracking in low alloy steels especially when these are heat-treated to tensile strengths of the order of

Fig. 2.29. a Fracture markings in steel. SEM micrograph of a cleavage fracture in a plate of low alloy steel. (Unpublished work by I. Eldror and B. Cina 1988a). *Bar* = 100 μm. **b** SEM micrograph of a fatigue fracture in the form of a thumb-nail zone in a spring made from stainless steel wire. (Unpublished work by I. Eldror and B. Cina 1988b). *Bar* = 500 μm. **c** SEM micrograph of the classical dimpled structure of a ductile failure in a casting made of stainless steel of composition to specification A.S.T.M. A 296. (Unpublished work by I. Eldror and S. Mehditash 1986). *Bar* = 50 μm. **d** SEM micrograph of fine striations on the fracture due to fatigue testing of an aircraft component made from an aluminium alloy of the 2124 type hardened by ageing at an elevated temperature. (Unpublished work by I. Eldror and B. Cina 1987). *Bar* = 10 μm. **e** SEM micrograph of an intergranular fracture due to stress corrosion of a hydraulic cylinder made from an aluminium alloy of the 7075-T6 type. (Cina and Kaatz 1976). *Bar* = 5 μm. **f** SEM micrograph of an intergranular fracture due to hydrogen embrittlement of a component of an aircraft brake made from a low alloy steel of the 4340 type electroplated with cadmium. (Unpublished work by I. Eldror and B. Cina 1989). *Bar* = 10 μm. **g** Schematic presentation of multiple chevron patterns in steel emanating from a crack origin. Each chevron points back (*arrows*) to the site of crack nucleation. (Hertzberg 1979). **h** Radial feathers (chevron markings) radiating from an internal flaw in a steel reinforcing bar 6 cm in diameter. (Courtesy of Roger Slutter, Lehigh University, from Hertzberg 1976)

1260 MPa or over. The hydrogen may be inherent in the material from its initial melting and casting, although these days this is rare, or it may be introduced to the steel during an electroplating operation, for example of cadmium or chromium, required for conferring resistance to corrosion. Failure results from applied triaxial stressing as, for example., at a threaded joint. A typical such failure (Fig. 2.29 f) is due to hydrogen embrittlement of a cadmium-plated component of an aircraft brake. The component was manufactured from a low alloy steel to specification S. A. E. 4340 heat treated to a tensile strength of 1520 MPa. An interesting feature of the latter failure is that fracture occurred at what were essentially the prior austenite grain boundaries of the material. The austenite phase of face-centred cubic F. C. C. crystalline structure was the allotropic form of the material when it was heat-treated at a temperature of about 900 °C. On quenching the material from 900 °C to room temperature, to harden it, the austenite phase would have transformed to martensite of body-centred tetragonal B. C. T. crystalline structure. Each austenite grain would have been replaced by possibly hundreds of martensite laths. The steel, however, still seems to "remember" its allotropic form at the elevated temperature and the hydrogen in the steel migrates to, and concentrates at the prior austenite grain boundaries, despite the fact that the austenite phase no longer exists. The increasing local gas pressure, due to the concentration of the hydrogen, combined with the applied triaxial stress, eventually results in cracking at those prior austenite grain boundaries.

Chevron or herringbone markings often characterize the fracture surfaces of metals (Fig. 2.29 g) and other materials (Kies et al. 1950). The morphology of these cleavage markings strongly resembles the straight plumes (Fig. 3.2 a) which commonly appear in rock exposures, although the latter are up to 10^5 times larger. Radial feathers that emerge from a single point on the fracture surface identify the origin of the cleavage fracture (Fig. 2.29 h). These have resemblance to radial plumes (Fig. 3.2 d) (see a further example of brittle fracture in Sect. 2.3.2.1).

Ductile Fracture. Ductile metals, when stressed in tension, fracture after having permanently extended in length and contracted in section. Both the extension and contraction are generally non-uniform, both being of greater extent towards the centre of the gauge length of the test specimen. The central portion of the fracture surface appears perpendicular to the direction of the tensile force, while the outer portion of the fracture for a specimen of circular section can form a cone inclined at an angle of approximately 45° with respect to the axis of the specimen. The central portion of the fracture represents tensile deformation, whereas the outer cone is due to shear, both forming the well-known cup and cone fracture. A cup and cone fracture that resulted from an abrupt tensile failure is shown in Fig. 2.41.

In metals showing ductility, plastic deformation often results in the development of microvoids of μm dimension, which form along particle-metal matrix interfaces. Under increasing stress, these microvoids coalesce and ultimately may join to form a crack. The microvoids appear on the surface of the crack as dimples. The latter may acquire circular, elongated and twisted shapes (Beachem 1975; Luyckx 1980). The shape of the dimple can help in giving an indication of the direction of the applied stress and of that of fracture propagation, relative stress magnitudes at the crack tip, and some idea of the fracture modes. The tips of elongated dimples usually point in the direction of fracture propagation. A rough dimpled structure results from a plastic deformation when a brittle crack suddenly transforms into a rapid ductile fracture (Fig. 2.29c).

When a flat specimen of mild steel is given a mirror-like polish and then tested either in tension or in compression, on reaching the yield point lines or stripes gradually appear on the polished surface inclined at about 45° to the direction of the applied stress and approximately coinciding with the planes of principal shear stress. These markings are known as Luders' bands, Luders' lines or as flow strain figures (Nadai 1950). The lines, which are several mm in width and seem to vary proportionately with the thickness of the specimen develop by the slip of layers of material stressed beyond the plastic limit of the steel. They appear as shallow grooves when failure occurs in tension and low ridges in compression. On a microscale these lines consist of very large groups of displaced atoms and even grains. Numerous slip bands, whose number increases with stress, appear on the surface of the grains. The width of the individual slip bands can be about 100 times smaller than the width of the Luders' lines. The orientation of the bands in the individual grains is determined by the crystallographic directions in the grains (see further reference to this subject in Sect. 3.2.1.3).

2.1.3 The Quantitative Mirror Plane

2.1.3.1 General

Following the spectacular experimental fracture results (De Freminville 1914) and the recognition that a fracture surface constitutes specific morphological features in a logical order (Preston 1931), further realization that the stippled perimeter of the smooth mirror surface defines its boundary (Figs. 2.30a, 2.31) (Smekal 1936, 1940; Terao 1953; Shand 1954) set the way for the estimation of fracture stress by fractography.

Smekal (1936) broke circular glass rods, about 1.4 mm in diameter in tension, and found that when the breaking load was divided by the remaining area outside the mirror, the nominal breaking stress was essentially constant for a fairly wide range of mirror sizes. Terao (1953) observed that the breaking stress varied quite consistently with the reciprocal of the square root of the mirror radius. Shand (1954, 1959) confirmed these relationships for glass and glass ceramics, while Kirchner and Gruver (1973) and Rice (1974) extended them to polycrystalline materials and single crystals.

Fractures in glass and ceramic bodies and in rocks propagate from pre-existing flaws. Those which are responsible for failure are the severest. In polycrystalline non-metallic materials they are usually taken to be the longest

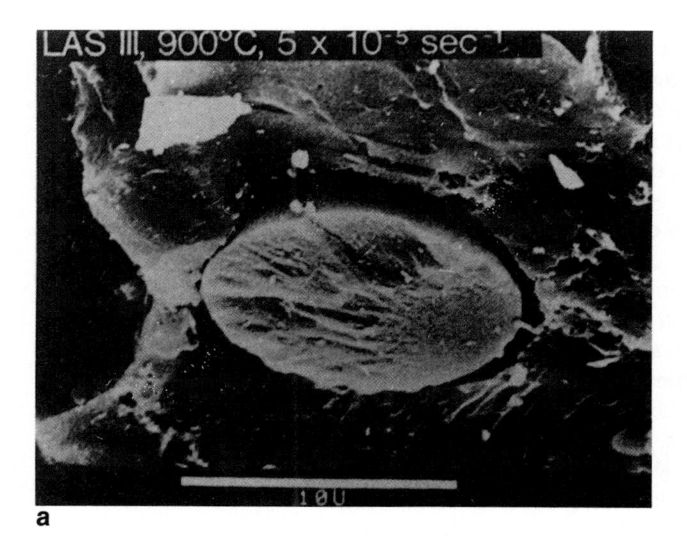

a

b

Fig. 2.30. a. Tensile fracture surface of a SiC fibre of about 10 μm diameter showing mirror, mist and hackle in a reinforced glass-ceramic composite. (Stewart et al. 1986). **b** Early curved flaw with concentric undulations *A* and radial scars within the flaw. Also radial fractures are shown on the mirror plane (*upper part* of picture). These markings reveal that the initial crack growth is controlled by a heterogeneous stress field. Length of flaw diameter is 0.9 mm. *B*-flaw to mirror transition curve at the tip of the critical flaw. (Bahat et al. 1982)

and straightest grain boundaries (Friedman and Logan 1970a). Such a grain boundary is designated as the initial flaw length $2c_i$, being distinct from the critical flaw length $2c_{cr}$ (Fig. 2.31). A fracture mirror plane extends from the critical flaw to three differently defined radii, r_m (the mirror-mist boundary); r (the mist-hackle boundary); and r_b (initiation of macroscopic crack branching) (Mecholsky and Freiman 1979). Many investigators (e.g. Levengood 1958) have obtained the semi-empirical formula:

$$\sigma_f r_j^{1/2} = A_j \ , \tag{2.7}$$

where σ_f is the fracture stress, r_j is the distance to a particular boundary, and A_j is the corresponding mirror constant. An important aspect of this equation is its validity for tensile, flexural and biaxial (tension) loading, and it is useful in the study of mixed modes which renders it applicable to the analysis of shaped bodies under complicated stress states (Mecholsky and Rice 1984). Naturally, there are some exceptions, and these include cases of large stress gradients and materials of very low or very high strengths.

Rice (1984) remarks that "while there is overall general similarity between the fractures on glass and polycrystalline fractures, there are a number of important differences". Decreasing strengths with increasing grain and pore sizes (and total porosity) reduce the density and often the clarity of fracture features. This effect, combined with increasing roughness with larger grain and pore sizes, typically makes the mirror, mist and even some hackle less discernable, in the extreme case obliterating all normally observed fracture features. In some ceramic materials these changes cause hackles to appear as distinct sharp ridges (Fig. 2.32). These ridges are termed here cuspate hackles. Increasing roughness of intergranular fracture accentuates these trends. Rice also mentions that other heterogeneities, for example, microcracking and second phases, can sufficiently perturb the crack front so as to make the various crack

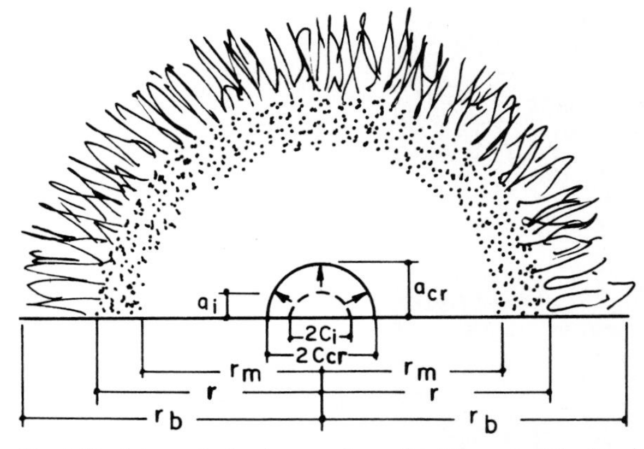

Fig. 2.31. Schematic fracture surface of brittle materials showing idealized initial flaw length $2c_{ci}$ and depth a_i, critical flaw length $2c_{cr}$ and depth a_{cr}. The three mirror radii, r_m (mirror-mist boundary), r (mist-hackle boundary) and r_b (initiation of microscopic crack branching) are shown as well. (After Mecholsky and Freiman 1979)

morphologies difficult, or impossible, to detect on polycrystalline fractures. "The only fracture feature that may be clearly observed (if the specimen is sufficiently large relative to the strength of the sample) is macroscopic crack branching"... "On the other hand, where a reasonable amount (25% or more transgranular failure) occurs, one can commonly identify the mirror, mist and hackle of polycrystalline features". The relevance of this description to rock investigators is shown in Sections 3.1.6.6 and 4.2.1.5.

Mecholsky et al. (1976) show that the elastic modulus, E, is proportional to the mirror constant, A, and probably to the critical fracture energy, γ, but that the latter is highly dependent on local microstructure. Good correlations between E and A were also found by Kirchner et al. (1976) for several hot-pressed and reaction-bonded ceramics. Kirchner and Gruver (1973) verified the relationship $\sigma_r^{1/2}$ = constant for alumina ceramics at room temperature and at elevated temperatures.

Kirchner and Gruver (1974) used fracture stress and measurements of mirror radius to estimate the reduction of residual surface stress resulting from annealing of flint glass and to estimate the residual stresses induced in steatite and silicon-carbide by quenching. Conway and Mecholsky (1989) use crack branching data for measuring near-surface residual stresses in tempered glass. Kirchner et al. (1975) used specific features of the fracture mirror to interpret impact fractures in brittle materials.

Gulati et al. (1986a, b) showed that in fractured glass ceramics there was a systematic decrease of the mirror radius (which corresponded with higher values of modulus of rapture) in specimens tested at the temperature of liquid

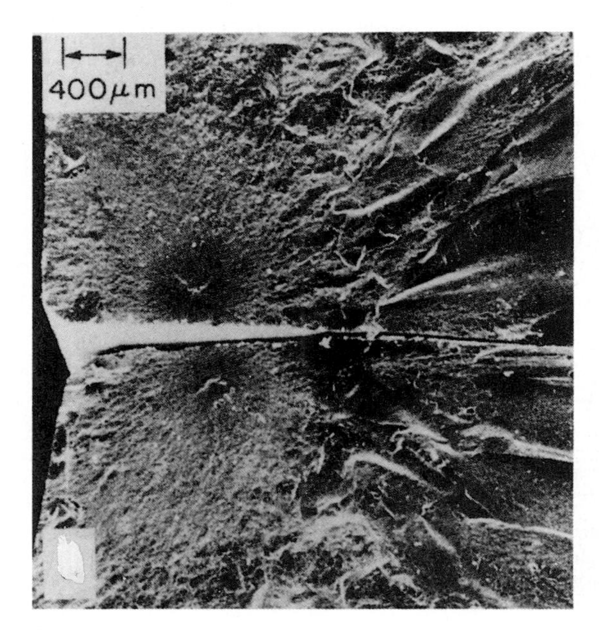

Fig. 2.32. The origin of a mirror fracture and distinct sharp hackle ridges which are termed cuspate hackles on two matching halves of a fracture of $BaTiO_3$. (Rice 1974)

nitrogen compared with those tested at room temperature and attributed this to a difference in susceptibility to stress corrosion.

2.1.3.2 The Shape of the Mirror Plane

The mist, hackle and branching boundaries in isotropic materials are circular when formed whether under uniform tension (Fig. 2.33 a) or at a corner (Fig. 2.33 b). The mirror boundary is seriously affected by stress gradients, secondary defects, edge effects, and the ratio of stress to specimen size. Shand (1954) showed that for a rectangular body in bending, due to the decrease in

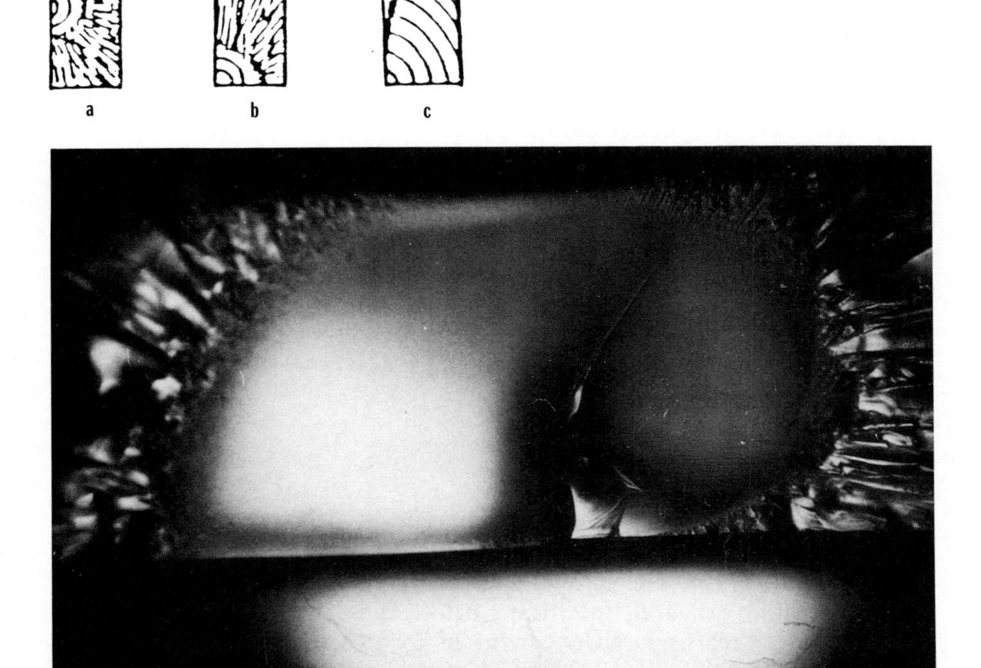

Fig. 2.33 a–d. Mirror shapes. **a** A circular mirror formed under uniform tensile stress. **b** A circular mirror in a corner fracture, in tension. **c** A non-circular mirror caused by bending (mirrors **a** to **c** from Orr 1972). **d** A bilateral asymmetrical mirror in a bottle fractured by internal pressure due to a biaxial stress. The thickness of the bottle wall (horizontal boundaries) is 3 mm. (Leonard 1979)

the tensile stress inwards from the surface, there is a transition from a semicircular to a semielliptical crack (Fig. 2.19), which is the reverse of that generally expected under uniform tension (Shand 1965). In bending, a mirror deviates from a circular shape also in a corner fracture (Fig. 2.33c). Stress gradients often result in asymmetric or non-circular mirrors with morphological features such as mist and hackle being absent from various parts of the fractured surface (Johnson and Holloway 1968). Bending may also result in bilateral asymmetric mirror (Fig. 2.33d) or even separate mirrors rather than a single one (Kirchner and Kirchner 1979). Hagan et al. (1979, Fig. 3), however, found that when fracture is rapid, for example at a deflection rate of 10 mm/min in four point bending, the mirror is almost semi-circular and the mist and hackle zones are elongated extending into the material rather than forming wedges.

In tempered (toughened) glass the mirror in the fracture surface is semi-elliptical (Shand 1967) when the fracture is slow or the fracture stress is low, but becomes more circular with increasing the rate of deflection or at high fracture stress (Hagan et al. 1979). Hagan et al. observed that in annealed glass the radius r_s of the mirror on the tensile surface is always less than the radius r_b of the mirror extending into the bulk of the sample (where r_b/r_s lies between one and two) but in thermally toughened glass this ratio is less than or equal to unity.

Shand (1967) also pointed out the effects of a low ratio of stress to specimen size, edge break and an elongated scratch on the mirror boundaries. According to Holloway (1986) for simple bending, complete arcs of mist will have radii $\leq t/6$, where t is the thickness of the specimen. The combined effects of the stress gradient and the surface edge on the shape of the mirror zone in glass were treated by Kirchner and Kirchner (1979) and by Bahat et al. (1982).

Comparatively little has been written on circular mirror planes as related to sub-surface origin in glass and ceramics (Fig. 2.34). A few examples of contributions related to this subject are those by Rice (1974), Bansal and Duckworth (1977) and Leonard (1979).

2.1.3.3 Repetition of Fracture Markings

De Freminville (1914, Fig. 36) and Preston (1926, Fig. 12) obtained mirrors with the repetition of mist, specifically before and after hackles. Shetty et al. (1980) predicted mathematically conditions for two separated mirror regions and mirror boundaries in a fractured surface. Ravi-Chandar and Knauss (1984c) produced a fracture surface in Homalite-100 which resulted from the interaction with parallel stress waves. This interaction resulted in the formation of alternating zones; mist, hackle, mist, hackle along the propagating crack (Fig. 2.23b). The second mist zone between the two hackle regions resulted from the compressive interaction with the crack tip. Rice (1984) suggests an additional possibility, that "repetition of mist and hackle means they continue on crack-branching surfaces".

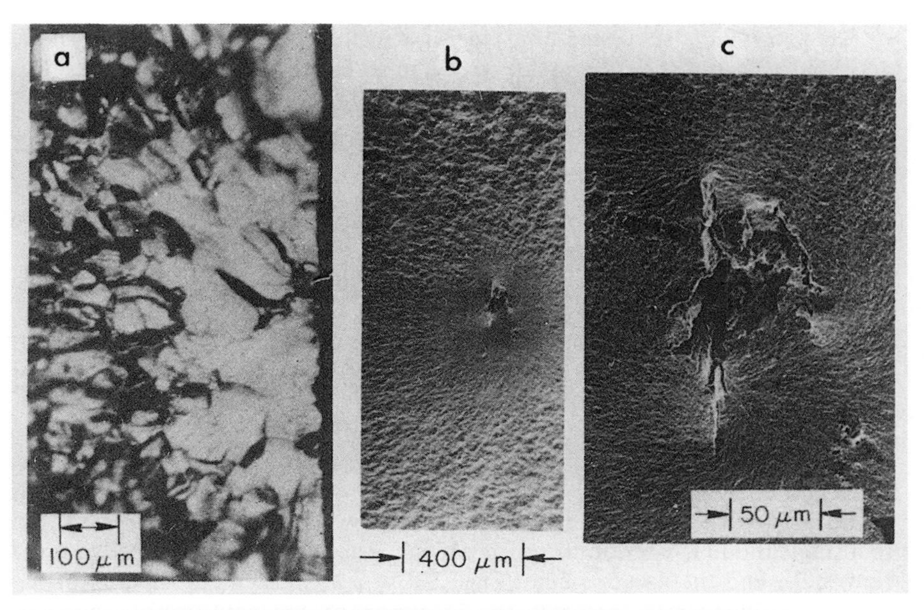

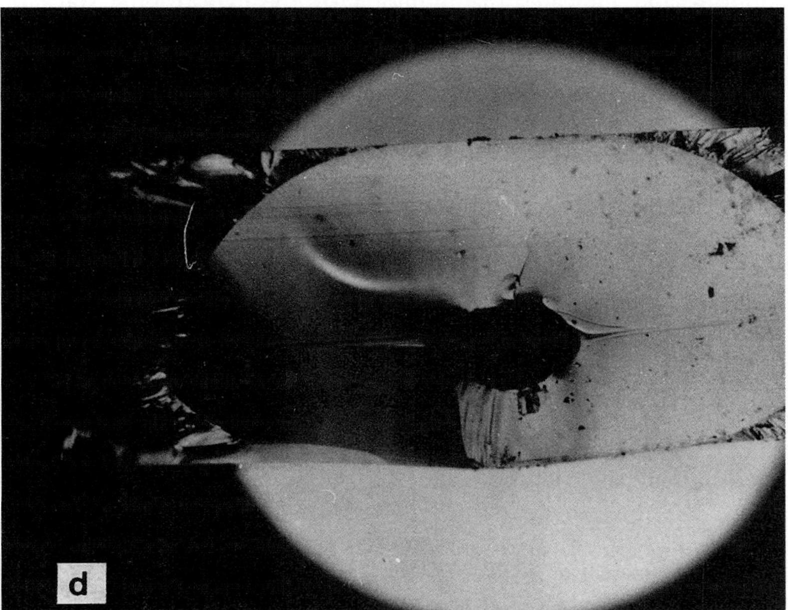

Fig. 2.34 a–d. Inner mirrors in ceramic and glass bodies. **a** Optical photograph of a clearer than average inner mirror in fully stabilized ZrO_2; note probable flaw origin (*arrowed*) $\sigma_f = 42\,000$ psi. **b** SEM of partially stabilized ZrO_2 which failed from a large internal pore. Failure stress at tensile surface (*right edge*) ~105000 psi, at pore ~85000 psi. **c** Higher magnification SEM of pore in **b**. Note the fine grainy fracture steps leading into sharp areas at top and bottom of pore, indicating fracture initiation from these areas. (Rice 1974). **d** Quasi-elliptical mirror developing from a subsurface crystal acting as a critical flaw, observed across the fractured wall of a glass bottle 3 mm thick. (Leonard 1979)

Shand (19967) suggested that in an asymmetrical mirror it is usually desirable to favour the shorter side to obtain an estimate of fracture stress [Eq. (2.7)].

2.1.3.4 Flaw Size and Shape

The term flaw shape should be distinguished from the term crack tip shape commonly associated with stress corrosion (Sect. 1.4.1.4). Doremus and Johnson (1978) suggested that the diameters of the flaw circles intersecting the final crack plane are not the depths of the crack-initiating flaw but result from the early stages of crack propagation. Common concentric sharp undulations in the curved flaw (Fig. 2.30b) indicate that there is an incremental growth of the flaw, $2c_i$, before the transition. These are analogous to beach marks (Sect. 2.1.2.6). The common occurrence of radial scars within the flaw indicates a heterogeneous field of mixed mode stresses on the curved surface, whereas radial scars on the mirror plane can be attributed to the inertia of mixed mode fracture propagating from the flaw surface prior to a transition to a single mode I stress on the mirror plane. The sharp transition between the flaw, $2c_{cr}$, and the mirror plane (Figs. 2.30b, 2.31) designates the loci of K_{Ic} stress conditions and indicates the initiation of the catastrophic stage of failure, namely, an abrupt transition to rapid fracture propagation.

Krohn and Hasselman (1971) showed the relation of flaw size to mirror in the fracture of glass, and Bansal (1977) demonstrated the importance of the overall flaw shape in the fracture behaviour, which led to a modified Griffith equation (Irwin 1962; Bansal 1976):

$$\sigma_f = Z/Y(2E\gamma)^{1/2}/c_{cr}^{1/2} = (Z/Y)(K_{Ic}/c_{cr}^{1/2}) \, , \tag{2.8}$$

where E and γ are the Young's modulus and fracture surface energy respectively, σ_f is the fracture stress at the origin, Z and Y are geometric constants, c_{cr} is the shorter radius of the flaw at the onset of catastrophic fracture (the flaw depth in the case of a surface flaw and the half depth in the case of a subsurface flaw) and $K_{Ic} = 2\gamma E)^{1/2}$. The combination of Eqs. (2.7) and (2.8) gives (Bansal and Duckworth 1977):

$$(Z/Y)(r/c_{cr})^{1/2} = A/K_{Ic} \, . \tag{2.9}$$

Krohn and Hasselman (1971) also suggested that the ratio between A (a mirror parameter) and K_{Ic} (a parameter associated with the flaw perimeter) is an absolute constant independent of type of material. Bansal (1977) found that this constant is 2.32 for soda lime glass.

The concept of a constant A/K_{Ic} ratio should be considered valid only when mechanisms that increase crack resistance with crack length (such as transformation-toughening) are excluded (Kirchner et al. 1981).

Bansal and Duckworth (1977) make several distinctions regarding the flaw dimensionless parameters: in the case of flaws much smaller than the dimensions of the specimen, $Y \simeq 2.0$ for a surface flaw and 1.77 for a subsurface flaw; Z, the flaw-shape parameter, is the product of two independent dimensionless parameters, Z_e and Z_d. The parameter Z_e varies with the flaw shape in the fracture plane (as shown in Fig. 2 by Bansal 1976). For failures from two-dimensional flaws with sharp tips, $Z_d \simeq 1$. In the case of three-dimensional flaws (e.g. pores and inclusions) of radius R, Z_d varies with the ratio $L/2R$, where L is the length of the sharp-tipped radial crack which extends from the flaw periphery in the fracture plane. It can be seen (from Fig. 7 by Evans and Tappin 1972) that Z_d, hence Z, becomes large when $L/2R$ is small, indicating that an inclusion or pore, per se, is not a severe flaw. It becomes severe only when a radial crack extends from it. Bansal and Duckworth (1977) show the difference between the initial flaw (in their case a subsurface pore) and the critical flaw in a fractured sintered Al_2O_3 (Fig. 2.35a). They also show how several sub-critical flaws may be combined into a critical one (Fig. 2.35b).

2.1.3.5 Mirror to Flaw Size Ratio

Mecholsky et al. (1974, 1976) found that the outer mirror r, to flaw size ratio is shown to vary about a value of 13:1. Thus the mirror constants are used to predict critical flaw sizes in glass and ceramic materials. The cases in which poorer correlation is obtained are those in which flaw sizes are smaller than the grain size, flaws where individual significant pores are surrounded by further porous regions or where severe microcracking exists. Contrary to these observations, Hagan et al. (1979), using flaws induced by indentations, found that the ratio of the radius of the mirror to the radius of the initial flaw size on the tensile surface varied from 55 to 8, depending on the initial flaw size, the rate of deflection and the ambient environment. On the other hand, Marshall et al. (1980) obtained about 5 for the latter ratio when testing artificially induced surface flaws in glass. Rice (1979) observed smaller mirror to flaw-size ratios in polycrystalline bodies compared with those in glass and offered an explanation (Rice 1984), based on the assumption that mirror sizes are linearly related to fracture energies. Accordingly, mirror-to-flaw-size (r_m/c_{cr}, Fig. 2.31) ratios should be proportional to the ratio of fracture energy for crack propagation (γ_p) before branching to that at the initiation of mist (γ_m):

$$(r_m/c_{cr})(\gamma_p/\gamma_m) = \text{ a constant } . \tag{2.10}$$

For glasses $\gamma_p/\gamma_m = 1/2$ (microbranching multiplies γ by 2) and $r_m/c_{cr} \simeq 12-13$. γ_p/γ_m for polycrystals varies from 3/4 to 20/21 because nucleation of mist involves low γ at grain boundaries and this gives $r_m/c_{cr} \sim 6$ to 8, for polycrystals in good agreement with experimental observations (Rice 1984). It follows that, due to the readiness of mist formation in polycrystals since γ is reduced, the onset of hackles is closer to the fracture origin (see comments on the width of the mist region in Sect. 2.1.2.4). This trend is intensified with the increase of grain size.

A poor distinction between c_i and c_{cr} (Figs. 2.31 and 2.35b) is possibly an important factor contributing to the above discrepancies in the ratio of mirror to flaw size.

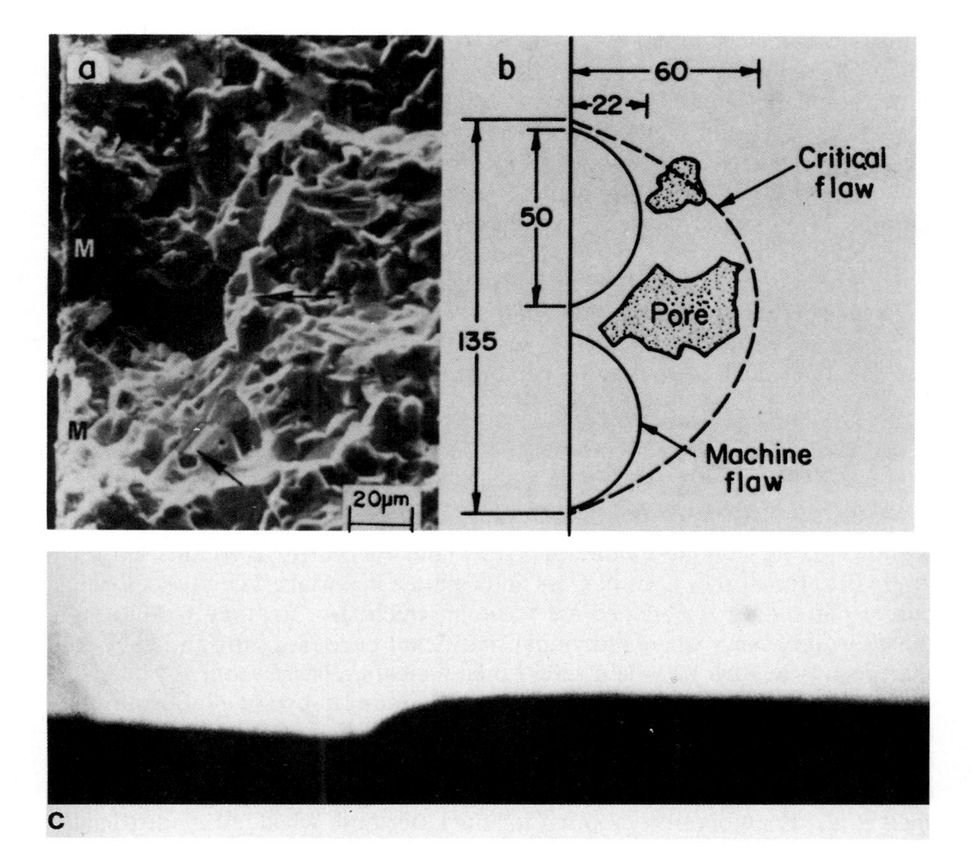

Fig. 2.35 a – c. Sintered Al_2O_3 tested in dry N_2. **a** SEM micrograph shows two "surface" semi-circular machine flaws at *M*, two pores (*dark areas*) one of which is quite large and *arrowed* critical flaw. **b** Schematic view of fracture initiation in **a** from the interlinking of two "surface" artificially machined flaws. Each flaw is ~20 μm deep in the tensile surface, and two sub-surface pores linked up to give the critical-flaw boundary which is ~60 μm deep and 135 μm wide. (**a, b** from Bansal and Duckworth 1977). **c** The critical flaw is at an angle to the mirror and produces a cusp at the approximately 1-mm-long tensile surface in a fractured glass bottle (the glass is *white* in picture). Such a situation often results in a bilateral asymmetric mirror (Fig. 2.33 d). (Leonard 1979)

Freiman et al. (1979) investigated glass failure under stresses of mixed mode (not stress gradients), resulting in asymmetric flaws and mirrors. They observed that similar ratios of mirror to flaw size characteristic of symmetric mirrors are also found on the larger zone of the asymmetrical mirror (note a conflicting comment by Shand in Sect. 2.1.3.3). Rice (1984) considers that irregular flaws, and especially machining flaws not normal to the stress axes, as well as inclusions and especially pores, can be important causes for the asymmetry of mirrors. Rice examined data by Freiman et al. (1979) and observed a systematic decrease in the ratio of mirror to flaw size for the mist and hackle boundaries as the angle of the flaw relative to the tensile stress axes (Fig. 2.35 c) decreased.

Rice (1984) also found that the ratio of the mirror to the pore size, where the pore serves as a flaw, is often much less than the values normally found for glasses failing from machining flaws (about 14 to 1). Rice attributed this to the bluntness of the pores compared with the sharpness of the more common flaws. Hence the ratio of mirror to flaw size is reduced by the bluntness of the flaw, that is, by $(\sigma_{\bar{a}}/\sigma_{\bar{p}})^2$, when $\sigma_{\bar{p}}$ is the failure stress from a pore plus crack, and $\sigma_{\bar{a}}$ is the failure stress from a flaw of the same size. Alternatively, multiplying the ratio of mirror to flaw size [Eq. (2.9)] for a more blunt flaw by $(\sigma_{\bar{p}}/\sigma_{\bar{a}})^2$ should give the ratio of mirror to flaw size for failure from a normal sharp flaw. The implication of Rice's work is that

$$r_j/c_{cr} = (A_j/A)^2(\sigma_{\bar{a}}/\sigma_{\bar{p}})^2 \quad (2.11)$$

if $A = \sigma_{\bar{a}} \cdot c_{cr}^{1/2}$ (Fig. 2.31). Rice (1984) found experimentally that realistically only upper limits to $\sigma_{\bar{p}}/\sigma_{\bar{a}}$ could be determined.

Flaws deviating from a single plane may lead to different parts of the flaw acting essentially independently (Rice 1984). A single flaw would act as two flaws with each generating its own portion of a mirror. Extreme distortions of the mirror are associated with pores. These effects seem to increase with the decrease of the average angle of the flaw relative to the tensile axis (Fig. 2.35 c), namely with the increase of shearing and the reduction of relative tension on the flaw.

Mirrors are not always confined to the main fracture surface. Rice (1984) shows various forms of mirrors on secondary fracture surfaces. These may result in a series of separate mirrors such as from multiple flaws in certain single crystals, or occasionally overlapping mirrors form even larger mirrors. Overlapping of secondary and primary mirrors and intersecting mirror geometries are not rare. Interactions between mirrors or between a mirror and a flaw may cause distortions, and an interaction of a crack with a pore may cause a deflection of the crack around the pore. These interactions may ultimately encourage the "early onset of mist and hackle" (Rice 1984).

2.1.4 Crack Branching

2.1.4.1 Branching in Nature

Ever since Leonardo da Vinci's interest in the branching relationships of plants, scientists have investigated these phenomena in nature. Perhaps the most famous studies include Murray's (1926, 1927) three-dimensional branching model of arteries and trees, and Woldenberg's (1969, 1970) branching investigation in human lungs and ruptured blocks of plastic, as well as Horton's (1945) two-dimensional model of the hierarchical network of streams. Numerical generations of branching (Leopold and Langbein 1962; Stevens 1974) have also contributed to this amazing subject.

Evidently, divisions occur by very slow processes such as branching in trees and by very rapid processes such as in a stroke of lightning. There arises the question of commonality in rules governing their formation.

Mechanical principles of crack branching were presented in Chapter 1. This chapter concerns various geometrical aspects of the branching of cracks.

2.1.4.2 General Aspects of the Geometry of Branching of Cracks

Symmetrical branching occurs under uniform uniaxial tensile conditions (Fig. 2.36a). When shear effects are influential, two branches can develop propagating at different velocities and angles (Fig. 2.36b) (Achenbach 1974).

Following results by Theocaris (1972) for asymmetric branching, Kitagawa et al. (1975) suggested that when one branch becomes longer than the other, the K_I value at the crack tip of the longer branch increases rapidly and that of the shorter branch decreases rapidly. In such branching which is sensitive to changes of K_I there will be a tendency for the preferential growth of one branch. Two branches of the same length will develop when the rate of crack growth is insensitive to the change in K_I. Insensitivity of the branching to changes in K_I will result from intensive microbranching that will cause small changes in K_I. This will enable the simultaneous growth of two branches at the same rate.

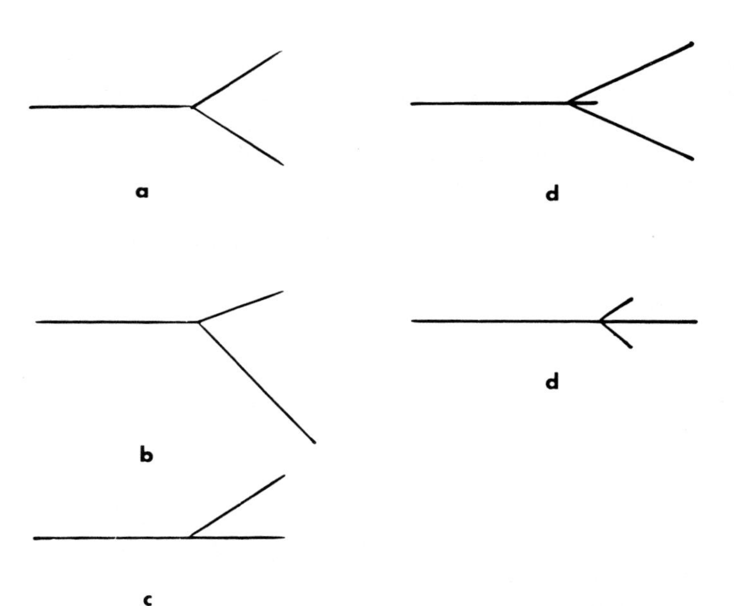

Fig. 2.36 a – d. Four modes of crack branching. **a** A symmetrical branching bifurcation due to uniform uniaxial tension. **b** An asymmetrical branching when tension is not uniform and shear is a significant factor. **c** Budding the parent crack continues while hiving off a new crack. **d** Two pitchfork branchings which combine bifurcation and budding, where three branches stem from the same location forking into a bifurcation which is bisected by the parent crack

Since the time of Leonardo da Vinci, people have noticed that crack branching may occur in more than one manner (MacCurdy 1955). The fracture may either bifurcate (Fig. 2.36a) or "budding" occurs when the parent fracture continues its propagation while hiving off a new crack (Fig. 2.36c). According to Rice (1984), the mode of branching appears to depend on the material and the type of test. He observed a strong preference for bifurcation (Fig. 2.36a) in a commercial crystallized glass tested biaxially, while similar testing of optical grade magnesium fluoride (MgF_2) much more commonly showed continuation of the parent fracture with the hiving off of a new crack (Fig. 2.36c). The cause of this difference has not been resolved. An additional type of branching is pitchfork, which combines bifurcation and budding (Figs. 2.36d).

Bottles fractured by internal pressure give a characteristic crack branching pattern (Sect. 2.3.1) but the detailed branching textures are somewhat variable. The branching modes shown in Fig. 2.36 can be recognized in Fig. 2.37. These include budding branches (at D in Fig. 2.37) pitchfork branches (at P in Fig. 2.37) and bifurcations (at B in Fig. 2.37). The budding branching probably reflects the greater local influence of shear stresses. The tendency for the bifurcation of the type shown in Fig. 2.36a along the $2r_b$ split as depicted in Fig. 2.43a is evidence for a local condition close to pure tension. The random distribution of B, D and P in Fig. 2.37 possibly implies that the branching mode is determined to a large extent by local conditions (angle between principal stress and critical flaw?).

Rice (1984) remarked that in flexural testing, the cracks in glass rods always branch vertically if branching occurs parallel to the direction of loading. This fracture direction is orthogonal to that of rectangular bars tested in flexure, which almost always branch horizontally in directions normal to the loading directions. The latter mode of fracture can perhaps be attributed to a gradient of tensile stress along the width of the bars.

Rebranching of cracks occurs at high failure stresses and can be investigated in large specimens. According to Rice (1984) rebranching can be expressed as:

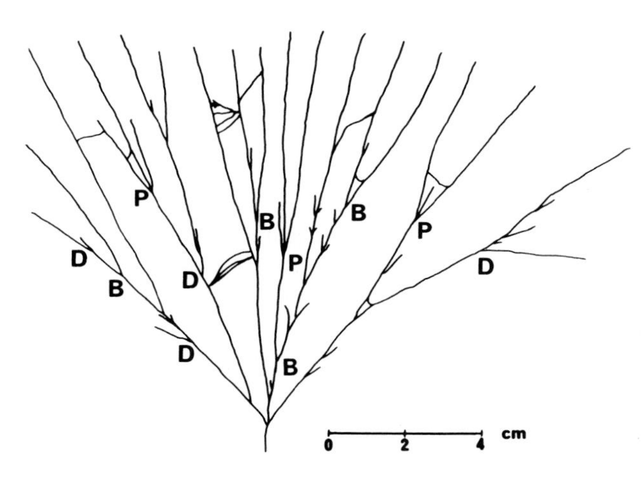

Fig. 2.37. An approximately 150° view of a glass bottle fractured by internal pressure showing budding *D* pitchforking *P* and bifurcation *B* cracks. Drawing of the individual cracks was carefully done on paper pasted on to the fractured bottle. The paper was then removed from the bottle and it now presents a two-dimensional fracture pattern

$$r_{b\acute{n}} \sim \acute{n} r_{bj} , \tag{2.12}$$

when new branches occur at multiples of the distance from the origin to the first branch (at least in the absence of large decreases in stress) and where $\acute{n}$ is the first, second, etc. stage of branching. Generally an increase in the distances for each subsequent set of branching is expected in the general situations of decreasing stress field (Rice 1984). Kirchner et al. (1981) investigate crack branching in transformation-toughened zirconia. They found an increase in K_b during the unstable propagation of the crack associated with the increase of the size of the transformation zone about the crack tip. This involved an increase in crack resistance which postponed crack branching until the crack lengthened.

2.1.4.3 The Branching Angle

According to Kobayashi and Ramulu (1985), theories used to predict the angles of crack branching obtaining under static loading conditions are the maximum circumferential stress theory (Sect. 1.1.5), the minimum strain energy density factor theory (Sih 1973), and the maximum strain energy release rate theory (Anderson 1969).

Preston (1931) made early observations that, on bending laths of glass to destruction, if a fracture advanced with any violence or with a jerk, it would cause "a sudden redistribution, both in amount and direction of the stresses at its head, the jerk is sufficient to split the front, to introduce hackle-texture, and thereafter, forking". . . . Under transverse bending (flexure) forking into multiple radiating branches will form a wedge of 45° (Fig. 2.38 a). The wedge attains a larger forking angle if there is a secondary tension at right angles to the primary one (Fig. 2.38 b). Hence, Preston (1935) stated that when the ratio of the secondary tension, f_y, to the primary tension, f_x, is $+1$, the forking angle is 180°. However, if a glass lath be twisted in pure torsion (Fig. 2.38 c), equal orthogonal tension, f_x, and compression, $-f_y$, develop in the glass ($f_y/f_x = -1$) (Fig. 2.39). "This amount of compression is not sufficient to suppress forking entirely and forks of about 15° angle are found".

Kalthoff (1973) produced a branching angle 2α of 28° between two cracks by tension (Fig. 2.40), a result that has approximately been obtained by others (Kobayashi and Ramulu 1985). Rice (1984) observed 2α of 44° in flexure which corresponded to the 45° fork angle found for the transverse bending of laths by Preston. Average values obtained by Rice (1984) for branching angles produced under biaxial conditions are between 90 and 156°. Congleton (1973) measured branching angles of about 30 to 40° in centre and edge-notched steel plates, and 70 to 80° in steel tubes burst under internal pressure. The latter results are close to the forking angles obtained by Preston (1935) for glass. The forking angle that Preston obtained in burst cylindrical glass bottles was close to 90° (when f_y is half f_x). Hence, for equivalent modes of brittle fracture, as determined by the relationship of principal stresses, branching angles for various materials such as glass, ceramics, metals and polymers are similar, and are related to Preston's forking angles.

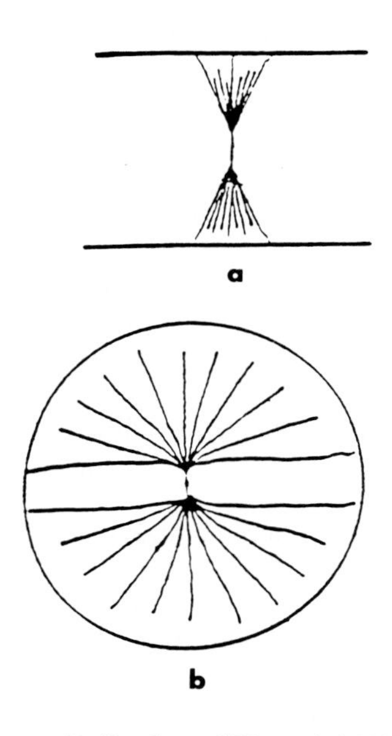

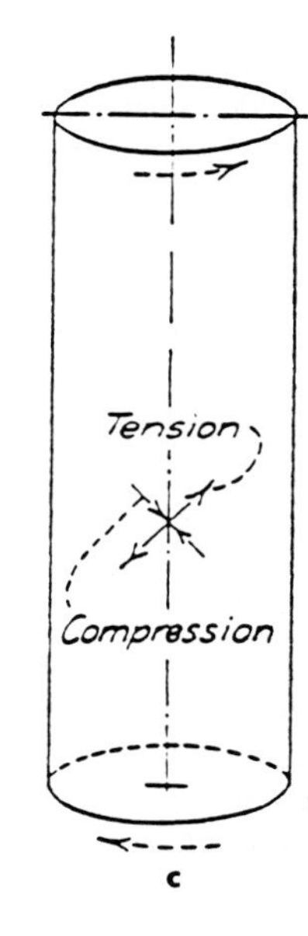

Fig. 2.38. a The fork which develops in the cross-bending of a lath includes several branchings of small angles, the forking angle being close to 45°. **b** Drumskin tension or isotropic tension ($f_y = f_x$) results in a 180° forking angle. **c** Pure torsion. (Preston 1935)

Bullock and Kaae (1979) fractured glassy carbon by a three-point bend technique. Failures in this material usually originated at interior spherical pores. Cracks branched and formed wedges with a forking angle of about 30°. Secondary cracks within the wedge developed secondary wedges with lower branching angles. Similarly, studies of branching in fractured glass bottles showed that, within the limit of the forking angle of 90° (Fig. 2.37) branching

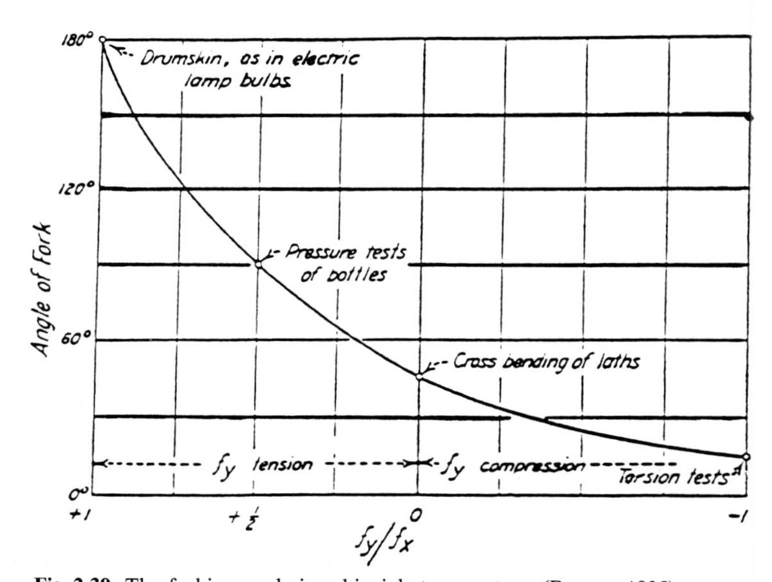

Fig. 2.39. The forking angle in a biaxial stress system. (Preston 1935)

angles are reduced at higher fracture stresses in order to accommodate more wedges in a given area (Bahat 1980b).

Kalthoff (1973) used the maximum circumferential stress theory for his numerical analysis of the crack branching angle 2α. He investigated the static post branching state of stress of symmetrically branched edged cracks under uniform uniaxial tension. He obtained 28° for the branching angle, which is in agreement with the angle which he measured in fracturing a polymeric glass. Under uniaxial tension, branching cracks are straight, but when $K_{II}/K_I > 0$ branching cracks of angles smaller than 28° repel each other and when $K_{II}/K_I < 0$ branching cracks of angles larger than 28° attract each other. Kalthoff showed that when branches interacted with each other, each crack deviated by an angle γ according to Erdogan and Sih (1963) (Fig. 2.40):

$$K_{II}/K_I = \sin\gamma/3\cos\gamma - 1 \ . \tag{2.13}$$

Kitagawa et al. (1975) confirmed Kalthoff's results by a numerical analysis and obtained a crack branching angle under uniaxial tension of 30 to 40°. Kitagawa et al. made a distinction between microbranching and macrobranching, and suggested that the branching angle for the former was larger than for the latter.

2.1.4.4 *Conic Branching*

Bullen et al. (1970) described an intriguing branching in heavily drawn chromium wires recrystallized at 950°. The wires, 0.25 mm in diameter, broke into three pieces due to an abrupt tensile failure (Fig. 2.41): the mating fracture surfaces were conical and formed two cup and cone fractures (Sect. 2.1.2.6). In the central cup there was a small central hole. This mated with small protrusions in the two conic fractures. "The cone angles, measured on a profile projector, were always near 120° at the apex; the cone was increasingly splayed as the diameter approached the wire diameter, until near the wire surface the frac-

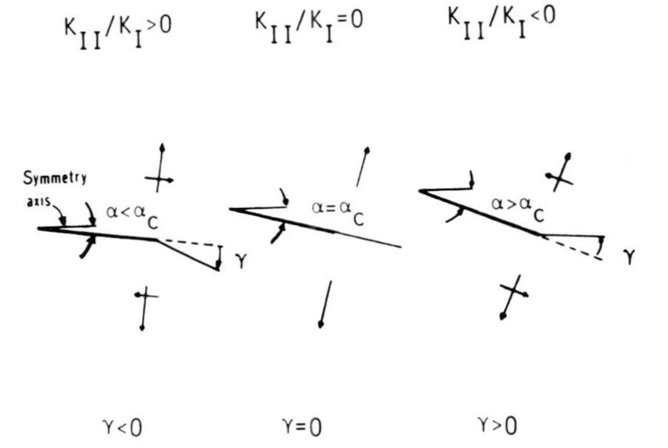

Fig. 2.40. Three possible propagation paths of a branch in a symmetric bifurcation. Only one branch (*thick line*) and the symmetric axis (*thin line*) are shown (*between curved arrows*). *Straight arrows* indicate near field stresses. Half the branching angle α is shown *between two curved arrows*, and γ is the deviation angle from the initial propagation path. (Kalthoff 1973)

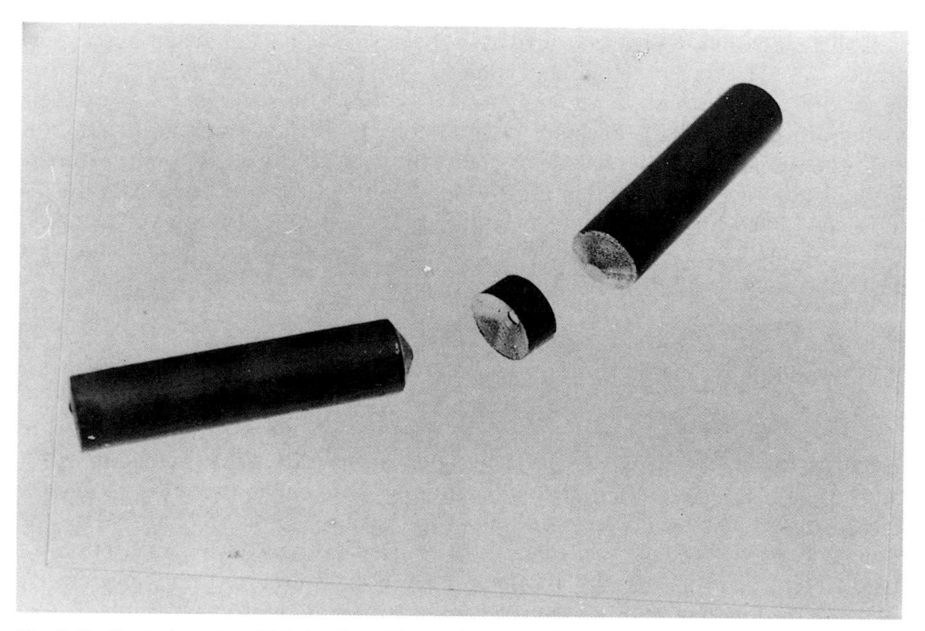

Fig. 2.41. Conical cracks which developed by an abrupt tensile failure of a chromium wire 0.25 mm in diameter. (Bullen et al. 1970)

ture surface was nearly normal to the wire axis". Fractography by SEM revealed that the hole in the cup appeared to be located randomly and that elongated flakes radiated from the periphery of the hole.

These chromium wires can be brought to failure at various fracture stresses yet with the same fracture morphologies as described above. Bullen et al. further reveal that the diameter of the hole in the cup decreases with increasing fracture stress in quite reasonable agreement with corresponding plots of mirror radius versus fracture stress (Congleton and Petch 1967). Hence, the implication is that the hole represents the mirror plane at whose periphery crack branching occurs three-dimensionally on two conical surfaces in analogy to the more common two-dimensional crack branching. Despite some resemblance to a ductile shear failure, evidence shows that the fracture is of brittle and presumably cleavage type. The authors consider that "the conditions at fracture are eminently favorable for crack-branching because of (1) the high stress level (20% of the theoretical), (2) the low effective surface energy for fracture, and (3) the highly preferred crystallographic texture which provides (111) planes of low effective surface energy in suitable orientations for propagation of the branched cracks".

Although the plot of variation of hole diameter versus fracture stress (Bullen et al. 1970, Fig. 8) is a strong argument in support of the conclusion that the two conical surfaces are formed by a crack-branching process analogous to that described by Congleton and Petch (1967), this conclusion still suffers from a few deficiencies: (1) No mirror plane has really been identified. (2) The radial elongated flakes shown by Bullen et al. (in their Figs. 3–6) maintain approximately unchanged dimensions regardless of their radial distance from the periphery of the hole. In normal crack branching, on the other hand, the radial hackles increase in size with distance from the mirror boundaries and the branching morphology generally changes accordingly. (3) The 120° cone angle is far too large for branching under tensile conditions. (4) The cones described by Bullen et al. may be explained to be a result of a Hertzian fracture (Sect. 1.1.7.3).

2.2 Terminology

The need for a unified nomenclature of fracture markings is well expressed in a discussion between two leading investigators in this field (Frechette 1984; Rice 1984). They represent two substantially different approaches to such a need for classification. Frechette (1984) states that "terms to characterize the markings on crack-generated surfaces of brittle materials have come to be used in different ways as a result of the variety of interests and backgrounds of workers in this field and the scattered nature of the literature. A single system of nomenclature which is explicit, universally applicable, and self-consistent should be helpful in aiding communication among fractographers".

In various of his publications (e.g. 1972), he defined the crack-generated surfaces of brittle materials primarily on the basis of the mechanisms that caused the fracture. Rice (1974, 1984), on the other hand, prefers to base his broad definitions upon physical characteristics such as shape, appearance, etc., bearing in mind that many of the fracture mechanisms are not yet fully understood or may even be quite controversial.

Rice's approach to the problem is adapted in this book. His nomenclature is a logical development which stems partly from the understanding of the progressive stages of early crack growth associated with the "foyer d'éclatement" (De Freminville 1907, 1914) and partly from the later concept of the quantitative mirror plane (Sect. 2.1.3). A distinction can be made between the more fundamental fracture markings which define the mirror plane and its periphery, and the lesser important ones. De Freminville (Sect. 2.1.1.1) provided, in fact, most of the elements required for the systematic identification of fractographic features. It is unfortunate, however, that Woodworth (1896) used the term hackle in too generalized a way and De Freminville did not quite distinguish between striae in the conchoidal zone within the mirror and their continuation as hackles in the tearing zone beyond the mirror. These two investigators were not yet familiar with the principles of fracture mechanics and the concept of the quantitative mirror plane.

Consequently, the term hackle has become quite controversial. Rice (1984) points out that the generalization of the term hackle and its division into various subcategories could be very confusing. The generalization of the term hackle (Frechette 1972) penetrated also to the geological literature. For instance, the sub-category twist hackle is being used by Kulander et al. (1979) as a substitute for both hackles (in their Figs. 14, 18) and for en echelons (in their Figs. 21, 29), while the term hackle represents plumes in their Figs. 30, 97, 107 and others. The classification is expected to help understanding the mechanism of fracture. This can be especially useful if assisted by the terminology of the mirror plane (Fig. 2.31). However, when the term hackle is being used in the definition of several different morphological features, Fig. 2.31 cannot be used. I therefore fully subscribe to Rice's suggestion that the terms mist, hackle and branching should exclusively be used for fracture beyond the mirror plane, to be distinguished from striae within the mirror plane, and even if all these markings are radial with respect to the origin, and the fracture velocities associated with these respective markings were below the terminal velocity (Figs. 1.26 and 1.29b).

Fracture steps, cleavage steps or simply steps, lances or river patterns are approximately synonymous terms. They represent elongated surfaces having a pronounced orientation, often around 90° with respect to the parent fracture or cleavage planes. They are termed striae on the mirror plane and they are the bridges between adjacent en echelon cracks close to the edges of the specimen. The bridges maintain a systematic orientation with respect to a parent fracture which generally excludes them from being termed hackles (Table 3.3).

Rice (1984) also expressed his concern about the term arrest line, which is sometimes used too loosely while being equated with rib marks. It can be questioned whether we really always know that a rib mark has been caused by true crack arrest. It is reasonable to regard many rib marks on surfaces of slow fractures as arrest lines. However, De Freminville (1907) fractured glass by impact and obtained two series of undulations (rib marks) as a consequence of both local stress and wave interaction. Again, it can be questioned whether the term arrest fits his rib marks. Hence, the term arrest line should be used with extra caution.

There is a host of fracture markings with a lesser diagnostic importance (Sect. 2.1.2.5). These include the parabola, hyperbola, ellipse and gull-wings, which are boundaries that develop when fracture surfaces meet at slight differences in depth. Also included are the fracture wedge or intersection scarp, which represent contacts of out of plane fractures. These features also serve to separate wet from not-wetted zones.

Fracture markings more familiar to metallurgists include fatigue striations which should be distinguished from striae in non-metals. The latter are generally much larger and have different orientations on the fracture surface. Beach marks in metals is a synonymous term to rib marks in non-metals.

2.3 Applied Fractography

Applied fractography is that branch of technology which concerns the analysis of the fracture features of products which have failed mechanically or by some other mechanism in actual service or in laboratory tests. Quite often the results of such analyses are brought up in court in product liability law suits. The impartial investigator is asked about the causes of failure and those bearing the responsibility for it, whether the manufacturer of the article or component, or the person who used the article. The result of a fractographic analysis can sometimes be frustrating when some questions are left unanswered. This can happen when certain fragments of the original article are missing, or even when all parts are found and the fracture morphology is clear, but the cause for failure is ambiguous. However, an analysis can be conclusive, with clear indications as to the location of the fracture origin or origins, of the directions and velocities of fracture propagation, and the modes of displacement and magnitudes of stress at failure. With a full picture at hand the investigator can define where the responsibility lies. Hertzberg (1976, p. 555) provides a "suggested check list of data desirable for complete failure analysis". This checklist can be helpful to those unfamiliar with the application of fractography to failure analysis in metals (see additional aspects of applied fractography in Sect. 3.2).

2.3.1 Fractography as a Tool of Fracture Diagnosis in Glass Bottles

Glass bottles may crack from several causes, getting hit, defective annealing, stones in the glass or similar gross inhomogeneities, small scale inhomogeneities (cordy, reamy, or lappy glass), thermal shock or excessive internal pressure (Preston 1932). Four major load conditions that may lead to fracture are elaborated in this section: (1) internal pressure (2) vertical load, (3) impact or diametrical squeezing and sliding and (4) thermal shock. Fracture under any of these conditions can happen unintentionally in service, and intentionally under controlled test conditions. Each of these conditions would characteristically start a fracture in a different weak region determined by the shape of the bottle and by the development of high stresses in the glass (Fig. 2.42). Accordingly, there seems to be some 15–20 characteristic crack patterns (Fig. 2.43), and the identification of a particular one can often indicate the cause and mechanism of fracture. The investigator can also determine at what stage of the bottle life the fracture occurred; e.g. during a certain stage in its production, when being filled with liquid, or while being handled by the consumer. In the following section the stress pattern in a bottle (Fig. 2.44) and the characteristic type of fracture which can develop for each of the four load conditions will be presented with some comparisons made between types of fracture that may occur similarly by two alternative such conditions, occasionally separate cracks at different locations propagating independently. Finally, most fractures initiate at the outside surface of a bottle, where interactions with the environment invariable lead to the development of many Griffith cracks (Sect. 1.1.2). The internal wall of the bottle is obviously more protected and an identification of a crack initiation on an internal wall is often quite diagnostic.

The level of sophistication reached in the utilization of fractography of glass bottles probably surpasses that in other engineering fields. The applica-

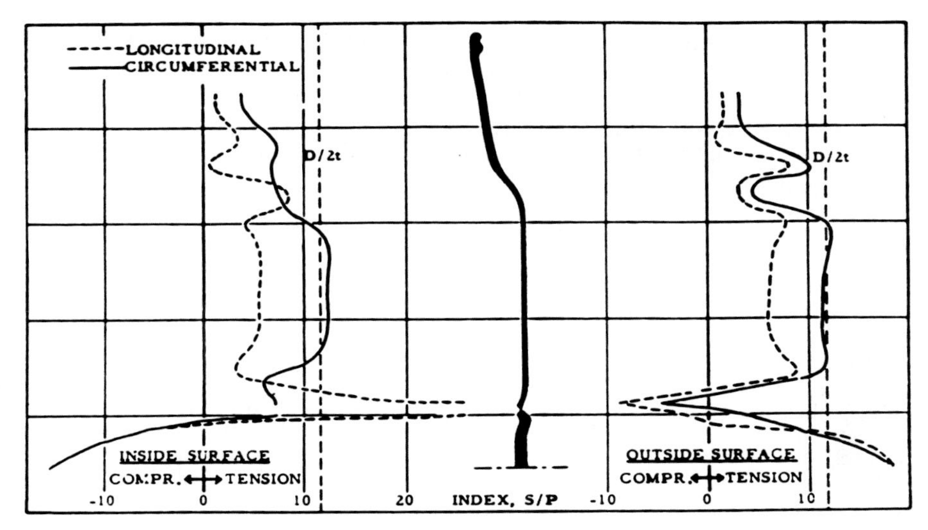

Fig. 2.42. The measured distribution of surface stress for a one-way beer bottle under internal pressure. The longitudinal and circumferential stresses on the inside and outside surfaces of the bottle are represented graphically in terms of the stress index *S/P*. Compression is indicated negatively. In the *centre* of each graph a vertical section through the wall of the bottle is shown schematically to relate the shape of the bottle to the stress curves. Note that the bottom section is graphically hinged at the outside heel and swung downward to a vertical position. The *vertical lines marked D/2t* on each graph indicate the theoretical average circumferential stress calculated for the cylindrical portion of the sidewall, where *D* is the inside diameter and *t* is the wall thickness. (Teague and Blau 1956)

tion of fractography in the analysis of jointing in rocks and related fields may never reach this level of sophistication but may nevertheless be challenging.

The present concise section on the application of fractography to glass bottles may serve in a limited way as an introduction to the wider field of the testing of glass containers, a subject with which the American Glass Research Inc. from Butler, Pennsylvania, U. S. A. has been involved during more than 60 years since it was founded by the legendary F. W. Preston.

2.3.1.1 Internal Pressure

Positive pressure in a closed bottle is exerted on its internal wall. It is commonly caused by CO_2 in carbonated water, wine or beer. This pressure increases with temperature and may vary up to about 10 kg/cm^2, but is commonly in the range 5.0 – 8.0 kg/cm^2. Internal pressure may also be caused when the liquid freezes or when the bottle is overfilled with the incompressible liquid. A negative pressure may develop when a hot liquid contracts as it cools down in the closed bottle.

Although the internal positive pressure is hydrostatic the stresses that develop in the bottle are not uniform, they depend on the geometry of the bottle.

High tensile stresses develop above the shoulder, along the cylindrical wall and particularly at the bottom (Fig. 2.42). A useful parameter in this context is the stress index S/P which is the stress σ per unit of applied pressure P.

The circumferential and axial stresses σ_{cir} and σ_{ax}, respectively, exerted on the cylindrical wall of the bottle are given by the two equations (Preston 1932; Teague and Blau 1956):

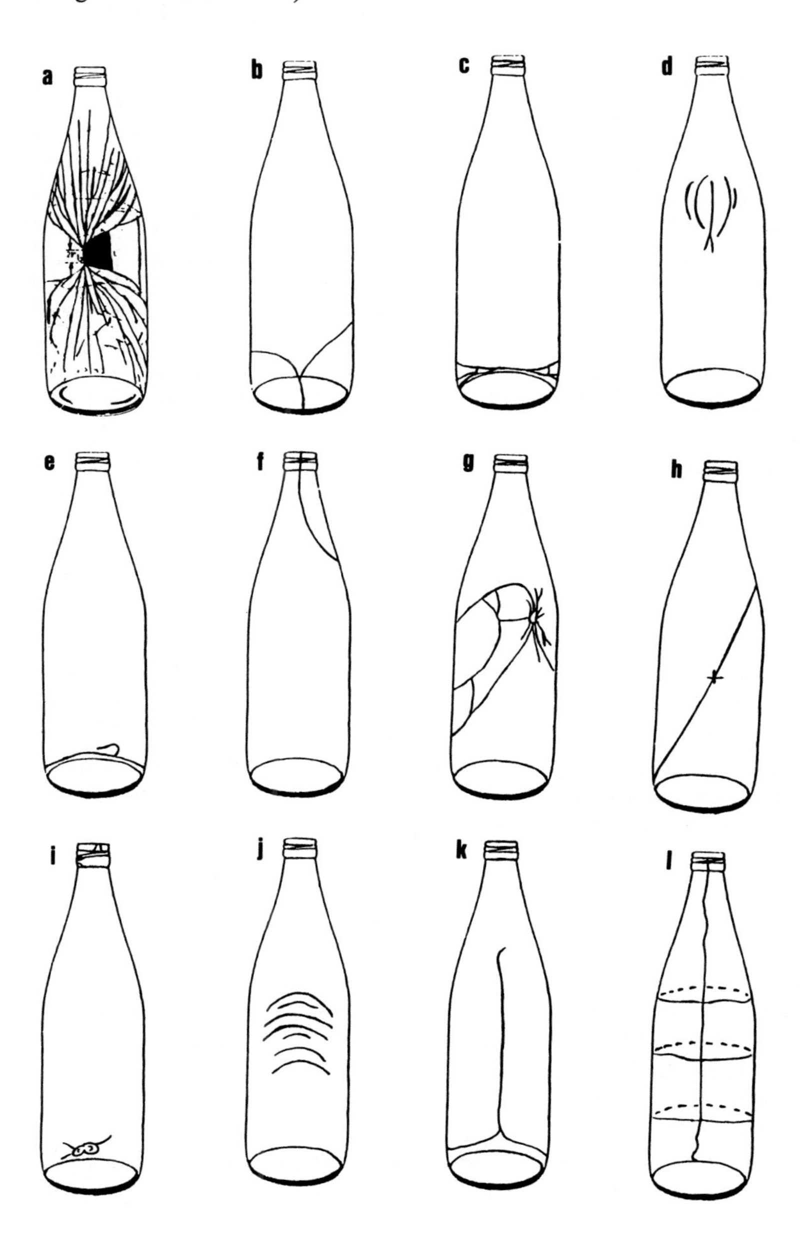

$$\sigma_{cir} = P \cdot D/t \tag{2.14}$$

$$\sigma_{ax} = 1/2\ \sigma_{cir} \ , \tag{2.15}$$

where P, D and t are the internal pressure, the inside diameter of the bottle and the wall thickness, respectively. Since the circumferential stress is double that of the axial stress, a vertical split is the most common form of fracture initiation due to internal pressure (Fig. 2.43 a). The vertical split displays a symmetrical mirror (Fig. 2.33 a) which is an indication of tensile fracture. Branching of the split will become more intensive with the increase in σ_{cir}. Any deviation of the perimeter of the bottle from a cylindrical form will increase the bending stress on the glass and fracture will initiate with an asymmetric mirror (Fig. 2.33 c, d). Accordingly, a vertical split with an asymmetric mirror may imply bending in a "straight" rather than a perfectly circular wall, which could have resulted from an uneven distribution of glass in the mould during production, with consequent local reduction in wall thickness. Most diagnostic are the two fragments on the sides of the vertical split. Each fragment contains one mirror. These fragments commonly have the shape of a rounded trapezoid or a saddle (Fig. 2.43 a).

The very pronounced inside bending and large thickness of the bottom in champagne wine bottles are complementary precautions against fracture in this highly stressed region. The inside bending performs better against the outside bending stress that is exerted by the internal pressure. These sensitive outside areas are also thus protected from scratching (see Fig. 2.44 a).

The bottle manufacturer must concentrate on producing a well-designed bearing surface in the peripheral circular region at the bottom. The angle between the vertical wall and the bent bottom determines the bending stress along the inside surface of the bottle above the bearing zone.

Fig. 2.43 a – l. Diagram of typical fracture patterns in cylindrical bottles. **a** A characteristic fracture due to internal pressure. The length of the vertical split (*between arrows*) is $2r_b$. This length is also the diameter of the mirror (Fig. 2.31). The vertical split together with the other vertical fracture *L* (and with the upper and lower crack branches) produce the saddle-shaped fragment (*black*) which is the most diagnostic piece of the fractured bottle. **b** A split due to a low internal pressure at the bearing which produces a D fracture at the bottom and branches towards the sidewall. **c** A horizontal fracture at the sidewall separates the bottom from the cylinder. If the fracture is intense it results in branching at the two tips. **d** Combined vertical split and concentric cracks caused by vertical load. **e** Fracture at the heel due to vertical pressure. **f** A vertical crack (check) at the top of the bottle (on the finish) is produced by vertical load while corking, and further shows a possible curving of the crack when the consumer opens the bottle. **g** Sidewall contact damage by a severe impact. Note the round hole at the point of impact, the hinge fracture at left, and the sub-horizontal leader crack between them. **h** A star crack is formed by a mild impact at a location which suffered an internal abrasion, it then continues as an extended inclined fracture. **i** A butterfly bruise due to a mild impact at the heel, a chip at the left side of finish by a sidewise or upward impact and a chip at the right by a downward impact. **j** Frictive damage manifested by crescent cracks which point their concave side downward results from a scratching downward movement of another body with a circular contact along the examined bottle. **k** A typical thermal shock inverted Y shaped fracture (shark's fin) at the heel. **l** A collar-split of a bottle caused by thermal shock

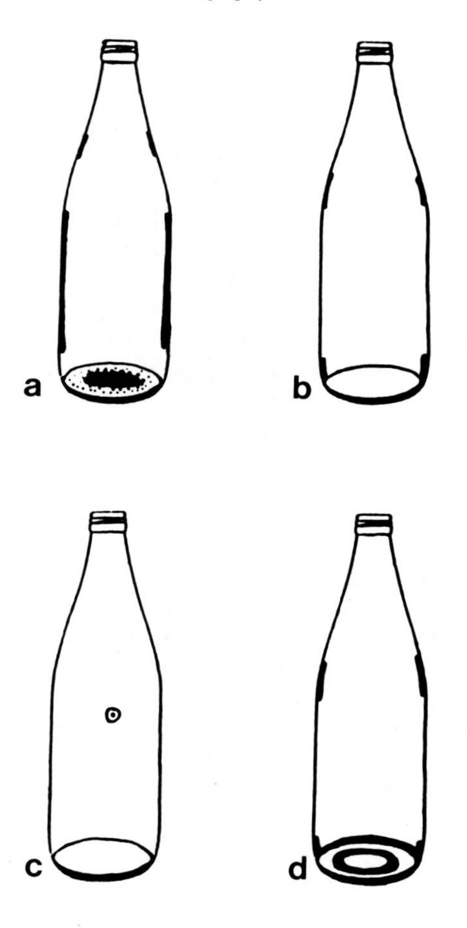

Fig. 2.44 a–d. A diagram comparing characteristic regions of high sensitivity in a bottle to the four major load conditions. *Heavy lines* designate regions of high stresses. **a** Internal pressure. **b** Vertical load. **c** Impact. **d** Thermal shock

Two regions where only low stresses are exerted on the glass are the shoulder and the heel, and the manufacturer often makes these regions slightly protrude from the outside vertical walls of beverage bottles. The consumer should therefore not be alarmed if he sees many cracks in these regions.

Fig. 2.45 a–e. Fracture patterns which were developed in bottles by internal pressure and thermal shock. **a** A split at the bearing due to high pressure. Branching occurs at the bottom and in the sidewall. The latter leads to additional branchings. **b** Two vertical splits caused by high pressure at the upper and lower parts of the cylinder instead of one at the centre. The lower branches of the upper split arrested at those branches which had earlier propagated upwards from the lower split. **c** A sub-horizontal split (*between H and M*) by internal pressure in a squashed ball bottle. **d** A very long vertical split (along most of the cylindrical part of the bottle) induced by a low internal pressure in a hot bath. *Scale* is 15 cm. **e** A fracture in a Pyrex vessel due to thermal shock. *Bar* = 5 cm. (Courtesy of Abraham H. Parola, the Ben Gurion University of the Negev). See text for explanation

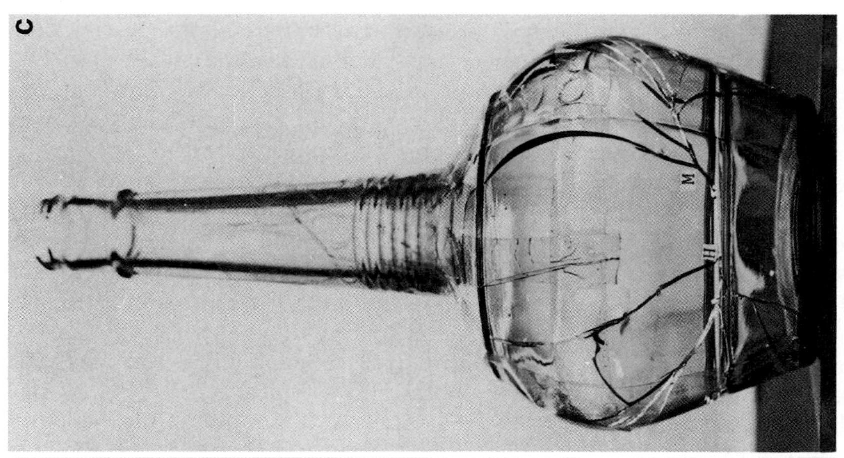
c

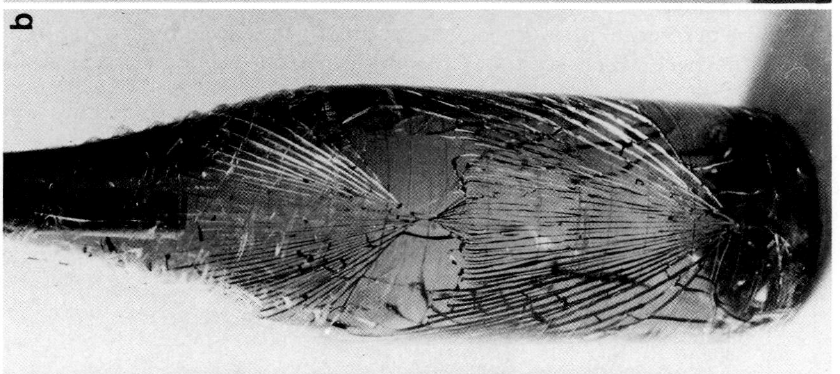
b

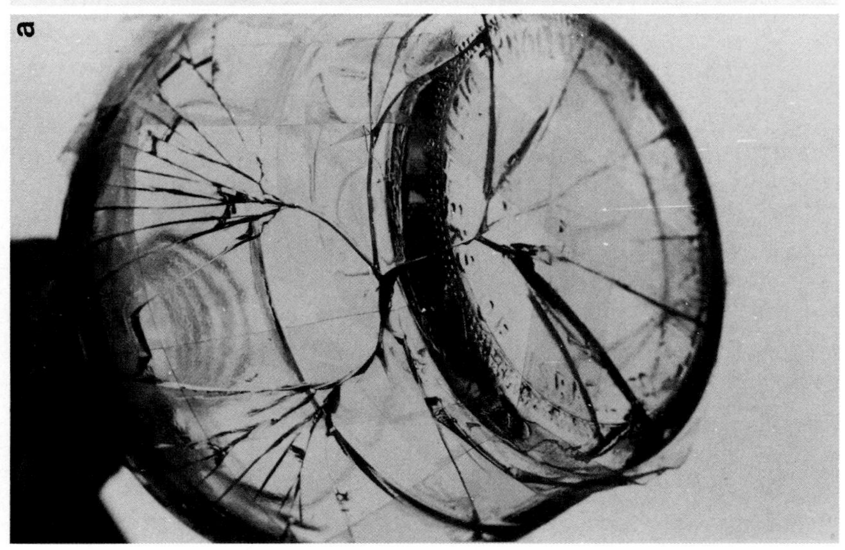
a

Fracture due to a high internal pressure may initiate at the bearing as a horizontal split. This split may then typically branch asymmetrically at its two edges: one branch will occur at the bottom and the other will "climb" the vertical wall (Fig. 2.45 a). Branching at the two tips of the split may start with straight fractures that then curve. Fractures which start at the bottom may be quite diversified (Fig. 2.46). When fracture initiates at the bottom centre due to a low internal pressure a single diametrical split will result in a typical D piece (Fig. 2.43 b). Such a fracture would be caused by inadequate glass thickness or by inside bottom bending, or by flaws introduced in printing. This kind of fracture would initiate with an asymmetrical mirror. A symmetrical branching that develops from a horizontal split initiated at the bottom centre is also possible (Fig. 2.46 d). A concentric fracture which initiates at the baffle (a small circular discontinuity at the bottom) is a typical failure (Fig. 2.46 b). An initiation at the baffle, however, may also result in a radial fracture propagation (Fig. 2.46 c).

Failure due to internal pressure may also be manifested by a horizontal fracture above the heel which separates the bottom from the bottle side walls

Fig. 2.45

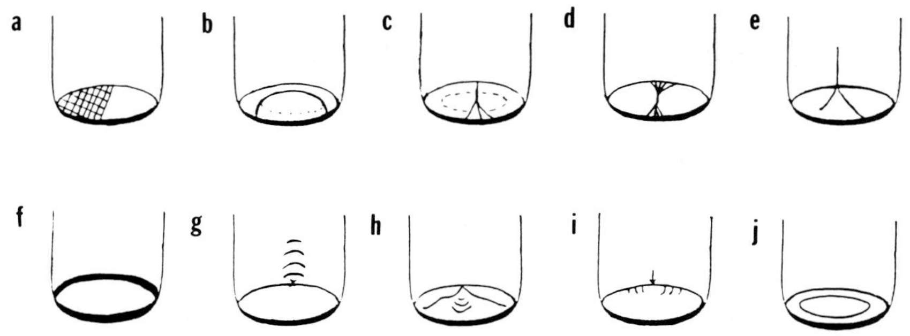

Fig. 2.46 a–j. Diagram of typical fracture patterns in bottle bottoms. Five fracture types due to internal pressure include: **a** D-shape fracture which initiated along the centre line at the bottom and then circulated along the bearing. **b** The fracture initiated at the baffle and followed it for quite a distance before deviating towards the bearing, where it ended its circulation. **c** The fracture at the baffle deviated towards the bearing trending sub-normally to it. **d** A split at the centre line that branched at its two tips in the bottom. **e** Branching at the bottom following fracture initiation split at the bearing. **f** A full circular fracture at the knuckle (the angle between the cylinder and the bottom). In this case the bottom has been completely removed. When this occurs due to a thermal shock, the surface of the sidewall part at the bottom appears smooth. **g** A vertical upward impact is identified by chipping and ripple marks on the side wall close to a bruise at the bottom. **h** A horizontal impact may be identified by the point of impact (on outside bruise) and cracks that spread out from it through the bottom. If there is chipping at the bottom, ripple marks will develop on the fracture surface. **i** When bottom removal results from vertical load, fracture initiation is identified at the outside knuckle (*arrow*) by the spreading of ripple marks and absence of the point of impact. **j** A low thermal shock may result in a circular baffle fracture with an infold kink

(Fig. 2.43 c). A single fracture caused by a low stress may imply that the fracture origin should be sought in the internal corner of the heel (Fig. 2.42). When a horizontal fracture is sub-divided into branches due to a high stress, the fracture origin should be sought in the outside wall of the bottle. A peripheral crack which separates the bottom from the bottle side walls characterizes particularly containers with square sections (Preston 1932).

Quite spectacular branching patterns may develop, albeit rarely, under high internal pressures when there are more than one fracture origins which initiate almost simultaneously (Fig. 2.45 b). This kind of failure may occur after an extended service life of the bottle when it becomes damaged at various points on its external surface (see also Bahat 1979 b, Fig. 2).

The fracture pattern of a bottle may change significantly with its design. For instance, a deviation from a cylinder to a squashed ball shape would cause the vertical split to be replaced by a sub-horizontal split (Fig. 2.45 c).

2.3.1.2 Vertical Load

A vertical load operates on a bottle during circulation. The tensile stresses developed are intensified in two locations, at the shoulder and at the heel. It is to be noted that these are locations where stresses due to internal pressure are low. The vertical wall becomes loaded in compression, but since both the shoulder and heel are located at the ends of "conical portions" of the bottle, they are significantly affected by bending. These distributions of stress are quite sensitive to small changes in the design of the bottle. It is recognized that circumferential stresses are greater than axial ones, therefore a vertical split which resembles that caused by internal pressure could initiate in the shoulder or in the heel under critical vertical loads. Such a fracture pattern would however be compounded by typical "looping" of concentric cracks around the vertical split (Fig. 2.43 d).

A fracture initiating at the heel and caused by a vertical load would be identified by a fracture origin at the outside corner of the bottle (Fig. 2.43 e). This origin is distinguishable from an origin at the baffle, which occasionally occurs due to internal pressure (Fig. 2.46 b). In both cases the bottom is likely to become separated from the rest of the bottle.

A vertical load may result in a single crack that starts at the top of the bottle (also termed finish) and propagates downwards when, for example, the bottle is being corked. This hidden crack may further propagate sideways when the consumer tries to open the bottle (Fig. 2.43 f).

2.3.1.3 Impact (Contact Stress) and Sliding

Impact. A bottle suffers from impacts on its side many times in its lifetime. This occurs during production when the bottle is still empty, during filling and while being transported after filling. Vertical impact occurs when the bottle is corked. Mould (1952) describes some of the various characteristic fracture patterns caused by impact. They result from a series of modes that develop in response to the impact. These stresses include localized contact and hinge stresses at the outside wall of the bottle and a localized flexural stress on the inside wall of the bottle opposite the location of contact (Fig. 2.47). These stresses often result in one, two or three fracture processes that initiate at different locations. The hinge fracture forms first, somewhat distant from the point of impact. The hinge usually forms as an elongated crack that may occur

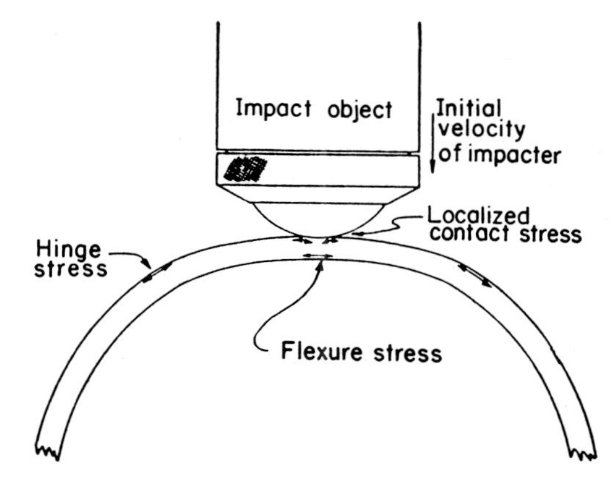

Fig. 2.47. Tensile stresses produced by impact on the middle side wall of a bottle. Only the three most important tensile stresses (those tending to break the bottle) are shown. Compressive stresses and secondary tensile stresses are omitted for clarity. Localized contact stresses result from the tendency of the impact to depress the outside surface at the point of contact. Flexural and hinge stresses are produced by general deformation of the bottle. (Mould 1952)

at any orientation. This, in turn, sends a leading crack (or cracks) to the location of contact. This is followed by a conical shaped circular crack (a Hertzian crack, see Sect. 1.1.7.3), which is caused by the localized stress and radial cracks resulting from the flexural stress. There is no undercutting between the hinge and the cone on the inside of the bottle. The radial cracks are often shown to precede the conical ones and they propagate mostly vertically and to a much lesser extent laterally. The number of the radial cracks increases with the intensity of the impact (Fig. 2.43 g). A very low intensity impact may result in a single sub-vertical crack, and the location of fracture initiation may be identified as a flexural star on the inside wall of the bottle (Fig. 2.43 h) or along the neck of the bottle. The press and blow process in the manufacture of bottles that involve contact of metal with the inside wall of the bottle could promote such a failure.

The contact stress increases with the rigidity of the glass. Therefore, conical cracks are more common at the heel than at the vertical cylindrical wall and they occur more frequently in cylindrical bottles than in square ones. A conical crack at the heel often appears in a complex butterfly shape (Fig. 2.43 i). A strong impact at the heel may even result in an undercutting of a butterfly piece of glass that falls into the bottle. Alternatively, a shallow conical crack may nucleate a later fracture when other loads create critical stresses. With the decrease of wall thickness and increase of bottle diameter, the rigidity of the glass decreases and the conical fracture is replaced by a flexural crack. The glass is quite rigid at its neck and an impact stress during corking may result in small cones at the top of the bottle.

Lateral and vertical impacts may result in the bottom coming away (Fig. 2.46 h, i). A horizontal squeeze test is often being used to simulate impact. In this test a uniaxial pressure is applied at the cylindrical wall of the bottle perpendicular to the axis of the bottle. This test produces a multi-stress effect which resembles the stress pattern in Fig. 2.47.

Damage Due to Friction and Scratching. A contact force which consists of a tangential component superposed on a normal component can result when a hard object, which does not have a sharp edge, slides over a bottle (Fig. 1.14 c). The frictional traction is proportional to the normal load. A typical result is a series of crescent-shaped surface cracks more or less regularly spaced which produce a "chatter track" (Preston 1952). Let us consider the case of frictional damage caused at a circular contact on the surface of the bottle. Here maximum tension occurs in the wake of the circle and maximum compression at its front. Consequently, the concave side of the surface crack should show the sliding direction of the hard object (Fig. 2.43 j). Note that on a fracture surface the convex side of undulations shows the direction of propagation of the fracture. When, however, the hard object that slides over the bottle has a sharp edge (Fig. 1.14 d), the effect is different. Elongated scratches accompanied by many elongated damaged areas develop instead of the crescent shape of surface cracks.

2.3.1.4 Thermal Shock

Most bottles are made of soda-lime-silica glass. The coefficient of thermal expansion of this glass is high ($\sim 85 \times 10^{-7}/°C$), consequently, any sudden change in temperature introduces considerable stresses in the glass. Generally below the temperature of the strain point (above a viscosity of $\sim 10^{14.6}$ poises), when the glass behaves as an elastic body the cold sides of bottles develop tensile stresses. A tensile thermally induced stress on the outside surface of the bottle results because the glass is a poor conductor of heat and when the hot bottle is suddenly brought into a cold environment, the cooler outside surface of the glass is forced to contract. However, the inner hotter glass does not immediately contract and it therefore results in a stretching of the cooler glass. The stress that results is termed skin stress. In bottles the heel and the bottom are the most sensitive locations to thermal gradients due to the large thickness in these areas. The angular geometry of the heel considerably increases its sensitivity to thermal shock due to the superposition of a bending stress on the skin stress. Thermal stresses may also cause fracture in the shoulder and in the vertical wall.

Characteristic fracture patterns are: (1) a fracture that starts at the heel and has the shape known as shark's fin (Fig. 2.43 k), (2) a semi-circular crack that starts at the baffle and reaches the bearing, either at a low angle or normal to it (Fig. 2.46 b) and (3) a slightly curved crack on the wall running from top to bottom of the bottle (Fig. 2.43 l). The latter crack would typically appear without branching and fracture markings, such as ribs or striae on the fracture surface, due to fracture initiation under low stresses. Quite often, when the crack approaches the base of the bottle it bends and runs peripherally round the bottom (Preston 1932).

A bottle may fracture if the annealing process carried out by the manufacturer has been inadequate and residual stresses are present above certain permissible levels. Such stresses, combined with the presence of stress concentrators such as a stone in the glass, would promote fracture under conditions of low stress. It should be recalled (Fig. 2.42) that at the inside surface of a bottle at the heel there is a higher tensile stress than at the outside surface.

Although the circumferential tension can be entirely relieved by the single vertical crack (Fig. 2.43 k), the axial tension cannot. A single circumferential crack around the middle will relieve the axial tension only for a short distance up and down from the crack. "To relieve this tension with any completeness, the bottle must split into a number of short lengths like so many collars" (Preston 1932). These circumferential cracks start at the primary axial cracks and run more or less at right angles to them around the bottle (Fig. 2.43).

The spectacular fracture pattern in Fig. 2.45 e developed when cold water was poured into the empty Pyrex vessel while being heated on a hot plate. Note the equally spaced parallel cracks whose tips reach the same height. The cracks twist into helices which maintain amazingly similar patterns. These uniform features seem to have responded to stresses controlled by the geometry of the vessel.

A fracture under combined stresses such as those caused by thermal shock and internal pressure may result in a mirror plane (Fig. 2.31) with a particularly large diameter (Fig. 2.45 d). A simulation of such a break may be easily carried out by applying a thermal shock as per ASTM C 149-77 on a filled bottle containing a carbonated liquid. The internal pressures that develop at 60° may be as high as $\sim 10\ \mathrm{kg/cm^2}$ depending on the concentration of CO_2 in the liquid.

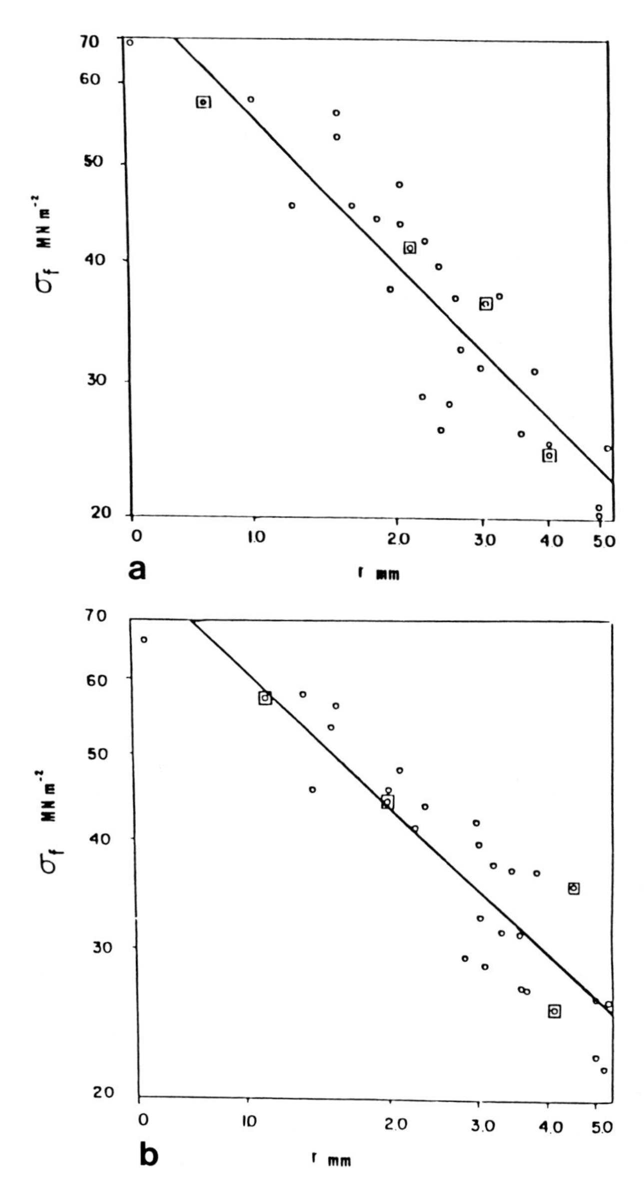

Fig. 2.48 a, b. A logarithmic plot of fracture stress σ_f (MN/m^2) versus mirror radius, r (mm) at the mirror-mist boundary for a number of glass bottles. Distinction is to be made between **a** measured results, and **b** calculated results (see Bahat et al. 1982). *Combined square and circle* represent an asymmetric mirror. (Leonard 1979, group III)

Such pressures, normally, would not cause failure in the bottle at room temperature. However, if the bottle is immersed in a hot bath, an intensified stress corrosion (Sect. 1.4.1) on the outside wall of the bottle results in failure under relatively low internal pressure. This is manifested by the unusually large diameter of the mirror.

2.3.1.5 Quantitative Determination of the Fracture Stress and Internal Pressure

The applicability of the empirical relationship $\sigma_f \cdot r^{1/2} = A$, where σ_f is the fracture stress, r the mirror radius and A the mirror constant [see Eq. (2.7)] to the analysis of fracture in bottles has been demonstrated by various investigators (e.g. Kerkhof 1975; Frechette and Michalske 1978; Bahat et al. 1982). A logarithmic plot of fracture stress versus mirror radius reasonably well fits a straight line with a negative slope (Fig. 2.48 a). Deviations of the slope from -0.5 (Mecholsky et al. 1974) have been examined by Leonard (1979) for various groups of bottles of different designs. Leonard finds a correlation between these deviations and the location of the mirror plane along the outside wall of the bottle. Leonard (1979) and Bahat et al. (1982) investigated the difference between the measured and calculated radius of the mirror. A somewhat better fit to a straight line in Fig. 2.48 b is obtained by using the calculated radius of the mirror. The fracture stress can be derived from the internal pressure, P, from Eq. (2.14). Normally, a fracture in a bottle from excessive internal pressure would result in a vertical mirror and the fracture stress would be σ_{cir}.

2.3.2 The Fractography of Metal Failures

This section is limited to examples of fracture diagnosis in metals where the fracture origin and other morphological features can be detected by the naked eye. Micro-fractography which applies microscopic markings such as fatigue striations (Sect. 2.1.2.6) is beyond the scope of this section.

Chevron markings are characteristic features on the fracture surfaces of steels for certain conditions of stressing. They often help to identify the fracture origin and the direction of fracture propagation (Fig. 2.29 g). Changes in the nature of the fracture surface can indicate changes in the fracture mechanism. Fatigue growth characterized by concentric beach marks may give place to a rougher fracture which is marked by a grainy texture of chevron markings (Fig. 2.49 a). This change also suggests a respective shift from a mixture of predominant mode I and mode II displacement to a mixture of predominant mode I and mode III (Sect. 1.1.7.2).

Two examples of quantitative fractography in metals are presented below (cited from Hertzberg 1976).

Fig. 2.49. a Two corner cracks emanating from a through-the-thickness hole, and showing bands and shear lips representing the growth of fatigue cracks. The depth of shear lips along the surface of the hole is about 0.8 mm. (Gran et al. 1971). **b** A flake (*dark circle*) due to excessive hydrogen which contributed to the fracture of a turbine rotor in Arizona. (Yukawa et al. 1969)

2.3.2.1 Example 1

In the case of the failure of a rotor generator forging in Arizona, the estimation of K_{Ic} was based on an excellent photograph of the fractured surface from Yukawa et al. (1969; Fig. 2.49b). An important cause of the failure was the formation of hydrogen flakes 2.5 to 3.8 cm in diameter as circular defects in the steel. Yukawa et al. determined the maximum tangential stress exerted on the borehole of the rotor to be 350 MPa at a speed of 3400 rpm at the time of fracture. Hertzberg used estimates by Sailors and Corten (1972) of the K_{Ic} range for the rotor material ($K_{Ic} = 34-59$ MPA $\sqrt{m}$), and by applying the modified Eq. (2.8):

$$K_{Ic} = 2/\pi \cdot \sigma \sqrt{\pi c} \ , \tag{2.16}$$

he obtained $34-59 = 2/\pi\ (350)\ (\sqrt{\pi \cdot c})$ and hence $c = 0.74$ to 2.2 cm (where c is the crack radius). Thus, the range of K_{Ic} which he used corresponded reasonably well with measured range of diameter of the hydrogen flakes.

The rotor had failed during a routine balancing test and the results were quite reliable. Hertzberg's assumption that the circular flake in Fig. 2.49b) designates the perimeter of the critical crack prior to catastrophic fracture is adopted here. One implication is that the echelon cracks which radiate from this perimeter developed at conditions of K_{Ic} or perhaps even at $K_I > K_{Ic}$ (see Sect. 4.2.1.7). One observes that the echelon cracks are mostly concentrated to the right and left sides of the circular crack, where they exhibit left stepping and right stepping modes, respectively. En echolon cracks are virtually absent from the zones above and below the almost perfectly circular crack in Fig. 2.49b. This suggests that the axis of the maximum principal tensile stress was normal to the plane of the fracture and the direction of minimum tension was up-down with respect to the photograph.

2.3.2.2 Example 2

Several additional concepts of fracture mechanics applicable to fractography were used by Hertzberg (1976) to estimate K_I values in a fatigue test which was conducted on a 1.78-cm-thick plate of (D6AC) steel which had been tempered to a yield strength σ_{ys} of 1500 MPa.

Fracture of the plate occurred when fatigue cracks which had developed on both sides of a drilled hole grew into a semi-circular configuration, as shown in Fig. 2.49a. One method of calculation was based on the measurement of the fatigue growth bands and on the power relationship that exists between the range of stress intensity factor ($\Delta K = K_{max} - K_{min}$) and the rate of fatigue crack growth da/dN (Fig. 1.29a). This relationship is of the form $da/dN = C(\Delta K)^2$, where C and n are constants. It was known that the last fatigue band was produced by fluctuations of 15 loads between stress levels of 137 and 895 MPa. Measuring this growth band to be 0.32 mm, the average crack growth rate was found to be

$$da/dN \approx 3.2 \times 10^{-4}/15 \approx 2.1 \times 10^{-5}\ \text{m/cyc} \ .$$

From a plot of ΔK versus da/dn by Carmen and Katlin (1966) the corresponding ΔK level was found to be about 77 MPa $\sqrt{m}$. The maximum K level was then given by

$$K_{max} = \Delta K(\sigma_{max}/\Delta\sigma) \ . \tag{2.17}$$

Hence, $K_{max} = 77 \cdot (895/758) = 91$ MPa $\sqrt{m}$. A similar calculation was made for the next to last band where $\Delta n = 2$, $\Delta a = 0.16$ mm, $\sigma_{min} = 138$ MPa and $\sigma_{max} = 992$ MPa and $da/dn \approx 1.6 \times 10^{-4}/2 \approx 8 \times 10^{-5}$ m/cyc. The ΔK level corresponding to this crack growth rate was found to be 82.5 MPa $\sqrt{m}$, and $K_{max} = 82.5 \cdot (992/854) = 95.8$ MPa $\sqrt{m}$.

The K_I value was also estimated by measurement of the depth h of the shear lip along the surface of the hole (Fig. 2.49a) and by using the relationship [Eq. (1.29)]:

$$h \approx 1/2\,\pi \cdot K^2/\sigma_{ys}^2 \ . \tag{2.18}$$

where σ_{ys} was the yield strength of the material, and h was measured as 0.8 mm. This gave $8 \times 10^{-4} \approx 1/2\,\pi(K/1500)^2$ and $K \approx 106$ MPa $\sqrt{m}$, which is reasonably close to the K_{max} results obtained by the previous method. Hertzberg also observed that the K_{Ic} for this material is 101 MPa $\sqrt{m}$, which implies that the fatigue crack possibly approached catastrophic conditions at its late stages when K_{max} reached the range 91 to 95.8 MPa $\sqrt{m}$.

2.3.3 Fractography in Polymethylmethacrylate

The morphology of the fracture surface polymethylmethacrylate, PMMA, varies when changes occur in the rate of loading (Zandman 1954) in temperature, orientation and molecular weight (Wolock et al. 1959; Doll 1975). Wolock et al. observed that as the molecular weight of the PMMA increases, parabolas (Sect. 2.1.2.5) become larger and less numerous per unit area around the mirror. Doll identified a jump in crack speed from about 300 to about 400 m/s on a fracture surface of PMMA with a molecular weight of 163000. The fracture surface was almost smooth and showed only parabolic markings at low crack speeds. A coarse fracture surface with rib-like features abruptly developed at the jump in speed. Doll considered that the fracture was essentially brittle at both the low and high velocities of fracture.

Rummel (1987) induced unstable fracture in a Plexiglas cube of dimensions $15 \times 15 \times 15$ cm, loaded biaxially in a triaxial press, when the principal horizontal far field stress was unequal to the vertical one, and fluid pressure was applied to a sealed-off borehole drilled generally parallel to the vertical far field stress. Typically in these experiments, a hydraulic oil with a viscosity of 32 cSt (32×10^{-6} m^2/s) was used to induce fracture and was injected at a pumping rate of 5 bars/s. In this hydraulically induced fracture, transitional markings (resembling Figs. 3.19 and 4.2a) of radial plumes superimposed on concentric undulations resulted. Rummel interpreted the concentric undulations as an indication of discrete events in the growth of the fracture. When he rotated the

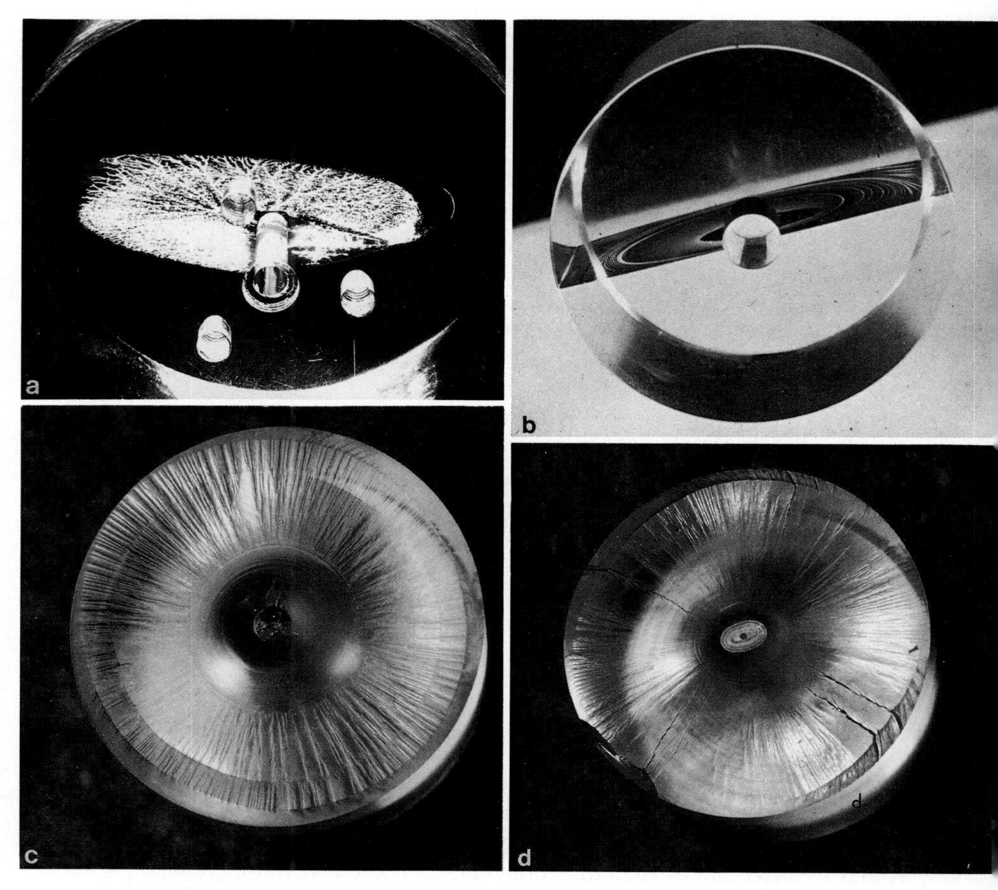

Fig. 2.50. Fracture markings in Perspex cylinders 42 mm high and 48 mm in diameter by hydraulic pressure. In **a, b** and **c** borehole and piston were circular, in **d** these sections were elliptical. Note radial zones of branching cracks in **d**. See explanation in text

direction of the external horizontal stress by $\pi/2$ the fracture also changed its direction of growth.

Four distinct fracture morphologies, obtainable by hydraulic means, are shown in Fig. 2.50. In each of the four experiments, grease was the hydraulic medium. Pressure was applied by a steel piston through an axial borehole drilled into Perspex cylinders 42 mm high and 48 mm in diameter. In the first three experiments, both borehole and piston were circular, whereas in the fourth experiment (Fig. 2.50d) these sections were elliptical.

Discrete plumes and concentric undulations were observed on flat axial fractures that grew parallel to the borehole as a result of slow application of the hydraulic pressure. The fracture with the plume marking (Fig. 2.50a) resulted from biaxial loading, when stress was applied normal to the cylinder axis in addition to the hydraulic axial pressure. The concentric undulations

(Fig. 2.50b) developed when no external loading was applied apart from the fluid pressure.

Two Hertzian cone fractures developed preferentially to axial flat fractures when the hydraulic pressure was applied rapidly (Bahat and Sharpe 1982, Sect. 1.1.7.3). A circular cone (Fig. 2.50c) resulted when the borehole and piston were circular in section. Note that circular markings developed before radial markings on the fracture surfaces.

The fracture surface which resulted from application of pressure through an elliptical borehole (Fig. 2.50d) developed in two stages. An elliptical cone fracture developed first. In this fracture the cone angle α (Fig. 1.14e) was larger along the small diameter of the ellipse than along the large one. Following fracture initiation, when the reflected stress waves propagated back from the cylindrical boundary towards the axis, they induced branching cracks that advanced from the periphery towards the centre, along four radial zones. The branching cracks identify zones along which the stress intensity factor was amplified. These zones seem to be radial to the cylinder at the loci produced where the elliptical section intersects a parallel elliptical section of the same size when rotated at 90°.

Chapter 3 Rock Fractography

3.1 Fracture Markings on Joint Surfaces

3.1.1 Early Studies

Woodworth (1895, 1896) set the stage for the science of fractography by astute observations of the morphology of joints in geologic exposures. He also noted the essential morphologic similarities between fracture markings in glass samples and rock outcrops. No one who has worked with glass as an industrial material can fail to be fascinated by its intriguing geometries and lustrous fracture features, so that the interest of the early glass experimentalists in fracture markings was no more than natural (e.g. De Freminville 1907, 1914; Preston 1926, 1931, Chap. 2). Rock fracture, by comparison, is drab and lacklustre, and it is therefore not surprising that only very few geologists were attracted to the subject of fracture morphologies on rock exposures during the long period that elapsed between the studies of Woodworth (1896) and those of Hodgson (1961 a, b) (see below). The systematic description of fracture markings in rocks begins with the trail-blazing investigations of Woodworth, followed by the elaboration on those studies by later observers. The present chapter deals mainly with fractography of sedimentary rocks, and only to a limited extent with fracture morphology of igneous rocks.

3.1.1.1 Woodworth's Fracture Marking Classification (According to Hodgson 1961 a)

Woodworth's work included a framework for classification. This framework has been presented in a concise version by Hodgson (1961 a), and it is essentially this review by Hodgson of Woodworth's work, slightly modified, which is presented in this section.

Woodworth's characterizations are based primarily on his observations of closely spaced parallel joints in argillites of the Mystic River quarries near Cambridge, Massachusetts. Most of the surface markings that were examined in detail are of small size, rarely exceeding 200 mm in length. They occur on joints that are perpendicular to the bedding and are generally elliptical in shape when occurring in a single layer, with elongation parallel to the bedding. Woodworth divides these planar joint surfaces into three major structural parts: (1) the joint plane, (2) the joint fringe and (3) the rim of conchoidal fracture. In addition, Woodworth describes the shoulder between the joint and the fringe, and discoid small joints which are quite distinct from the flat joints.

Joint Plane. The joint plane commonly presents a rough, granular surface on which appear faint rides or rays. These diverge outward from a focal area or point of minimal relief and form a plume-like or radial pattern (Fig. 3.1 a, b). The divergent lines are comparable to the percussion rays of flint chips (well known to prehistorical archaeologists) and indicate the direction of fracture propagation. Woodworth (1895) describes the structure of the joint plane as follows: "These apparent rays are simply cross fractures between thin laminae of rocks formed by minute planes of fracture, and some of the typical joint planes are made up of combinations of these small joint planes and the cross fractures". Slight variations in the size of these small joint planes and the cross fractures produce the above-described faint ridges.

The most common fracture marking is the plume or plumose structure which consists of a focal area from which faint ridges diverge bilaterally outward toward the boundaries of the joint plane. Generally, but not invariably, the pattern becomes increasingly distinct and coarse toward the periphery of the joint plane. Woodworth (1896, p. 168) describes one example of plumose structure where two roughly parallel plumes occur on a single joint face, noting that "the separation of the rock may begin at points separated by appreciable distances and the fronts of the growing fracture may travel towards each other and pass on different levels".

Fringe. The joint plane is commonly surrounded by a fringe of small en echelon joints (Fig. 3.1 a, b, c). Woodworth (1896, p. 169) refers to these joints as "border planes" or "b-planes" and describes them as follows: "These b-planes, where their relation to the joint plane is most clearly shown, spring out from the edge of the joint plane with which they are at first parallel, and proceeding outward turn gradually so as to be inclined from 5 to 25 degrees or more to that plane". The intervals between the b-planes are usually formed by rough, irregular fractures at about right angles to the b-planes. These are called "cross-fractures" or "c-fractures". By far the majority of these c-fractures are straight. Woodworth notes one example, however, where the c-fractures of the fringe are curved and merge into one another, forming a continuous scalloped surface. These are described as "curved lateral" c-fractures.

Plumose structures are not restricted to the main joint plane but occur also on the b-planes (Fig. 3.1 c).

Rim of Conchoidal Fracture. Woodworth (1896, p. 173) describes the rim of conchoidal fracture ase follows: "Outside of the fringe, or bordering the joint plane when the fringe is wanting, there may be developed a rim of conchoidal fracture. Structurally the rim is a compound surface of fracture. It tends to be inclined to the joint plane and to the joint fringe". This description is especially applicable to discoid joints (see below).

Shoulder. Woodworth (1896, p. 170) notes the existence of a shoulder or curved edge to some joint planes where they join the fringe (d-d in Fig. 3.1 b). He describes the shoulder as follows: "Even when the surfaces of the fringe

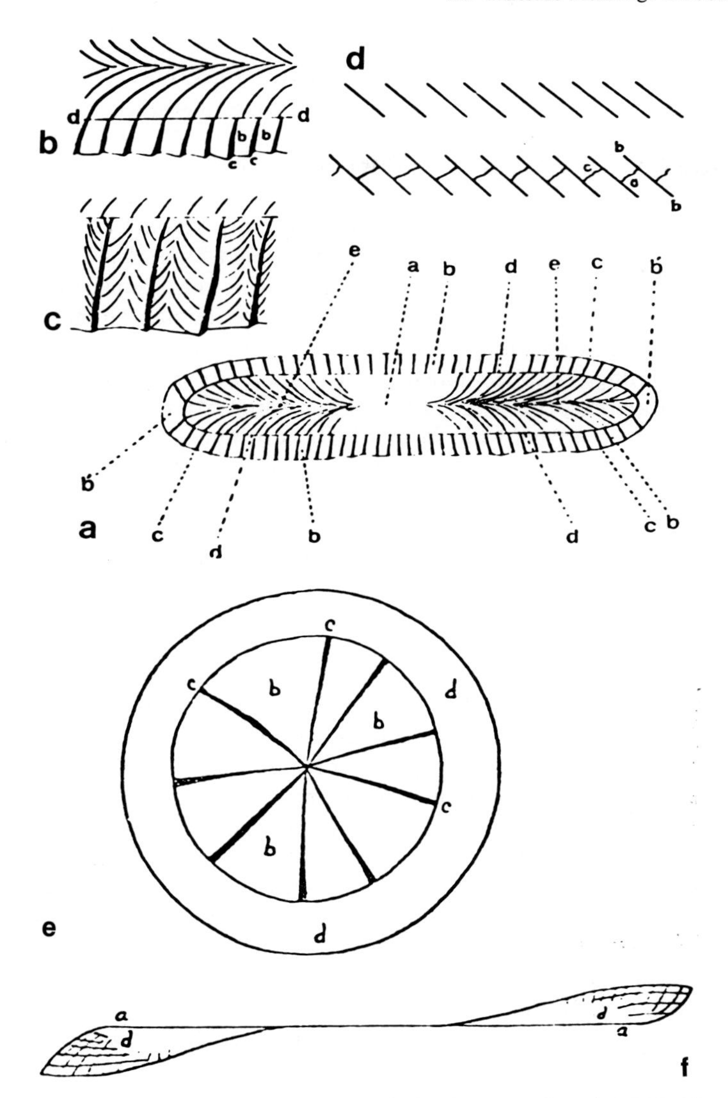

Fig. 3.1 a–f. Fracture surface morphologies from Woodworth (1896). Note a distinction between the **a** : **b** : etc. designations of the different figures and the *a, b* etc. designations with which Woodworth describes the various parts of a particular morphology. *Terms in parentheses* are added to Woodworth's description. **a** Idealized scheme of elliptical joint in a stratified rock: *a* Cental area of the main joint surface; *b* border planes of the fringe (echelon cracks); *b'* distal border planes; *c* cross fractures (steps); *d* edge of joint plane (shoulder) and inner margin of the fringe; *e* axis of feather fracture (plume), parallel to the stratification. **b** Border planes *b* and cross fractures *c* in a fringe that parallels the axis of the feather fracture. **c** Distribution of feather fractures on border planes. **d** Relationship between border planes (*above*) and border planes with cross fractures (*below*). **e** A plan of discoid joint includes *b* imbrication planes dipping largely in one direction around the centre; *c* cross fractures; and *d* rim of conchoidal fracture. **f** Cross-section of discoid joint, showing antipodal sections of the curved rim of conchoidal fracture: *a* joint plane; *d* conchoidal fracture surface

and the joint plane are parallel, there is sometimes an offset or shoulder of hackly fracture, separating the two planes". This structure, which is very prominent on some joints, is nevertheless not noted in his classification.

Discoids. Woodworth (1896, p. 175) describes groups of small, warped discoid joints on the surface of one of a set of large joints intruded by diabase dikes in the Mystic River quarries. The surface markings of these joints are radial rather than plumose: "At a number of localities in the argillaceous strata of the Boston Basin, but nowhere more clearly shown than in the Somerville district, are small discoid fractures having a diameter varying from half an inch to two inches. The fractured surface may be divided into two fields, one forming a narrow rim surrounding the other. The large central area (Fig. 3.1 e) consists of b-planes and c-fractures, the latter radiating outwards from the centre so as to expose wedge-shaped areas of the former". He continues: "The outer rim is conchoidal in its habit of fracture. The warped surface thus formed may in some specimens be likened to the rim of a soft hat curled up behind and pulled down in front. This rim of conchoidal fracture is in other cases curved upward like the edge of a saucer, so that there is a close resemblance to the ball-and-socket jointing of the prismatic columns of igneous rocks" (Fig. 3.1 f).

Thus Hodgson's citation from Woodworth's classification actually describes the essential fractographic elements on a joint surfaces. These include: (1) the fracture origin; (2) faint ridges that are the most common markings, and which appear in various styles, such as plumes; and (3) the rim of conchoidal fracture, all these on the joint plane; (4) b-planes and (5) c-fractures both on the fringe that either merge or do not merge into one another; (6) the shoulder that separates the joint plane and the fringe; and finally, Woodworth also introduces the (7) discoid joints.

Some markings that are described in Chapter 2 for technical materials have not yet been encountered or described in rock exposures or geological samples, though they may well be present. Fringes appear to be more diversified in fractured rock than in fractured synthetic materials, which makes this feature all the more intriguing to the geologist.

Many contributions made by investigators that followed Hodgson dealt with analogies between fracture surfaces in natural rock, and those simulated in technical materials (see Chap. 2).

3.1.2 Plumes and Related Structures

Ridges develop in various styles and geometries. They often occur as curved or sinuous lines called barbs. Series of barbs in a feather like structure (Figs. 3.2 a – c) have been described under roughly synonymic terms such as feather-fractures (which are quite distinct from feather-fractures which are associated with faults, see Friedman and Logan 1970 b) or hackle markings (the latter term is not accepted here for reasons given in Sect. 2.2). Barbs also occur

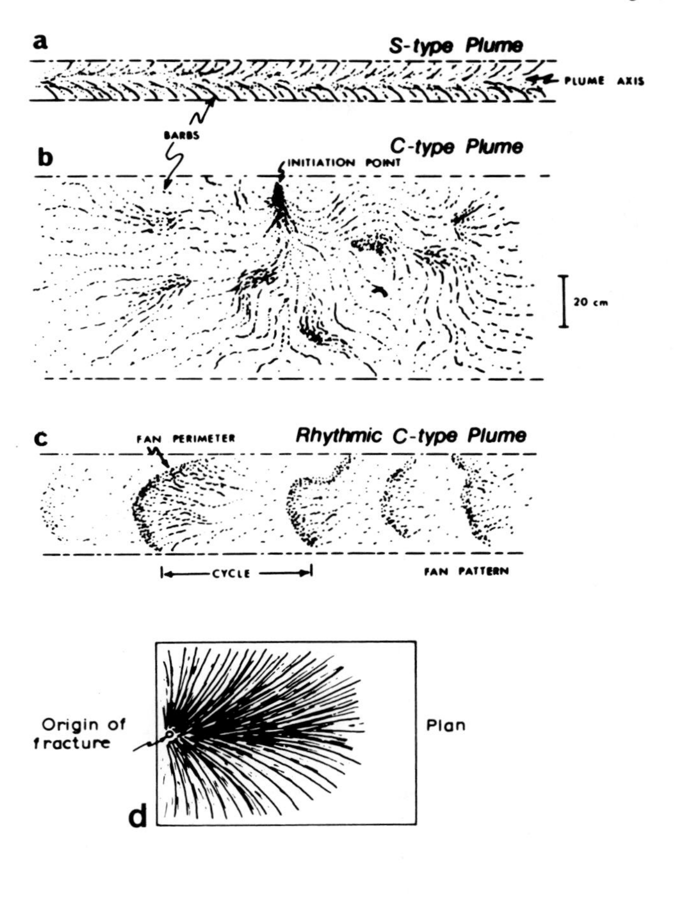

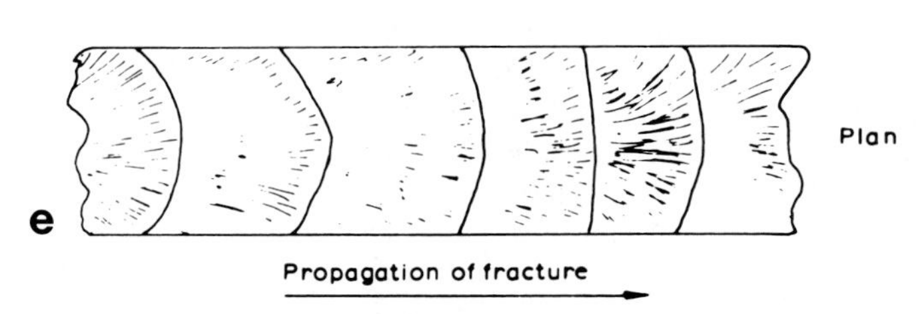

Fig. 3.2 a–g. Various plume patterns observed in siltstones of the Appalachian Plateau. (After Bahat and Engelder 1984). **a** Straight plume. **b** Curving plume. **c** Rhythmic plume. Barbs on each pattern are fine lines within the plume that mark the local direction of fracture propagation. *Fan perimeters* in **c** designate loci of arrest lines. The perimeters convex toward the direction of fracture propagation and the barbs may be partly branched due to lagging embayments (cf. Fig. 2.27 b). **d** Barbs fanning out from a common origin. **e** Combination of barbs and rib markings (**d, e** from Syme Gash 1971). **f** A spiral plume in limestone from the Franconian Alb. (Ernstson and Schinker 1986). **g** A bifurcating plume (from an Upper Cretaceous chalk in the Paris Basin) which starts as a single plume on the *right side* and branches towards the *left* into separate plumes in parallel two different joints (see photograph in Fig. 5.10 d)

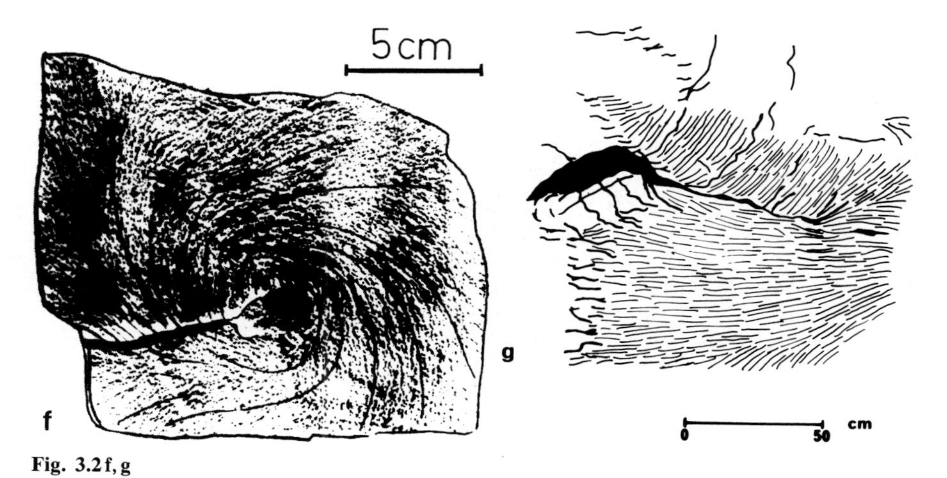

Fig. 3.2 f, g

as long radial straight ridges, or alternatively as periodic arc series of short radial straight barbs (see next section).

Straight barbs resemble striae, described in Chapter 2 (for details see Figs. 2.3, 2.4, 2.5h–l, 2.11 and 2.25). Plume classification provides the basis for further characterizations.

3.1.2.1 Classification of Plume Morphologies

The classification adopted here is a combination of Syme Gash (1971) and Bahat and Engelder (1984). Bahat and Engelder suggest a classification based on three plume end-member geometries, as occurring on joints from Devonian siltstones of the Appalachian Plateau: the straight or S-type plume (Fig. 3.3a, b); the curving or C-type plume; and the rhythmic C-type plume (Fig. 3.2a–c). The type B1 plume of Syme Gash in which the ridges radiate in a fan shape from a common origin (Fig. 3.2d) is considered here as a fourth end-member. The spiral plume (Ernstson and Schinker 1986, Fig. 3.2f) and the bifurcating plume (Fig. 3.2g) are rather rare types (the latter should be distinguished from barb branching, Fig. 3.7, see below).

The S-type plume is very similar to Syme Gash's chevron (or herringbone) type. The rhythmic C-type plume seemingly resembles Syme Gash's combination of plumes (which he actually calls hackles) and concentric rib markings (see below), but in fact they are not quite the same. Concentric rib markings are emphasized in the latter (Fig. 3.2e), whereas in the rhythmic C-type plume (Fig. 3.2c) the plume relief gradually intensifies and then periodically the plume disappears, designating crack arrest without producing continuous rib-like concentric ridges (see details in Fig. 3.7). The curving C-type plume (Fig. 3.2b) does not appear in Syme Gash's classification. Hence, the plumes are divided into straight, curving, rhythmic fan-like, radial, spiral and bifurcating types.

Fig. 3.3 a, b. Vertical straight plume in a fractured volcanic rock from the East African Rift in Kenya indicating upward fracture propagation. **a** Photograph. **b** Drawing

Less common than the pure plume geometries are plume combinations that consist of specific arrangements of several plume types. Two intriguiging plume combinations are the chaotic plumes (Fig. 3.4a), which consist of disoriented S-type plumes (Ernstson and Schinker 1986) and the vicinal plumes (Fig. 3.4b), which consist of two or more neighbouring plumes in a radial arrangement around a common origin and which maintain approximately the same strike.

3.1.2.2 Review of Observations on Plumes

Parker (1942), who examined Devonian siltstones and shales from the Appalachian Plateau in New York, notes that plumes are rare on strike joints (joints that are parallel or sub-parallel to the fold axis) but common on cross-fold joints (joints approximately perpendicular to the fold axis). He observed that the pronouncement of plumes changes with lithology. Plume axes run parallel to the bedding, usually separately on each layer. Some of the plumes are straight and others are sinuous, pointing in opposite directions. In addition, Parker notes that some of the wavy plumes turn up or down (but these should be distinguished from strictly vertical plumes, Fig. 3.3), and a few of these are seen to cross-layer boundaries. Parker remarks that fracture mor-

Fig. 3.4 a, b. Plume combinations. **a** Chaotic plumes. (Ernstson and Schinker 1986). **b** Vicinal plumes, markings *A* and *B* initiated from a common centre (*arrow*). *A* propagated downward and *B* propagated upward (see photograph in Fig. 5.24)

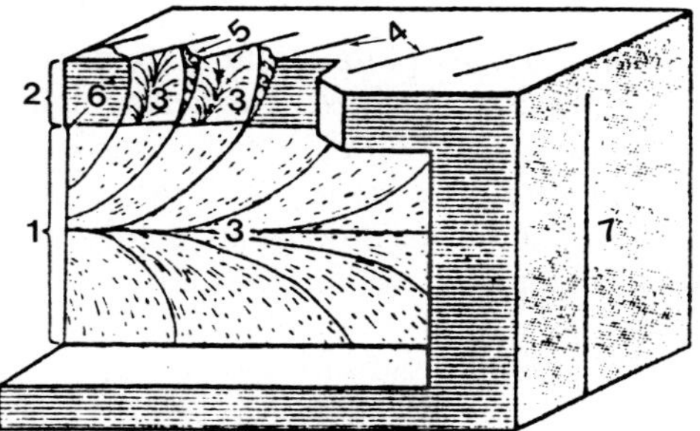

Fig. 3.5. Schematic block diagram showing primary surface markings of a systematic joint. (Hodgson 1961 a). *1* main joint face; *2* fringe; *3* plumose structures (plumes); *4* F-joints (b-planes); *5* cross-fractures (c-fractures); *6* shoulder; *7* trace of main joint face

phologies on opposite walls of a joint correspond, indicating that they constitute the fracture surface itfself, and not a subsequent coating.

Raggatt (1954) describes radial fracture markings on closely spaced, vertical to near-vertical joint surfaces in brownish black lignitic siltstones. The radial diameters commonly range from about 20 cm to about 60 cm.

Hodgson (1961 a) distinguishes between joints in systematic and non-systematic categories: fracture markings are well displayed on systematic joints, while in non-systematic joints surface markings are irregular (non-oriented) or of nondescript character. In a schematic block diagram, he shows

Fig. 3.6 a, b. Transitional markings. **a** Fracture marking on a horizontal joint on the ceiling of an artificial, bell-shaped cavern excavated in Middle Eocene chalks near Beit Govrin. Note the concentric ribs, a sub-horizontal plume at the *lower part* of picture, and radial hackles on the *outer periphery* of the distal ribs (see drawing in Fig. 4.4). **b** Superposition of rib markings and plumes on vertical joints in the Lower Eocene chalks near Beer Sheva. *Bar* = 70 cm

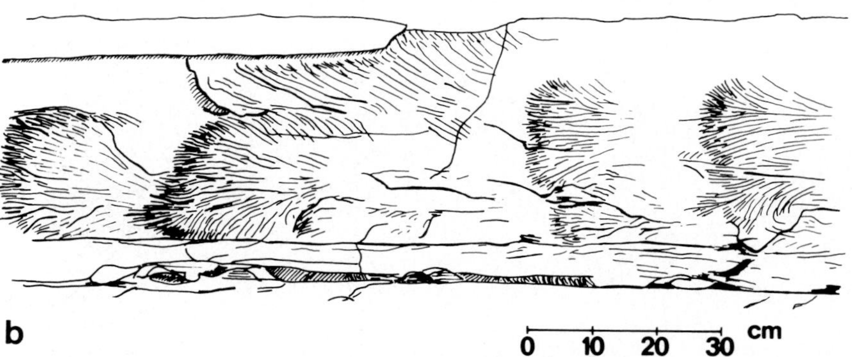

Fig. 3.7 a, b. Cyclic fracture propagation in a siltstone from the Appalachian Plateau represented by rhythmic plumes propagating in opposite directions in mineral precipitates along two adjacent beds. The fan perimeters occasionally display barb branching: they are indented at branching points, which form smaller fans. **a** Photograph, note 15 cm scale. **b** Drawing

some of the primary surface features of a systematic joint (Fig. 3.5). This diagram, which is a considerable improvement on Woodworth's diagram (Fig. 3.1 a), has guided geologists for a good number of years.

Roberts (1961), who investigated fracture markings in rocks of varying lithologies, found that plumes (fearther markings) occur most frequently in fine-grained rocks, and flaggy beds normally have well-developed plumes and fringes. He noted the resemblance of the herringbone pattern in cleavage fractures of mild steel (Sect. 2.1.2.6) to the plume patterns found in the joints of natural rocks. He also comments on outward-radiating plumes that develop as a result of blasting in quarries.

Cegla and Dzulyaski (1967) show plumes on surfaces of disjunctive cracks developed in semi-consolidated Recent and Pleistocene silts and in experimental material (see Sect. 3.2.1.2).

Bahat (1979a, 1980a) adapts the terminology of fracture surface morphology from the ceramic literature (Rice 1974). In so doing, he discriminates between striations (preferably termed striae) that occur closer to the fracture origin and which presumably reflect relatively low stress intensities, from hackles that develop closer to the outer boundaries and which possibly reflect greater stress intensities (Fig. 3.6a). The plumes are identified with striae (Sect. 2.1.2.2).

The ubiquity of fracture markings on vertical joints has been established by previous authors. Bahat (1980c), on the other hand, shows that arched undulations may be crossed by delicate plumes and also marked by coarse hackles on horizontal joints (sheets) (Fig. 3.6a).

Bahat and Engelder (1984) demonstrate various styles of plume initiation, starting mostly at layer boundaries, but also at random locations within the layer. Plume axes are commonly curved near initiation points and straighten along the direction of fracture. Distinct C-type plumes on adjacent beds

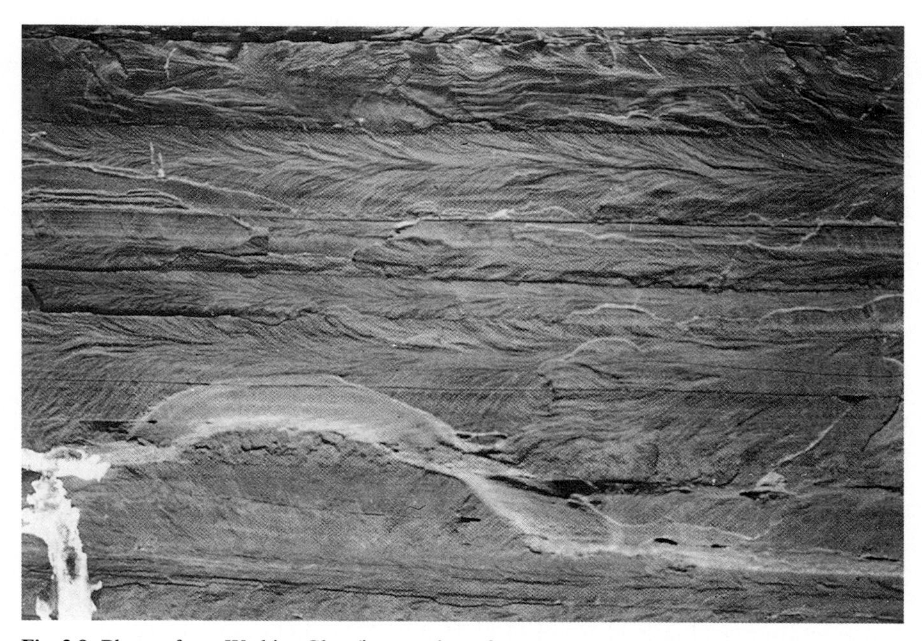

Fig. 3.8. Plumes from Watkins Glen (intersection of routes 414 and 79 west), New York. Approximately straight plumes, with slight wavering and barb curving. Note that the plumes are generally confined to individual beds, but crossing of the layer boundaries is also observed (see drawing in Fig. 4.5)

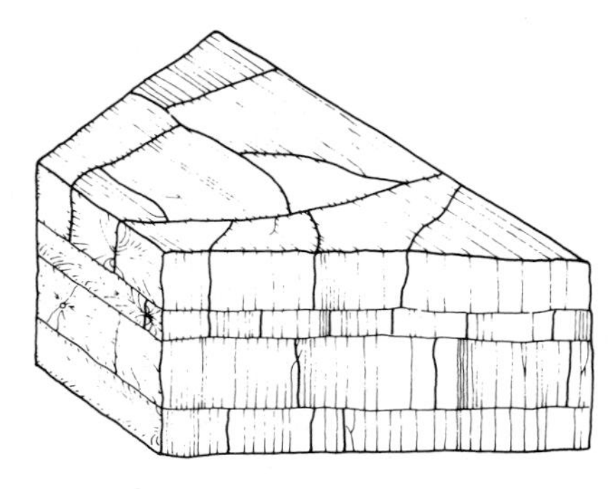

Fig. 3.9. Closely spaced jointing with plumes in Malm Zeta limestones in the Southern Franconian Alb. (Ernstson and Schinker 1986)

generally show entirely different patterns, indicating that they fractured independently (Fig. 3.7).

Various combinations of three plume types (Figs. 3.2a–c and 3.8) are distinguished on joints in thick siltstone beds west of Watkins Glen, New York. Of particular interest are the various propagation styles of the barbs. In approximately straight plumes the barbs generally curve from the central horizontal axis towards the upper and lower layer boundaries, sometimes with a wave or two (Figs. 3.8 and 4.17a). No branching occurs. In rhythmic (Fig. 3.7) and curving (Fig. 3.2b) plumes, barb branchings are visible, resembling the curving and branching of vein systems in leaves, and implying discontinuities in the fracture front (see also Sect. 3.1.2.7). Barb branching occurs on the same fracture surface, unlike a bifurcating plume (Figs. 3.2g, 5.10d), which splits along two separate fracture surfaces. Kies et al. (1950) suggest a more generalized concept for the former discontinuity. The fracture front is roughly convex towards the direction of propagation, but in detail it connects many lagging embayments (see Fig. 2.27b). This description fits quite well the fronts of rhythmic plumes (Fig. 3.7) where barb branching occurs due to lagging embayments (Fig. 3.2c).

Ernstson and Schinker (1986) describe a set of vertical joints, striking $110° \pm 5°$ in Malm Zeta limestones (on the southern Franconian Alb). The joints are distinctive by reason of their close spacing, generally between 1 and 20 mm (Fig. 3.9). In addition, a multitude of parallel surfaces of reduced cohesivity occur between the open joints. These are recognizable by discolorations and by signs of corrosive action. Plumes, generally of the S style, occur on all surfaces and are typical of the 110° set. They do not appear on adjacent older or younger joints which are of different orientation. The 110° joints with their plumes are not deflected in the vicinity of older joints. However, where joints interfere with each other, no plumes are found on the diverted joint surface. The same combination of closely spaced jointing, directional persistence and fracture marking is also found in Middle Triassic limestones (Wellenkalk and Muschelkalk) but not in other lithologies.

Closely spaced jointing or microfracturing is also described by Tuttle (1949), Wise (1964) and others. Spacing between planes in these structures varies from several mm down to several μm, and the planes maintain a remarkably uniform orientation over hundreds of square km. Tuttle observed these planes in quartz which behaves like an isotropic material with respect to jointing.

3.1.2.3 Dimensions of Plumes, Ridges and Barbs

Information on the sizes or ridges, barbs and plumes is rather scarce. Parker (1942) found that the plumes on cross-fold joints in Devonian siltstones in New York consist of a series of minute, irregular ridges, 6 mm or less in width, with a relief of about 1.5 mm. Roberts (1961) observed that in fine-grained rocks and flaggy beds ridges are closely spaced, with intervals of 0.25 to 0.50 mm. Bahat (1987 a), describing joints in Eocene chalks of the Beer Sheva syncline, notes that the straight axial plumes (Fig. 3.10) of the Lower Eocene are coarse and their barbs are 0.5 cm to several centimetres wide. In the Middle Eocene, on the other hand, the straight axial plumes are delicate and their barbs less than 0.5 cm wide (Fig. 3.11).

3.1.2.4 Ridge Microstructure Close to the Surface of the Joint

Woodworth (1896) made the important observation that the faint ridges on the joint plane are minute laminae formed by microcracks that enter the rock at

Fig. 3.10. A coarse straight plume in Lower Eocene chalks near Beer Sheva. This is a bilateral asymmetrić plume whose long axis (127 cm) is confined within an elliptical arrest line. The limb that propagates from the origin *toward the left* is 96 cm long. The limb that propagates *toward the right* is about three times shorter. *Bar* on the right = 58 cm (see drawing in Fig. 4.1 c)

small angles and which in turn are cut three-dimensionally by other fractures. Such laminae are not detached from the joint plane but remain attached to the rock at various points.

Syme Gash (1971) studied the microstructure of ridges in the median, chevron and fringe zones of a plume (Fig. 3.13) in limestone. Thin sections across the ridges reveal that the microcracks that form grooves in the median and chevron zones are approximately normal to the joint surface, have greater depth than width, and often penetrate into the rock along a line of fracture (Fig. 3.12a). In the fringe zone the groove profiles are notably different. They are about three times greater than in the median and chevron zones, and their microcrack continuations penetrate the rock to a depth of about 0.05 mm at angles from under 10° to about 30°, causing inclined and even overhanging ridges along the grooves (Fig. 3.12b). Syme Gash also notes similarities between fringe profiles and microcracks in the chevron marking of fractured steel described by Tipper (1957), indicating that bedding plays no controlling role.

The behaviour of "sub-surface cracks" has been treated by various material scientists from different points of view (see Sects. 2.1.2.4 and 2.1.2.5, observations by Preston 1926; Kies et al. 1950; Johnson and Holloway 1968).

3.1.2.5 Zonation of the Joint Plane and Angular Relations of Plumes

Syme Gash (1971) divides the joint surface marked by a straight plume (which he calls chevron structure) into three zones: the median zone along the plume

Fig. 3.11. A delicate straight plume and a fringe of echelons along the lower layer boundary in Middle Eocene chalks near Beer Sheva. *Bar* = 15 cm

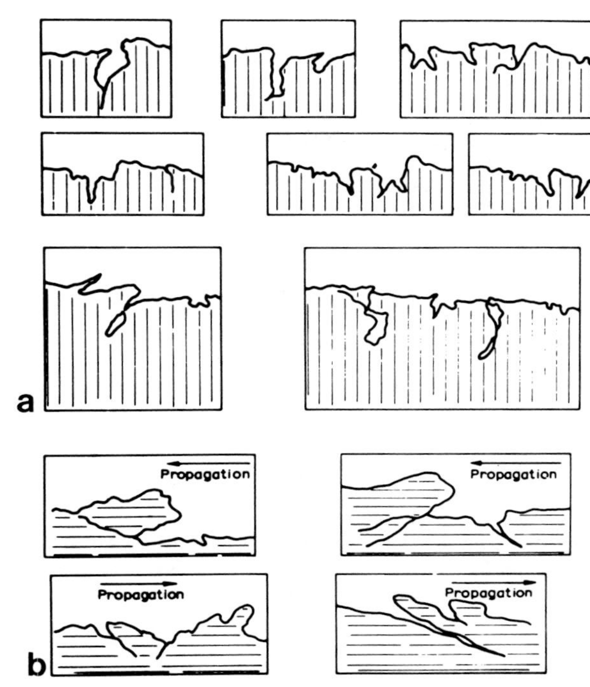

Fig. 3.12 a, b. Cross-sections of ridges in plumes. **a** The chevron and median zones (see zonation in Fig. 3.13). **b** The fringe zone. *Width* of a rectangle sketch in **b** is about 0.07 mm

axis, the two chevron zones on both sides of the median zone (still on the main joint plane) and two fringe zones outside the two shoulders of the main joint face (Fig. 3.13). This is essentially an elaboration of Hodgson's diagram (Fig. 3.5). He observes that the median angle m produced between the plume axis and the curving barbs at the median zone, often indistinct, varies over a wide range of angles and is not diagnostic. On the other hand, the interface angle i produced between the barbs and the boundary of the main joint plane varies with distance from the origin but is constant for any one plume specimen and is considered to be a diagnostic parameter for the analysis of plumes. Syme Gash distinguishes three ranges of interface angles (Fig. 3.14): A from under 30° to over 60°, with a distinct peak at 45° for joints without fringe zones in sedimentary rocks; B from under 50° to over 80° with a peak at about 60° for sedimentary rocks, and C from under 60° to over 90° with a peak at about 80° for metals that show fringe zones. It appears that in group A the free surface (or high reflectance surface) is associated with lower interface angles than the other two groups at which fringe zones are usually present. Syme Gash's results are comparable to those of Roberts (1961). Roberts finds that the barbs form a constant angle of 40 to 45° with the axis, but on approaching the axis the angle decreases to 30–35°, while at the joint fringe it increases to 70–75°. For further treatment of this subject, see Section 4.2.1.2.

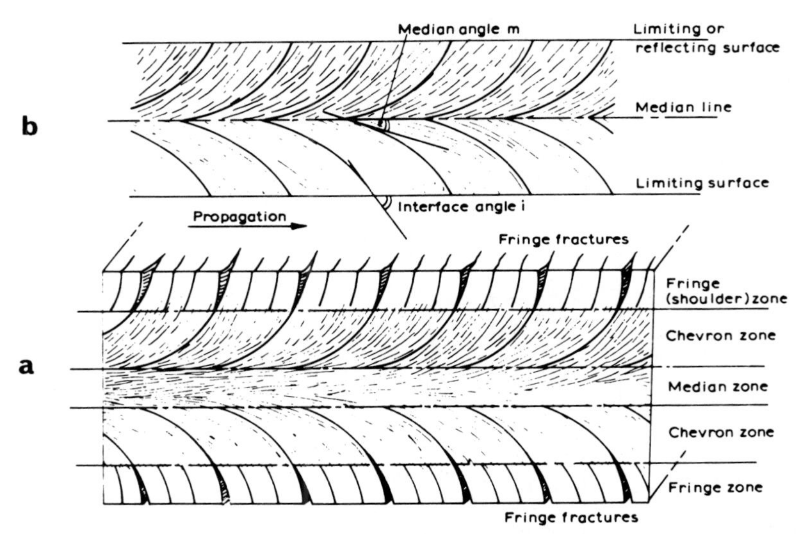

Fig. 3.13. a Subdivision of a joint plane into median zone, chevron zones and fringe zones. The fringe zone is separated from the joint plane by the shoulder. **b** The corresponding median angle *m* and interface angle *i* on a joint plane which has a clear layer boundary (a reflecting surface)

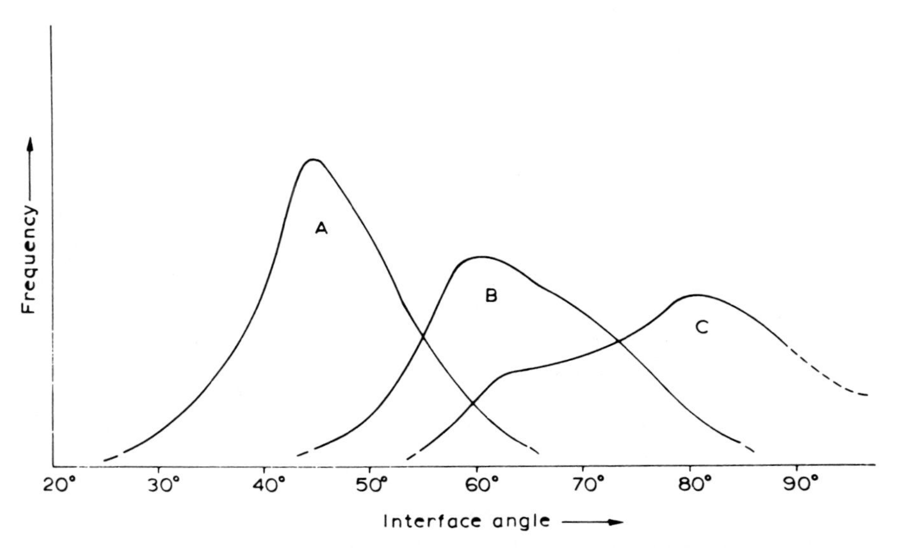

Fig. 3.14. Frequency distribution diagram of the interface angle *i* for markings in different materials. (Data from multiple specimens, photographs and reports). *A* interface angles for markings in sedimentary rocks, where the interface is a free, or high-reflectance, surface; *B* as for *A*, but where the interface is an imperfect reflector; *C* interface angles for markings in metals (usually at a free surface). A fringe zone is usually present in *B* and *C*, and absent in *A*

In straight plumes (that occur in rocks and technical materials) the barbs are curved with their concave side facing the layer boundaries and the convex side pointing towards the layer median.

3.1.2.6 A Model of Plume Generation by Stress Waves

Syme Gash (1971) sugggests a model of plume generation by stress waves. He divides the main fracture into C-T zones where pulses of compressive and reflected tensile stresses intersect, and T-T zones where pulses of tensile stresses intersect (Fig. 3.15). Accordingly, microfractures – usually on grain boundaries – are caused by the passage of interfering incident and reflected stress waves emitted from the running parent fracture. They cause a zone of weakness which predetermines the main fracture path, and enables it to propagate at much lower stresses than are required for initiation. As the fracture propagates, it incorporates these preformed flaws, which are expressed as surface markings. When the greatest tensile principal stress (negative) is greater than, or equal to, the tensile strength of the material, tensile microfracturing will result perpendicular to it. When the greatest tensile principal stress is less than the tensile strength of the material, either the stress difference ($\sigma_1 - \sigma_3$) is low and there is not microfracture initiation, or the stress difference is high and microfracturing will result at an angle to the σ_1 direction. The size of this angle depends on the magnitude of the stress difference and the least principal stress σ_3 for given conditions depicted by various parabolic Mohr envelopes. Hence in metals, which have very high tensile strengths, it is unlikely that the

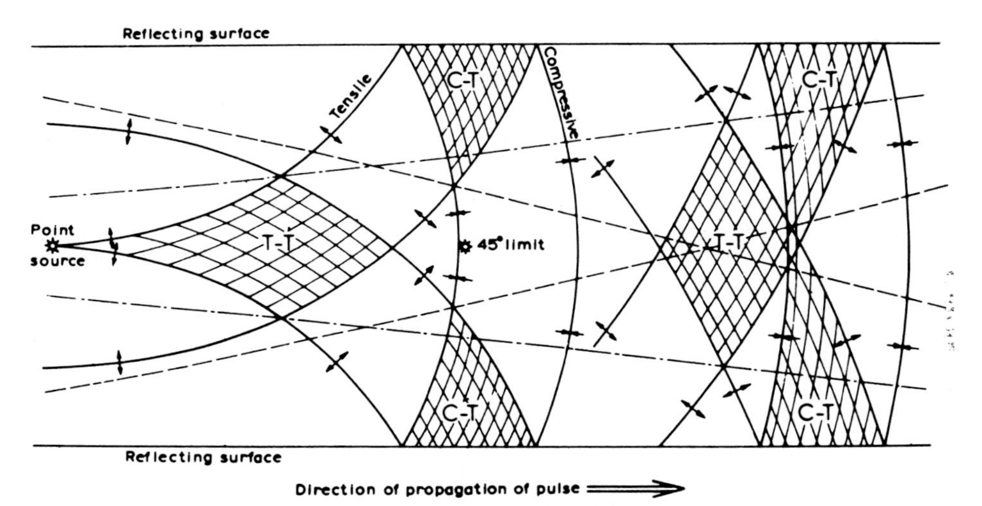

Fig. 3.15. Diagram showing the path of an advancing compressive pulse with a reflected tensile pulse in its wake. *C-T* zone of intersecting compressive and tensile stresses. *T-T* zone of intersecting tensile stresses. *Dashed line* shows the limits of C-T zone. *Dot-dashed line* designates limits of T-T zone. The zones resulting from pulse duration = 0.3 · t (separation of reflectors)/Vl (velocity of primary wave) · s (Figs. 3.12, 3.13, 3.14 and 3.15 from Syme Gash 1971)

greatest tensile principal stress will be greater than the tensile strength. Shear microfracturing would occur in a C-T zone due to stress difference, but not due to tension in a T-T zone. In rocks with low tensile strengths, the greatest tensile principal stress could easily exceed the tensile strength. Therefore tensile microfracturing would occur in the T-T zone and also in the C-T zone adjacent to a good reflector (for a high σ_3 value). Shear microfracturing would occur in the C-T zone adjacent to a poor reflector (for a low σ_3 value); the lower the value of σ_3 (negative) the higher the angle of the shear microfracture to the σ_1 direction.

Syme Gash considers that fan-like plumes (Fig. 3.2d) are the result of single sporadic progressive shock sources, while straight plumes (the chevron type, Fig. 3.2a) result from running fractures with an advancing shock source, either multiple or continuous. The straight plumes are indicative of high-energy, fast-running fractures with a running shock source.

Syme Gash suggests that a running fracture is dependent on the energy-to-thickness ratio. If the stratum or test plate is below a critical thickness, a running fracture with straight plume will result. A fracture will run if the reflectors are sufficiently close for the reflected shear stresses to reach the fracture apex before it is arrested. If the fracture has sufficient energy to pass the angle of total internal reflection (36° for a rock with a Poisson ratio 0.25) it will run to completion. If the stratum is too thick, fan-like plumes will result from the limited shock source. Fracturing in metals proceeds at much higher energy/thickness ratios than in rock under natural conditions.

Syme Gash concludes that plumes are indicative of shear fracture but could possibly develop on fractures perpendicular to tensile fractures as a subsequent event, or on complex tensile fractures.

In general, shear microfractures with high angles of p wave reflection form the fringe zone, while tensile microfractures form the chevron and medium zones. A poor reflector will produce wide fringe zones and high interface angles, while a good reflector will have narrow fringe zones, if any, with tensile microfracturing parallel to the σ_1 directions and hence low interface angles.

Ernstson and Schinker (1986) disagree with the two main theses of Syme Gash's (1971) theory, which argues that fracture markings are the product of elastic wave interference. They present various arguments against the first thesis which states that plumes are the image of microcracks formed in the tensile field of interfering elastic waves within the perimeter of the spreading fracture front, and suggest that Syme Gash's model may explain the development of simple plumes, but cannot account for complicated marking morphologies like spiral plumes in the vicinity of hollow spaces or on strongly curved joint surfaces. Furthermore, they argue that Syme Gash's theory does not explain bilateral plumes, plumes of transitional styles and other peculiarities.

With regard to Syme Gash's second thesis that plumes are overall diagnostic of shear fracture, Ernstson and Schinker state that plumes appear clearly and systematically in tensile fractures.

The hypothesis of microfracturing induced by stress wave still needs to be proven. Particularly so because straight plumes are believed to grow in nature also (if not exclusively) in a sub-critical slow process, in sedimentary rocks due to rhythmic increase and decrease of stresses (Figs. 3.2c and 3.7a), in columnar joints formed in basalts due to periodic thermal gradients (Fig. 3.50), and in mud cracks due to stress gradients that accompany gradual drying (Figs. 3.17d and 4.18).

3.1.2.7 Lateral and Vertical Relationships of the Plume with the Stratum

It is generally accepted that a joint in an unconstrained medium propagates along a circular front even if the areas of initial flaw or critical flaw (Fig. 2.2d) are elliptical. The argument for this is that the stress intensity factor (Sect. 1.1.4) is higher at the smaller dimension of the flaw (Irwin 1962). This tends to increase the small dimension at a faster rate than the long dimension, which results in a decrease of ellipticity. In unconstrained rocks, circular annular fracture markings are fairly common (Fig. 4.2a). However, stress differences in rocks parallel to joint surfaces may cause deviations from circular perimeters. Fracture morphology in layered rocks often displays radial geometry close to the origin of the fracture (Figs. 3.16, 3.17b) and rapidly transforms into a bilateral plume that propagates horizontally, parallel to the layer boundaries (Fig. 3.17a). The vertical propagation (being constrained by the layer boundaries) is very limited. There is an additional constraint that appears to be fairly common: the joint does not propagate really bilaterally, rather, it advances in one direction while in the other direction it is usually soon arrested. This type of partial bilateral propagation (Fig. 3.10) should be distin-

Fig. 3.16. Radial plume with a circular perimeter 20 cm across starts to grow to a horizontal plume, shown within an elliptical perimeter. *Pen* below the plume shows scale

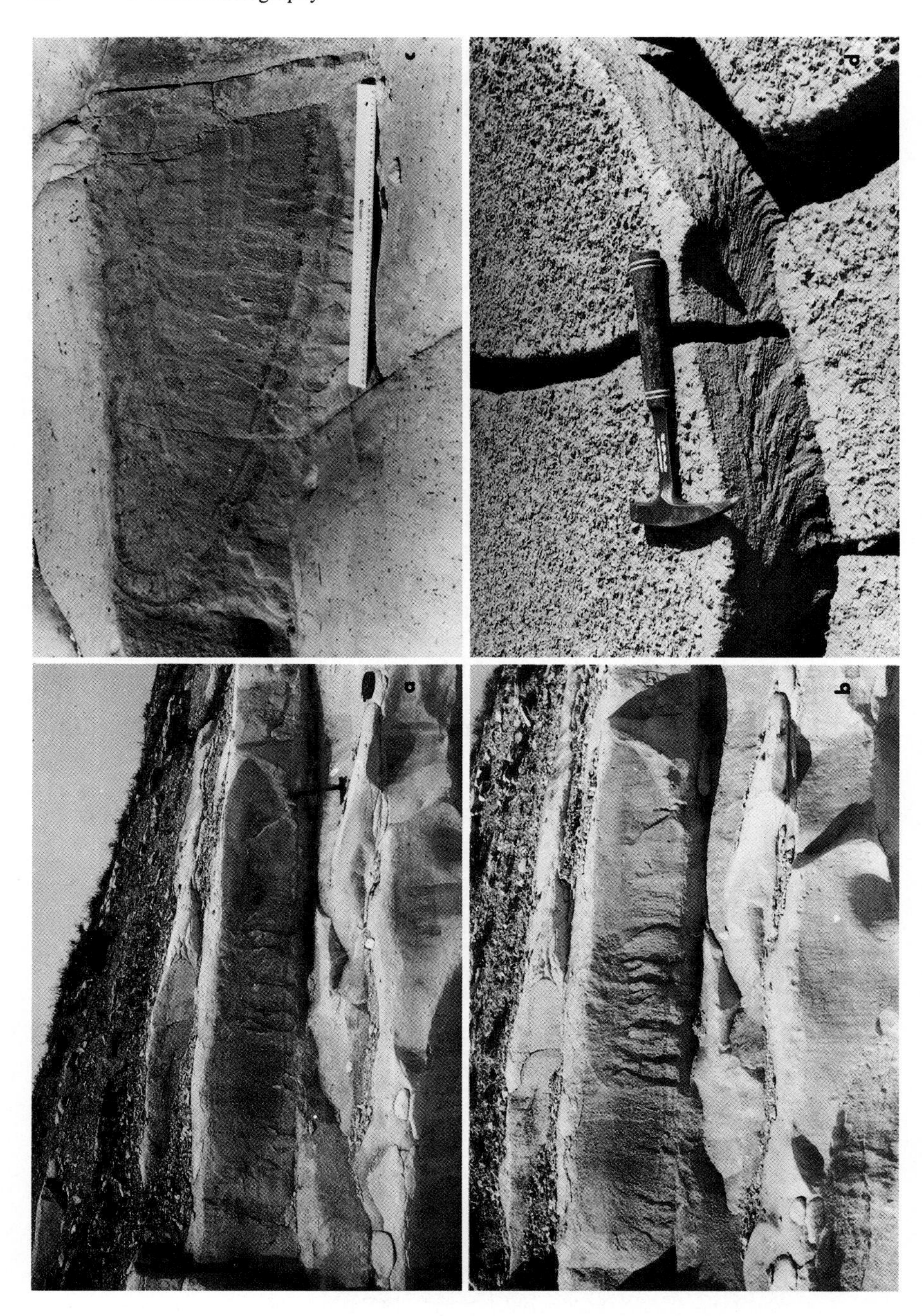
a
b
c
d

guished from a unilateral plume whose fracture origin is close to a free surface (previous joint) and which can propagate only away from the surface but cannot cross it (Fig. 3.17c). This type of plume is obviously distinct from a portion of a plume that visibly propagates in one direction only, but in fact it is a remnant of an originally bilateral plume whose other portion has been erased.

Distinctive bilateral and unilateral plumes are known from joints in Devonian siltstones on the Appalachian Plateau. Here one can distinguish between purely unilateral plumes, and single branches of bilateral plumes which initiate at layer boundaries, diverge at the middle of the layer, and then propagate horizontally in opposite directions. Particularly spectacular is a single plume 48 m long which propagated from SE to NW (at centre of Fig. 3.18). The length of its propagation path from NW to SE is not exactly known because the location of plume initiation is not certain (see question mark at initiation 6 in Fig. 3.18). However, the branch from NW to SE could not be longer than about 2 m before interacting with another plume which propagated from SE to NW (see Bahat and Engelder 1984, Fig. 4).

Apart from lateral plume asymmetry parallel to the bedding there is also asymmetry perpendicular to the plume axis. Woodworth (1896) discusses the plume axis "where a straight line is not usually in the middle of the stratum". Such a plume reflects a stress gradient normal to the plume which results in its propagation closer to one boundary. In many mud cracks the plume has barbs that initiate close to the surface and curve only downward (Fig. 3.17d). In columnar joints, on the other hand, the barbs curve mostly upward from an axis in a region of strong stress towards a more ductile zone with diminished stresses (Fig. 3.50). It is quite possible for a plume to initiate at the layer boundary and to propagate close to the middle of the layer (Fig. 3.18), preserving symmetry in relation to the plume axis.

3.1.3 Rib Markings

Woodworth's (1896) data on joint fracture morphology were obtained from argillitic rocks in which rib markings were not considerably developed. This led him to comment that the rim of conchoidal fracture, under which he included also rib markings, appeared to be restricted to discoid joints (see below).

Hodgson (1961a, b) investigated joints in sandstones on which rib markings (also variously termed as undulations, ripples, curvilinear ridges with conchoidal appearance, and arcuate ridges) are well developed. He constructed a schematic block diagram of the joint surface showing the morphologic rela-

Fig. 3.17a–d. Growth styles of plumes. **a** A bilateral plume propagating both left and right from the origin. **b** The same plume is at **a**, note distinction between an early semi-elliptical joint at the centre and the final bilateral joint (see drawings in Fig. 4.1a, b and Sect. 3.1.5.1). **c** A unilateral plume. At the *right end* of the 50-cm scale there is a joint normal to the page. The plume initiated close to this free surface and propagated towards the left but did not cross the free surface. **d** A mud crack. The horizontal plume axis is nearer the upper margin of the plume

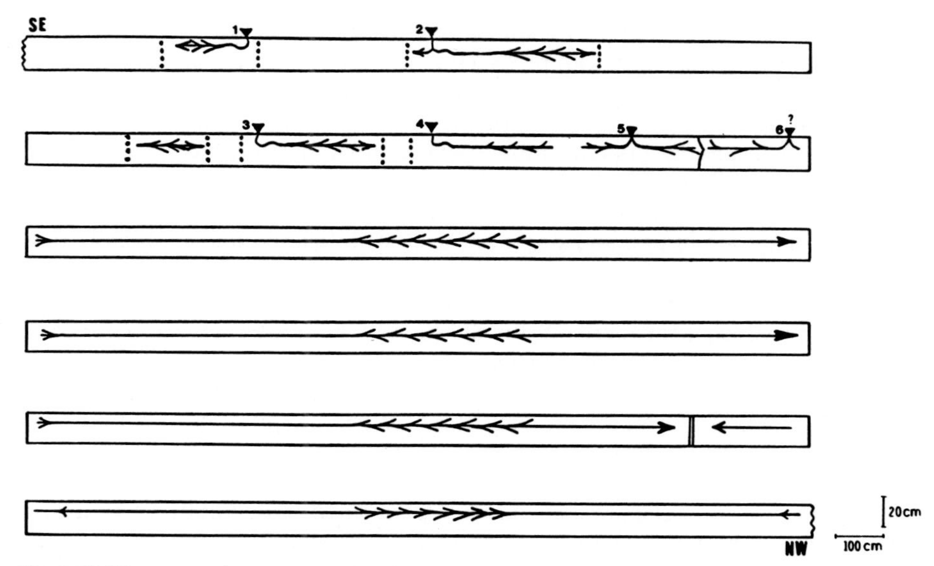

Fig. 3.18. Diagrammatic summary of various surface features of a joint from Watkins Glen, New York. A single joint surface, 100 m long, is presented in six successive sections from SE to NW. *Triangles* mark the points of plume initiation. The plumes that are closer to the SE end that initiate at points *1, 3* and *4* are unilateral, and those that initiate at points *2, 5* and *6* ar bilateral. However, the plume that starts at point *6* is notably asymmetrical. The branch that propagates from SE to NW is about 25 times longer than the branch that propagates from NW to SE. Note the slight curviness of plumes along the 30 cm or so, between the initiation point and the section where they straighten parallel to the bedding. Note also the plume at the bottom that initiated at some point beyond the NW end of the joint and propagated towards the SE. *Blank regions* (*between vertical dotted lines*) designate eroded areas. The *vertical double straight lines* mark a node which reflects an interaction between two joints that propagated in opposite directions on a slightly non-coplanar surface. (Bahat and Engelder 1984, Fig. 5). Note differences in vertical and horizontal scales

tionship between plumes and rib markings (Fig. 3.19). Hodgson observed that the plume barbs impinge radially on the concentric rib markings, expressing a directional relationship which resembles the orthogonality that was identified by De Freminville and Preston between the rib marks and striae on glass fractures (Sect. 2.1.2.3).

3.1.3.1 Classification of Rib Markings

Rib markings may be grouped according to either strongly arcuate or divergent, slightly arcuate geometries. Syme Gash (1971) classifies the arcuate rib markings into two types. Type 1 consists of arcs of approximately similar curvature but not concentric to the origin of fracture (Fig. 3.20a). A series of sub-parallel parabolic undulations (demonstrated by Kulander et al. 1979, Fig. 33) falls into this group. Type 2 includes patterns of arcuate to hemicircular ridges which are concentric to the origin of fracture (Figs. 3.20b and 3.21).

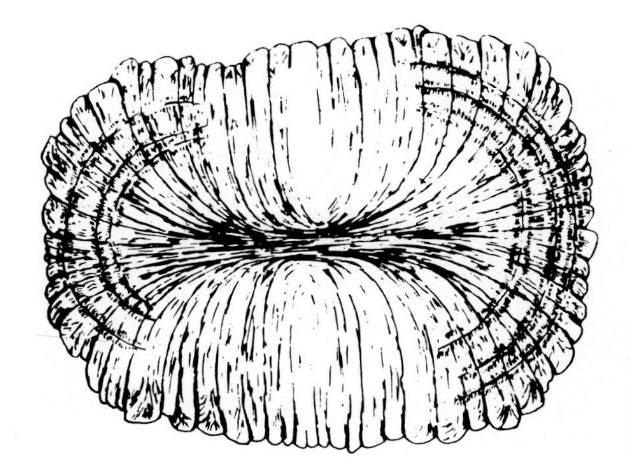

Fig. 3.19. Sketch of orthogonal crossing of plume and rib markings as displayed on a complete face of a systematic joint. (Hodgson 1961b)

Bankwitz (1966, Fig. 5) presents a scheme of concentric rib markings on a discoid (ring) joint (Sect. 3.1.7), and Bahat (1987a) elaborates on concentric series of circular rib markings (varying in number from 1 to about 15) in annular flat geometries (Fig. 3.22). Such concentric circular undulations may be classified as Type 3. In some cases they trend towards ellipticity (Fig. 3.19), usually with the long axis parallel to the bedding.

Rib markings of Type 3 indicate crack initiation within the layer. In Type 2, the fracture probably initiated at a surface such as an earlier joint or a layer boundary.

Another style of arcuate rib markings is classed as Type 4. These have an apparent centre but they are sequential, or they may deviate from circularity and appear as asymmetric, quasi-spiral series or as individuals, occasionally interfering with each other (Fig. 3.23).

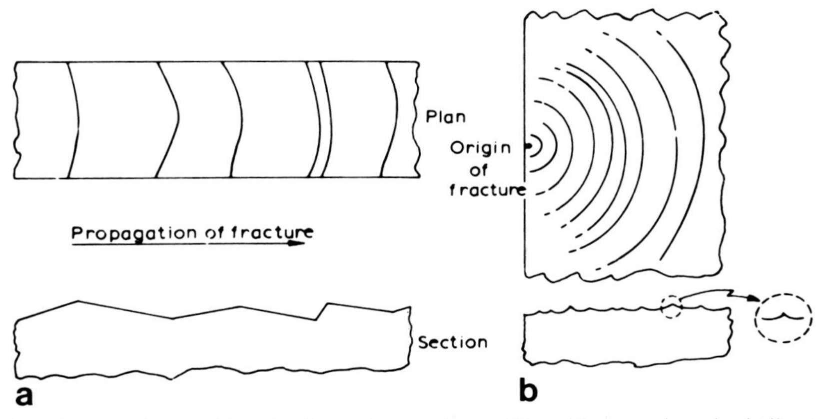

Fig. 3.20 a, b. Rib markings in plan and in section. **a** Ribs with approximately similar curvature, but without a common origin. **b** Semi-circular concentric ribs with a common origin. Corresponding cross-sections of the ribs are shown below. (Syme Gash 1971)

Fig. 3.21. Semi-circular undulations on 327 °C joint. (Outcrop 20, Bahat 1987a). The *bar* in the *lower part* of the picture is 35 cm. Note that the fracture markings are particularly abundant along the upper layer (see drawing in Fig. 4.8)

Divergent undulations are markings without a center. They stem from an origin or an imaginary origin and diverge fanwise (Figs. 3.24 and 5.21). Occasionally, truncated sub-circular markings may resemble divergent structures (Fig. 3.25). Strongly arcuate and divergent rib markings appear on both flat and curved surfaces.

3.1.3.2 Rib Markings in Cross-Section

Solomon and Hill (1962) noted that the crests of the undulations on a joint fringe are rounded and slightly asymmetrical. Bankwitz (1966, Figs. 5 and 29) shows schematically that rib markings on discoid joints may be either rounded or sharp. Syme Gash (1971) correlates arcuate rib markings that are non-concentric to the origin (Type 1) with a profile formed by the junction between non-parallel surfaces (Fig. 3.20a) and Type 2 rib markings with a cuspate profile (Fig. 3.20b). Barton (1983) and Kulander and Dean (1985) make similar profile distinctions without correlating them to particular types of rib markings.

A section examination of concentric circular rib markings on two matching sides of a single-layer joint in chalk reveals an approximately symmetric rounded section (Fig. 3.26a). Wavy round cross-sections with considerable asymmetries seem to be typical of many circular rib markings (Fig. 3.26b). Note

Fig. 3.22. Circular rib markings in Senonian chalk near the town of Arad, southern Israel. Note the ellipsoidal shape of the inner initial crack and first rib

that in Fig. 3.26b the slope that faces the fracture origin is considerably longer than the one facing the fracture end and its slope is gentler. Nevertheless, rounded section with reverse asymmetry is observed on an impact fracture in chert (Bahat 1977b), with the shorter and steeper slope facing the fracture origin (Fig. 3.26c). On fracture surfaces in limestone concretions, presumably caused by a natural thermal process (Sect. 3.1.9.3), the undulations are also asymmetric and the section appears irregular (Fig. 3.26d). Table 3.1 shows ridge height h, wave length l, and steepnesses (h/l) for various sections in Fig. 3.26. Unexpectedly, the steepnesses of three rib markings that differ in lithology, size, shape and presumed fracture conditions are quite similar.

Discoid joints from flint clays (Sect. 3.1.7.2) display the full spectrum from symmetric rounded through various asymmetries to cuspate profiles. A single discoid may reveal a wide variety of adjacent profiles of different wave length (Fig. 3.26e). Furthermore, in discoids one finds all gradations from near-flat

Fig. 3.23. Arcuate rib markings with an apparent common centre – but asymmetric and non-circular – on a joint surface in Palaeozoic sandstone, Wadi Tweibeh, northern Sinai (see drawing in Fig. 4.2b)

joints to conic joints (Fig. 3.43), bearing in mind that a matching of a "positive" (convex) conic discoid is a "negative" (concave) conic discoid. Generally, rounded profiles occur on flat surfaces, and cuspate profiles connect surfaces from different levels. As discoids commonly occur in groups, surfaces of various morphologies often appear laterally adjacent to each other.

The profile in Fig. 3.26a resembles swells in the open sea (Press and Siever 1982, Fig. 10.9). The section in Fig. 3.26b is similar to the asymmetric ripples which form in experimental sand beds under water currents at various flow velocities. Cuspate crests in discoids (Fig. 3.26e) resemble the sharp crests of oscillation ripples that are formed by the back-and-forth movement of waves (Fig. 3.27). The geometry of the sand bed is determined by three main variables: the flow velocity, the grain size of the sediment, and the depth of the flow (Press and Siever 1982, pp. 159–160). The resemblances between the morphologies of fracture surfaces in brittle materials to surface patterns of low-cohesion sand dunes is impressive: even the fine structure of delicate ripples that climb over larger ones at the backs of dunes has its analogy in frac-

Fig. 3.24. Divergent rib markings that stem from an origin and fan out on a curved horizontal joint surface in Middle Eocene chalks. Joint on the ceiling of an artificial, bell-shaped cavern near Beit Govrin. *Bar* = 20 cm

Fig. 3.25. Sub-circular rib markings which are truncated by a layer boundary and appear to be divergent. Lower Eocene chalks near Beer Sheva. *Bar* = 1.0 m

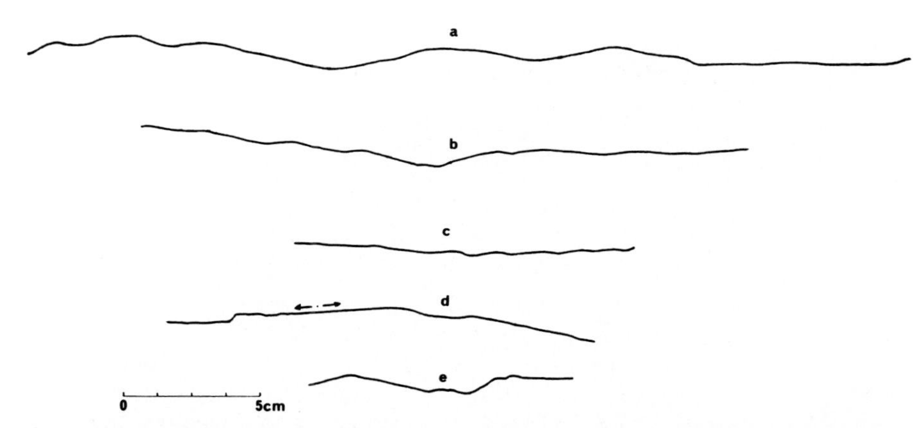

Fig. 3.26 a–e. Profiles of rib markings drawn on sections cut from gypsum casts of rock surfaces. **a** Single-layer joint, Lower Eocene chalk, near Beer Sheva. **b** Turonian limestone from the Judean Desert. **c** Artificially impacted chert (prehistoric tool) (Bahat 1977b). **d** Presumably thermal fracture in a limestone concretion from Tsin Valley. **e** Discoid joint in Ramon flint clay. Propagation direction in **a–c** and **e** *from right to left*. Bilateral propagation in **d** is shown by *two arrows* diverging from the point of initiation

ture markings of technologic materials (Sect. 2.1.2.6) and rocks (e.g. Bahat 1977b, Fig. 3f; Fig. 3.26e). What are the mechanisms that form joint surfaces with various profiles like these presented in Fig. 3.26? A systematic investigation of this question, taking into account the causes for undulation profiles in other materials (e.g. Fig. 2.20) should be a welcome scientific endeavour.

3.1.4 Combined Markings of Plumes and Ribs

Joint surface markings often show superpositions of straight or fan-like plumes and circular rib markings (Fig. 3.19). On single-layer joints (joints that

Table 3.1. Wavelength, height and steepness of cross-sections from rib markings

Section	Parameters	Measured rib markings							Mean
		1	2	3	4	5	6	7	
a	Wavelength l (mm)	48	50	58					52 ±4.3
	Height h (mm)	4.5	3	4					3.8 ±0.6
	Steepness h/l	0.09	0.06	0.07					0.07±0.1
b	Wavelength l (mm)	27.69	21.69	24.46	26.3	18.46			23.72±3.31
	Height h (mm)	2.76	1.15	1.38	1.84	0.92			1.16±0.65
	Steepness h/l	0.1	0.05	0.06	0.07	0.05			0.07±0.02
c	Wavelength l (mm)	11.53	12.92	11.53	13.84	13.84	12.92	13.84	12.92±0.96
	Height h (mm)	0.46	0.92	0.46	0.92	0.46	1.15	0.92	0.76±0.27
	Steepness h/l	0.02	0.07	0.04	0.07	0.03	0.09	0.07	0.06±0.02

a, b and c are respective sections from Fig. 3.26. Measurements were carried out on enlarged sections. Unpaired numbers are results below the curve and paired numbers are results above it.

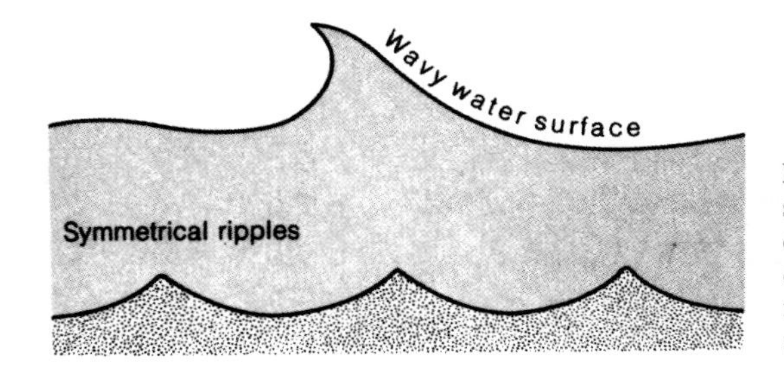

Fig. 3.27. Oscillation ripples are symmetrical concave ridges produced by wave movement in the overlying water column. They commonly have sharper crests than current ripples. (Press and Siever 1974)

are limited by layer boundaries) in the Lower Eocene chalks around Beer Sheva the surface markings display a full spectrum, between coarse axial plumes and concentric ribs (Fig. 3.6b). Characteristically, the joint surface has a single weak (or hardly noticeable) circular undulation and a delicate straight plume that radiates from the centre of the undulation. Occasionally, the rib marking becomes intense and the plume is very weak. Alternatively, the barbs are emphasized, the circular undulation changes into an elliptical shape and becomes less pronounced, and the transition into an axial marking becomes more emphasized. There are also transitional markings in which both the axial and the undulatory elements are strongly developed.

Perfectly annular undulations that imply pure tension, superposed by radial striae, are displayed on a joint surface in a Senonian chalk in the Judean Desert (Fig. 4.2a). Deviations from rib annularity are attributed to deviations from pure tension fracture conditions.

Plumes and circular rib markings superpose each other on a joint surface when their independent origins coincide. The orthogonal crossing of barbs and ribs at their various intersections indicates that the two markings were caused by the same process.

3.1.5 Affinities of Specific Joint Markings to Certain Joint Directions

Bahat and Engelder (1984) noted that specific plume end-members (Fig. 3.2) occur in certain sets of cross-fold joints. The more easterly striking cross-fold joints 345° (N 15° W), which cut thin siltstone beds embedded in thicker shale formations, are marked by the straight S-type plumes. The more westerly striking cross-fold joints (335°), cutting thick siltstone beds, are predominantly marked by either the curving or the rhythmic C-type plumes. Joints in this orientation are confined to the thicker siltstone beds and do not extend downward or upward into the adjacent shale beds.

3.1.5.1 Distribution Counts of Fracture Markings

In an attempt to quantify the preferential distribution of specific joint markings in Lower Eocene chalks near Beer Sheva, detailed counts were made at five outcrops (Table 3.2). The counting procedure presents problems. The identification and classification of joint markings depends on the discrimination

Table 3.2. Orientation of four types of fracture markings in Eocene chalks around Beer Sheva

1 Station number and syncline	2 Set azimuth (°) (mean and standard deviation)	3 Fracture markings type	4 Number of measurements (mean and standard deviation for station 20)	5 Percent in each station (rounded)	6 Formation
3 Shephela	309 ± 3 (27)	S	33	49	Mor
	309 ± 3 (27)	t	6	9	
	328 ± 6 (36)	u	20	30	
	328 ± 6 (36)	t	8	12	
	59 ± 5 (22)	u	2		
4 Shephela	309 ± 5 (25)	S	4	8	Mor
	309 ± 5 (25)	u	2	4	
	326 ± 4 (33)	u	36	75	
	326 ± 4 (33)	t	6	13	
	61 ± 7 (16)	u	4		
14 Beer Sheva	345 ± 7 (20)	S	28	90	Mor
	345	t	3	10	
20 Beer Sheva	329 ± 4 (40)	u	71 ± 9	90	Mor
	329 ± 4 (40)	t	5 ± 2	6	
	329 ± 4 (40)	S	2 ± 1	3	
	341 ± 1 (3)	t	1	1	
	51 ± 7 (40)	u	6		
40 Beer Sheva	35	d.s.	34	100	Horsha
	28	d.s.	36	100	

Key: S – coarse straight; d.s. – delicate straight; u – undulatory; t – transitional. Fracture markings on strike joints (059, 061, 051) in the Mor Formation are present for comparison with cross-fold fractures, but they are excluded from percent calculations. In column 2 numbers in parentheses indicate number of strike measurements. For station numbers see map in Fig. 5.29.

and experience of the investigator. Vertical joint marks are most clearly observed on wet surfaces (immediately after rain), and at a low angle overhead illumination (e.g. full summer sunlight around noon). Since the definition of a marking depends on contrasts between the shaded and the brighter parts, some joints have morning and others have afternoon preferences. The joint shown in Fig. 3.17a and b was photographed at 10.00 and 14.00 h, respectively, and the results are significantly different. A counting test was carried out around noon on a certain outcrop by the writer and two students. The differences between the results obtained by the three persons are indicated as standard deviations with respect to the (average) counts given in Table 3.2 (column 4). The differences of ±13% with respect to the most common marking type is not considered to be significant, and does not call for reservations regarding the main observations.

3.1.5.2 Preferential Distribution of Joint Markings

The distribution of joint markings according to particular arrangements is highly noticeable in Lower Eocene chalks around Beer Sheva. The first preferential trend is the general persistence of a certain joint surface morphology along a given joint orientation. There are always minor exceptions to this rule, but these amount to no more than 10–20% in each of the five outcrops examined. The second preference is for a specific joint marking to appear in abundance along a particular chalk layer, and to appear seldom or not at all in other layers (Fig. 3.28). It appears that in greyish white clay-rich chalks, joint markings are less common than in less clayey white chalks (see Sect. 5.2.2.3).

Thirdly, there appears to be a regional distribution factor. The cross-fold joints occur in three sets, 309 (N 51° W), 328 and 344°. The results in Table 3.2, which were observed on 225 joints, show that about 56% of the markings are circular rib markings that occur on the 328° set (Fig. 3.21), while about 29% of the markings are coarse axial plumes that occur along the 309 and 344° sets (Fig. 3.10). This correspondence between certain joint markings and sets of a particular orientation is displayed by 85% of the fracture markings on cross-fold joints. Twelve percent of the markings on cross-fold joints are transitional (Fig. 3.6 b), and about 3% belong to other categories.

Fig. 3.28. Straight plumes on 309° joints appear selectively along certain chalk layers (marked by *vertical arrows*) but not in others. Circular rib markings on 328° joints (*between horizontal arrows*) occur in adjacent layers, and these can be distinguished by their low length/height ratios. Bed thickness ranges from 60 to 90 cm. (Station 3 from Bahat and Grossmann 1988)

These results agree to some extent with observations made by Hancock and Al-Kadhi (1982). Of the 1100 joint orientations measured by these authors in horizontal strata of the central Arabian graben and trough system (Qaradan quadrangle), 41% of the joints were classified as extension fractures (analogous to above set 328°) while 16% were classified as hybrid members of two conjugate systems (Hancock 1985) (analogous to sets 309 and 344°).

3.1.6 The Fringe

According to Woodworth (1895, 1896), the fringe is the surface that surrounds the joint plane (Fig. 3.1). Some joint types, such as flat joints with dominant circular rib markings (Fig. 3.22), usually lack such fringes. Many joints with dominant horizontal plume markings also have no fringes. Among these are the cross-fold joints in the Appalachian Plateau (Bahat and Engelder 1984) (Fig. 3.7).

Other joints marked by plumes may have prominent fringe zones (Figs. 3.1 b, 3.11). In discoid joints (see below) fringes often are much more prominent than the joint plane (Fig. 3.29). Moreover, joint fringes that develop under different geological conditions may show totally different morphologies. Some are very simple, and others are quite complex. The implication is that

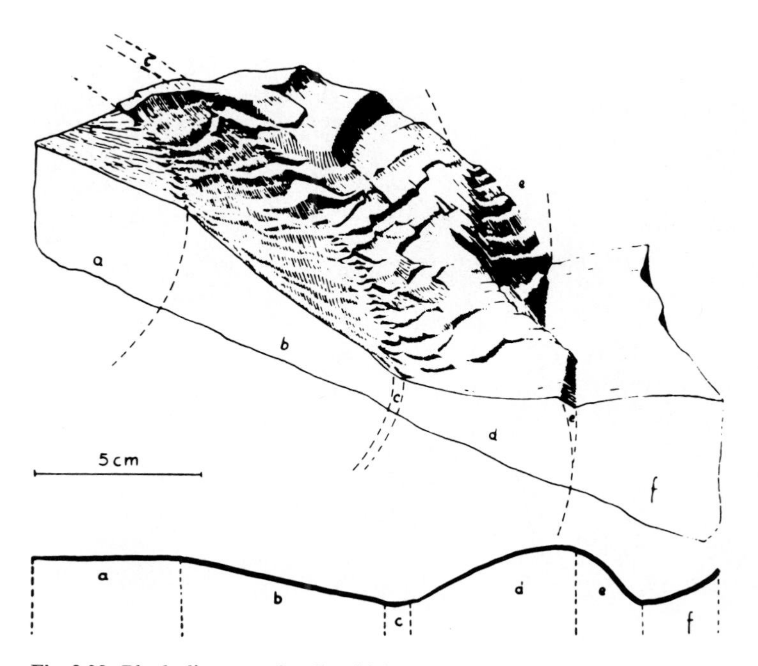

Fig. 3.29. Block diagram of a discoid fracture in graywacke. The parent fracture is segment *a* in the left-hand corner. The other segments *b*–*f* are components of a large fringe representing several propagation stages. The boundary between *a* and *b*, as well as segments *c* and *e* appear to be steeper than segments *b*, *d* and *f*. (Bankwitz 1965)

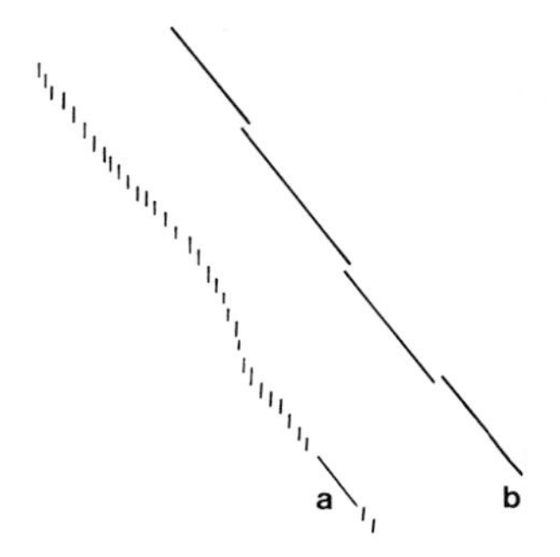

Fig. 3.30 a, b. Two styles of en echelon cracks in plan. **a** Segmentation of a large fracture into small parallel cracks oriented obliquely to the large fracture. **b** A series of parallel fractures, some of them arranged en echelon

investigation of the joint fringe may turn out to be a useful tool of tectonophysics.

Two basic fractographic elements of the fringe (according to Woodworth), are the b-planes and the c-fractures, which are also known as the en echelon cracks and the steps, respectively (see also sections on striae and steps in Sect. 2.1.2.2). A third element of the fringe are the various types of hackles and cuspate hackles. These markings, however, are notably less common on natural joints than are echelons and steps.

En echelon cracks occur in various strain environments, including dilational terrains as well as shear zones (such as great wrench faults). They appear in different styles and range in scale from microns to kilometres. Echelon fringe markings have been obtained in various materials (Sect. 2.1.2.2). The present study is limited to dilational joint fringes associated with Type a joints (Fig. 3.30), which are the product of a mixed mode fracturing, and does not concern Type b, which reflects extensional stresses only. The following descriptions deal with several aspects of en echelon fractures: (1) general geometry,

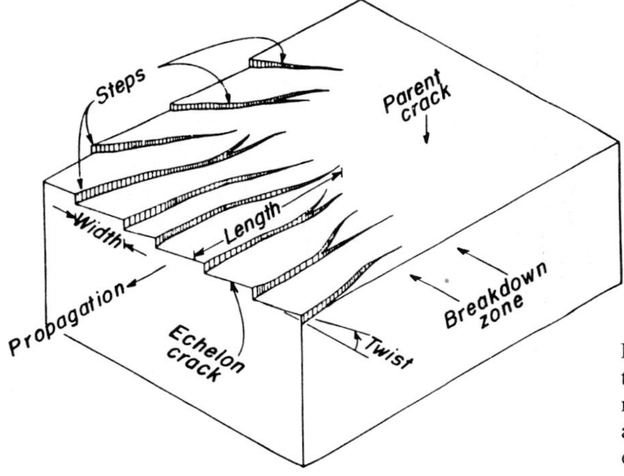

Fig. 3.31. Block diagram illustrating the geometry and mutual relations between parent crack and en echelon cracks. (Pollard et al. 1982)

(2) breakdown (parting) styles and root zones, (3) breakdown mechanism, (4) en echelon linkage, (5) subdivisions of echelon bridges into secondary and tertiary bridges and (6) a case of ripple ellipticity in the fringe (7) stepping.

3.1.6.1 En Echelon Cracking

General Echelon Geometry. The three idealized block diagrams in Fig. 3.31 (after Pollard et al. 1982, modified from Lutton 1971) and Fig. 3.32 (after Pollard et al. 1982) illustrate the geometric relationship between the parent fracture and the en echelon cracks as well as their relations with applied stresses.

The difference $W - w = 2m$ in Fig. 3.33 expresses the extent of echelon overlapping. When there is no overlapping, $2m = 0$. For a given 2d, 2m increases and w decreases with the twist angle ω since $w/2d = \cos\omega$. Steps often start with $2m = 0$ on the joint plane in a gradual breakdown (Fig. 3.34), or at the shoulder in a discontinuous parting (see next section). 2m increases gradually as the steps grow. The overlap of W increases from ~6.7% to ~14.2% with the increase of ω from 20 to 30° (Fig. 3.33). Pollard et al. (1982) show that:

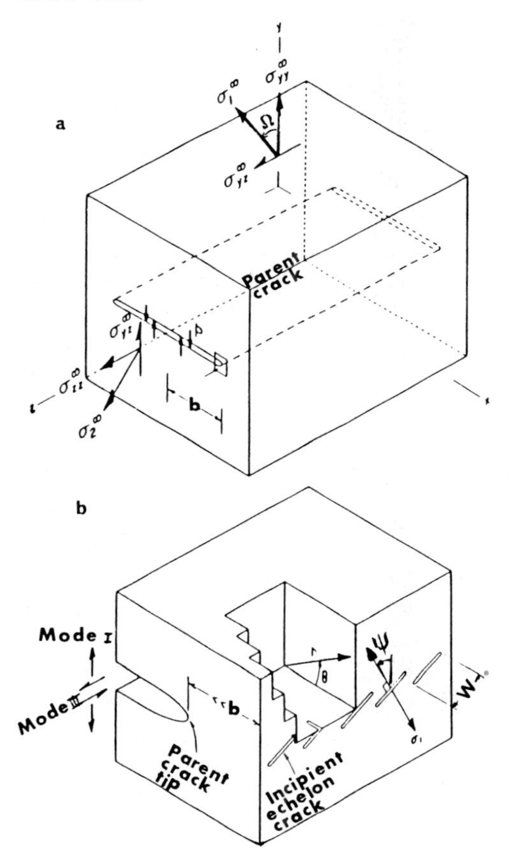

Fig. 3.32 a, b. Block diagrams illustrating relations among applied stresses and cracks. **a** Parent crack of width 2b and infinite length in z is loaded by pressure P and remote principal stresses σ_1, σ_2. Rotation Ω of principal stresses about x introduces stresses σ_{yy} and σ_{yz} acting on the crack plane. **b** Small element cut from near parent-crack tip (*small arrowed square* in **a**) with incipient echelon cracks of width W whose normal, the maximum local tension σ_1 making angle ψ to the y axis. Polar coordinates centred at parent-crack tip are r, θ. *Heavy arrows* indicate relative motion of parent crack walls when subject to mode I and mode III deformation. (After Pollard et al. 1982)

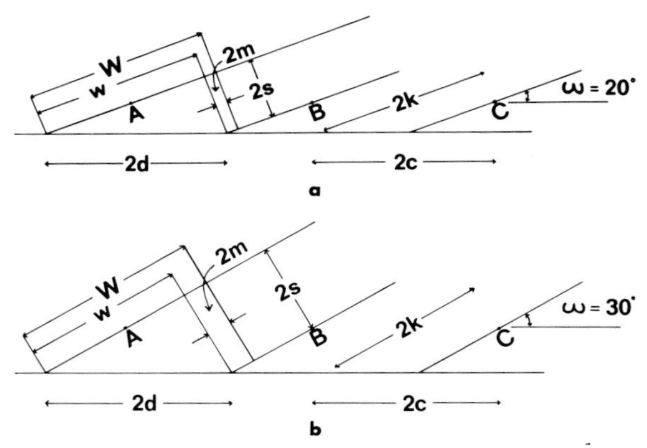

Fig. 3.33 a, b. Schematic sections of a discontinuous fringe normal to en echelon crack lengths. *A, B* and *C* are three en echelon cracks; *W* crack width; *w* part of the width which is not overlapped; *2m* the overlapped portion of the width, when $W = w + 2m$; *2s* step, the normal distance between two adjacent cracks; *2k* the distance between centres of two adjacent cracks parallel to cracks; *2c* the distance between centres of two adjacent cracks along the shoulder; *2d* the distance between initiation points of two adjacent cracks along the shoulder, and ω the twist angle between the shoulder and en echelon cracks. **a** twist angle 20°; **b** twist angle 30°. A comparison between **a** and **b** shows that overlapping increases with ω. At low ω, $W \cong w$

Fig. 3.34. Two joints in the Middle Eocene chalks near Beer Sheva become parallel to each other after the joint *at right* initiated on the surface of the other (above the 30-cm scale) bent towards the reader, and then continued straight. This is revealed by the delicate plumes on the two joints that initiated at the centre of picture and propagated away from each other in opposite directions. Note the gradual (continuous) breakdown to en echelon cracks below the parent joints

$$s = c \sin \omega \ , \quad \text{and} \tag{3.1 a}$$

$$m = W/2 - k = W/2 - c \cos \omega \ , \tag{3.1 b}$$

where ω is the twist angle, $2c$ and $2k$ are different distances between centres, and $2s$ is a step.

En echelon cracks occur both in straight rows and in radial geometries (Figs. 3.30a and 3.29) (see also Fig. 2.8).

Breakdown Styles and Root Zones. Two breakdown styles are distinguishable, gradual (continuous) and shoulder (discontinuous). In layered rocks the gradual breakdown starts on the joint plane at various distances from the layer boundaries and the steps have no shoulder as a starting line. Such breakdowns often display fringes with steps of quite different shapes and sizes (Fig. 3.34).

On one continuous breakdown, Bankwitz shows plumes that initiate on the parent joint and continue to propagate along the echelon crack. Note that in Hodgson's diagram (Fig. 3.5) echelons appear in a discontinuous style and the plumes on them (marked 3) initiate at the fringe of the parent joint (not on the parent joint itself).

The discontinuous breakdown starts along a shoulder at the edge of the parent joint from where the steps approach the layer boundary. When the shoulder is straight and parallels the adjacent layer boundary, steps of a given fringe would reach approximately the same maximum size (Fig. 3.35). However, a discontinuous breakdown with arcuate shoulders will have steps of different sizes (Fig. 3.36).

A discontinuous breakdown is often characterized by a root zone (Bankwitz 1966). This zone is a narrow strip of small echelon cracks along the joint

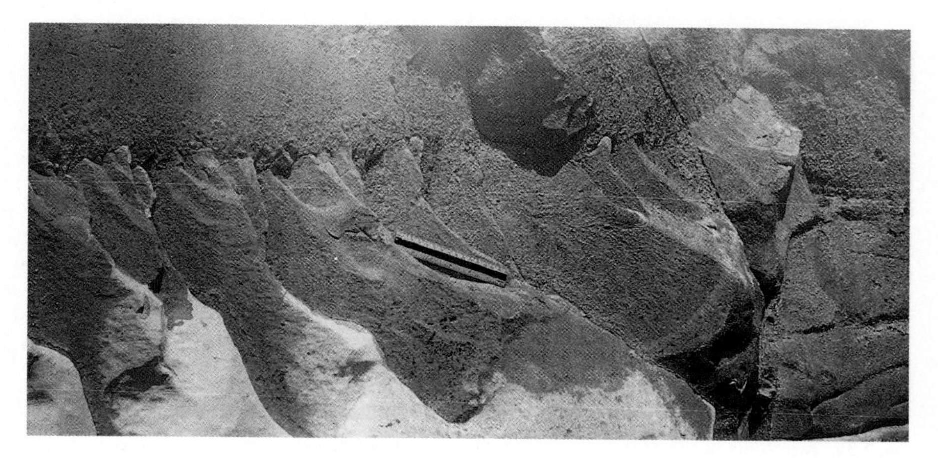

Fig. 3.35. A vertical joint in a chalk bed from the Middle Eocene near Beer Sheva, with a straight shoulder, a root zone along the shoulder and en echelon cracks in a straight row along the shoulder, displaying discontinuous breakdown. Note that radial plumes on the echelon surfaces initiate at a considerable distance from the shoulder. *Scale* = 15 cm

shoulder which are replaced by larger echelon cracks as the fringe develops (Fig. 3.35). Root zones are less common in continuous breakdowns.

Field observations of adjacent continuous and discontinuous breakdown styles in the same rock layer (Bahat 1986a) suggest that local conditions of fracturing are important in determining the style of breakdown. Figure 3.17b suggests that the gradual breakdown is associated with the early growth stage of the joint. The en echelon cracks seem to be mostly linked with the centre of this joint and not with its entire length (Fig. 3.17a).

Breakdown Mechanism. Pollard et al. (1982) examined the mechanism of a fracture breakdown into echelon cracks and this section basically is cited from their work. They adapt Sommer's (1969) explanation (Sect. 2.1.2.2) but add the internal pressure P (see Sect. 1.6.2) to the normal stress σ_{yy}^{∞} acting on the fracture plane (tensile stress is positive) (Fig. 3.32). Pollard et al. consider an elastic body subject to remote principal stresses σ_1^{∞} and σ_2^{∞} acting in the yz plane, and a fracture of width 2b subject to pressure P lying in the xz plane with straight tips parallel to z (length and width of parent fracture by Pollard et al. are reversed in the present text, to be consistent with other descriptions). Breakdown into echelon segments is independent of the third remote principal stress, which acts in the propagation direction x. A uniform rotation Ω of the remote stresses about the x axis introduces a normal stress σ_{yy}^{∞} and a shear stress σ_{yz}^{∞} acting in the fracture plane. According to Pollard et al., breakdown of the parent fracture depends upon Ω and the applied stress ratio

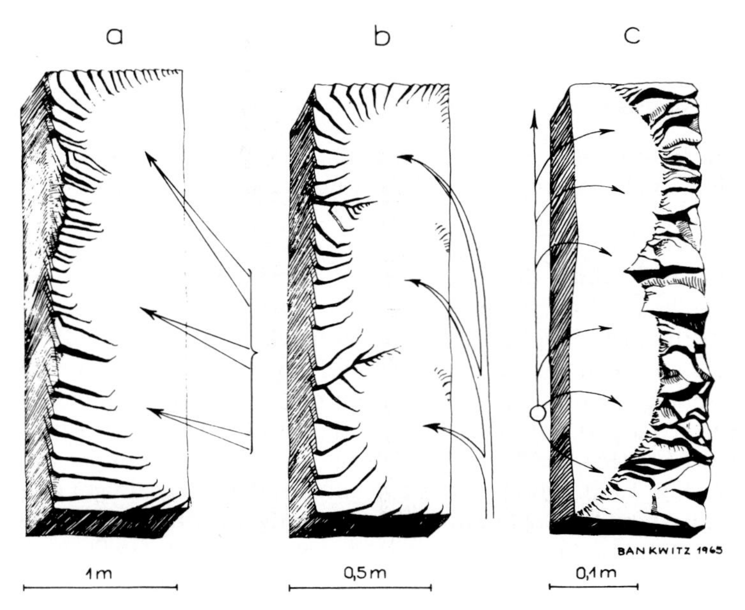

Fig. 3.36a–c. Schematic drawing of arcade edged joints with their fringes; *arrows* indicate directions of crack propagation. **a** Single joint plane. **b** Three separate joint propagations. **c** Two separate arced joints, note a somewhat disturbed fringe. (Bankwitz 1966)

$$R = (2P + \sigma_1^\infty + \sigma_2^\infty)/(\sigma_1^\infty - \sigma_2^\infty) \ , \quad P > -\sigma_1^\infty \ . \tag{3.2}$$

The inequality relating P and σ_1^∞ is required for the parent fracture to be open. The numerator in Eq. (3.2) is the difference between the internal pressure and the remote mean stress or pressure, $-(\sigma_1^\infty + \sigma_2^\infty)/2$. The denominator in Eq. (3.2) is the remote maximum shear, $(\sigma_1^\infty - \sigma_2^\infty)/2$.

Near the crack tip the stress field is proportional to the stress intensity factors. Pollard et al. resolve that the remote normal stress and internal pressure contribute to a mode I stress intensity, $K_I = (\sigma_{yy}^\infty + P)\sqrt{\pi b}$, and the remote shear stress introduces a mode III stress intensity, $K_{III} = \sigma_{yz}^\infty \sqrt{\pi b}$. The tendency for breakdown increases as the ratio of stress intensities K_{III}/K_I differs from zero. Writing σ_{yy}^∞ and σ_{yz}^∞ in terms of Ω and the remote principal stresses by resolving stress in the xz plane, Pollard et al. obtain

$$K_{III}/K_I = (\sin 2\Omega)/(R + \cos 2\Omega) \ , \quad R + \cos 2\Omega > 0 \ . \tag{3.3}$$

Inequality of R and $\cos 2\Omega$ is required for the parent fracture to remain open. For a particular rotation angle, the stress intensity ratio increases as the stress ratio R decreases. For $R \cong 1$, which is reasonably close to the range of R in nature, $K_{III}/K_I \cong \tan \Omega$, and Pollard et al. find that relatively small rotations of the principal stresses could initiate breakdown.

For complementary studies of echelon crack mechanisms, see references to Preston (1962), Sommer (1967, 1969), Mukai et al. (1978), Tschegg (1983) and Section 2.1.2.2.

3.1.6.2 Interaction and Linkage of En Echelons in Joint Fringes

En echelon crack linkage may be examined from the following aspects: (1) geometry; (2) dilational deformation; (3) sequence; and (4) subdivision of echelon bridges.

Geometry. Woodworth (1896) describes two geometries of en echelon (b-planes) linkage: "These cross-fractures are of two types. In most cases, the oblique plates or bridges of rock between b-planes break across about midway of their length and form straight medial c-fractures. Another mode of fracture arises from the continuation of the rupture from the end of one b-plane, in a curved surface to the next adjacent b-plane" (Fig. 3.1). These are two tip-to-plane linkages. Note that the "straight" linkage c is not quite straight but is slightly twisted (Fig. 3.1 d). Woodworth's two end members, the straight and the curved, have also been distinguished by other investigators. De Freminville (1914) observed partly straight and partly curved steps in experimentally fractured bitumen (see Fig. 2.4). Preston (1931) shows in detail the curved linkage in glass (Fig. 2.5). Swain et al. (1974) add to the above end members a theoretical third type, the tip-to-tip (asymptotic) linkage. See also Section 5.9.1 on joint linkage.

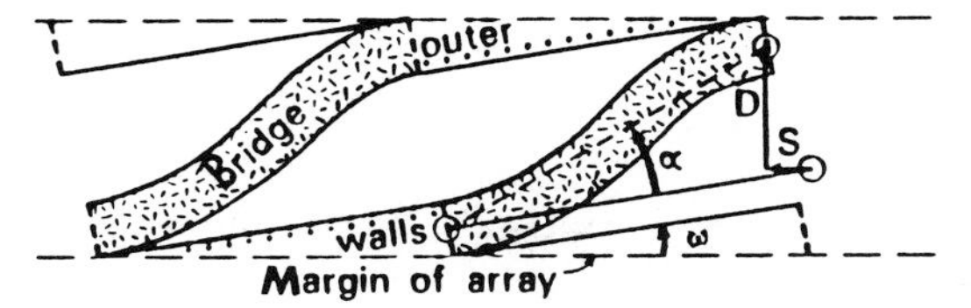

Fig. 3.37. En echelon fractures seen in profile of a shoulder breakdown. The dilatation angle α is formed by the straight line joining the ends of an undeformed bridge and the staight line joining its ends after deformation; S is the component of displacement parallel to the margin of an array (the joint shoulder); D the component of displacement at right angles to the margin of an array, and ω is the twist angle. (Nicholson and Ejiofor 1987)

Dilational Deformation. Nicholson and Pollard (1985) and Nicholson and Ejiofor (1987) show that sigmoidal, mineral-filled en echelon arrays do not necessarily form through ductile simple shear (Ramsay 1967), but they may well be the consequence of dilation. They present a cross-sectional model normal to the echelon lengths which enables to measure the amounts of lateral motion shear S, and perpendicular (dilatational) motion D, which are related to one another in proportion to the angle of dilation α (Fig. 3.37). Gradually, as α increases during extension, the host rock included between the en echelon cracks which they call bridge is strained. The bridge partly bends, so that each bridge wall is divided into a rectilinear (outer) part and a curved (inner) part. Advanced dilation may ultimately lead to bridge failure (not shown in Fig. 3.37).

Sequence. Woodworth (1896) observed that steps (c fractures) develop after en echelon cracks (b-planes). The succession of these fractures is often quite evident (Fig. 3.1 d). Bankwitz (1966) notices this order of events, but he also identifies the reverse sequence of steps preceding the echelons. This implies that the two fracture elements may also develop contemporaneously.

Subdivisions of Echelon Bridges. Kulander et al. (1979) show secondary en echelon crackings, and Kulander and Dean (1985) present a block diagram showing primary, secondary and tertiary en echelon cracks (which they term twist hackles) along the layer boundary (Fig. 3.38). This division corresponds to Preston's (1931) postulation of ad infinitum cracking (Sect. 2.1.2.2).

3.1.6.3 Ripple Ellipticity in the Fringe

Solomon and Hill (1962) describe fracture markings on vertical joints in thick-bedded, fine-grained homogeneous graywacke. Their illustration (in their Plate 2) suggests that the markings are normal to the bedding. The authors describe slightly asymmetric rib marks (ripples) in a quasi-concentric arrangement about 30 cm in size within the fringe zone which occurs above a vertical joint 2 m long or longer (its edge is not seen in their picture). These undulations present no more than 150° of arcs which are convex upward. Their wavelengths

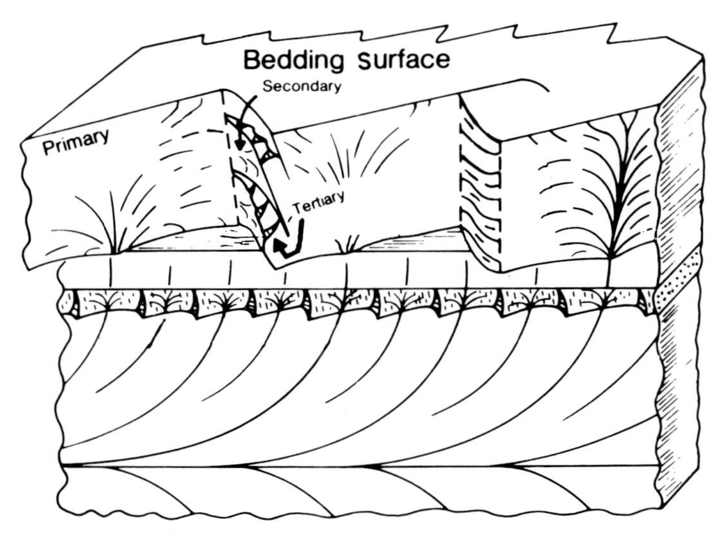

Fig. 3.38. Block diagram showing primary, secondary and tertiary en echelon cracks in fringe. (After Kulander and Dean 1985)

vary from a few millimeters to 3 cm, their amplitudes average 1–2 mm, and in a few cases exceed 2–3 mm. Solomon and Hill note that the wavelengths of the rib sections measured vertically are greater than those measured horizontally, (their pattern resembles the one shown in Fig. 3.47a) as if the ripples have "travelled" farther and faster upward than sideways (that is, faster across the fringe than along its length).

3.1.6.4 Determination of Stepping

The sense of stepping has to be the same, whether determined in plan or in vertical section. It is assumed that the echelon segments propagate away from the shoulder of the parent joint (parent crack in Fig. 3.31). The right-stepping in Fig. 3.31 is determined when the observer looks from the shoulder along the length (the propagation direction) of a step. The sequence of "shingles" then appears to rise toward the observer's right (if it were a stairway, it would ascend to the right). The same sense of stepping is shown when the observer steps on the vertical face of the block diagram along its upper edge looking towards the parent joint. Right-stepping is determined whether the observer advances from right to left or from left to right. This can also be verified in Fig. 3.5.

3.1.6.5 Influences of Remote and Local Stresses on Echelon Orientation in the Fringe

Bahat (1986a) investigated fracture markings in Middle Eocene stratified chalks from a syncline near Beer Sheva. Measurements were taken along a

950-m creek bank. At the eastern end of the outcrop considerably greater azimuth variations are observed for the main joint planes than for the en echelon cracks at their fringes. The mean azimuth of the joints is NO3° W±12° (63) (i.e. 63 measurements) and N23° W±6° (91) for the en echelon cracks. This suggests that the azimuths of the en echelon cracks are not influenced by the orientation of the individual joints, but rather by a remote stress which acts more uniformly than could be expected from the spread of orientations of the joints. Moreover, in joints with bilateral plumes, the en echelons maintain a single orientation along a continuous breakdown (Fig. 3.39), indicating independence of en echelon azimuth from the local stress on the surface of the parent joint and propagation directions of the joint.

A gradual rotation of 31° of the joint strikes occurs between the two ends of the investigated strip along the creek bank. The en echelon cracks at the fringes of these joints rotate in covariance with the joints (Fig. 3.40), supporting the suggestion that the orientation of these cracks is controlled by the areal tectonics (see also Sect. 5.2.4).

On the other hand, there are indications that the style of en echelon propagation is at least partly influenced by local (rather than remote) conditions. At the western end of the investigated strip, the transition from the main joint plumes to the en echelon cracks at the fringes is both discontinuous and continuous in adjacent locations along the same layer, the former being the more

Fig. 3.39. A close-up view of Fig. 3.17a showing a rhythmic relationship between vertical en echelon cracks and horizontal segments of rib marks (*dark lines*) in a continuous breakdown

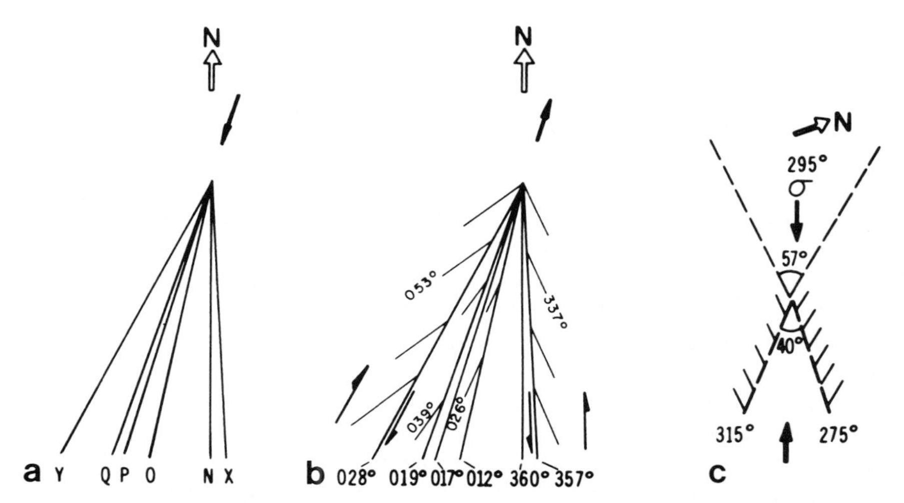

Fig. 3.40 a–c. Orientation of joints and en echelon sets. **a** Joint rotation in six adjacent outcrops (Bahat 1986a, outcrop *Y*), and implied direction of compressive stress. **b** Azimuths of rotated joints and en echelon cracks in six outcrops. Directions of shear stresses and elongation are derived from the orientation of the en echelon cracks (outcrops *X* and *Y*). **c** Two conjugate en echelon sets and the inferred compression direction. (After Shainin 1950)

common. Further support for local control is derived from the fact that en echelon cracks usually increase in size with bed thickness, implying that at least the size of the cracks is determined by local stress behavior of the stratum.

Size parameters at another outcrop near Beer Sheva (Bahat 1986a, Tables 1–5) show that the orientation of small and large en echelon cracks in individual root zones is generally similar, suggesting either simultaneous or sequential cracking. Rhythmic relationships between the vertical en echelon cracks and horizontal segments of a rib marking (Fig. 3.39) seem to reflect simultaneous arrest in the straight fracture fronts of the many downward-propagating en echelon cracks. It is less likely that the cracks grew at different times, and yet were all arrested at the same height.

It was also found that en echelon cracks (like plumes) display a preference for particular chalk beds and are rarer in adjacent ones. With some exceptions, most en echelon cracks are wider than the width of the interlaid steps. The same investigation (Bahat 1986a) also found a rough relationship between length of the crack and some of its other parameters (see Fig. 4.21).

3.1.6.6 Hackles and Cuspate Hackles

In the present study the terms hackles and cuspate hackles are used to designate markings which appear only at the fringe of a smooth joint, and which signify high local stress conditions. Plumes and rib markings, on the other hand, are mostly confined to the smooth central surface of the joint and they imply relatively low local stress conditions (see Sect. 2.2). En echelon cracks seem to serve an intermediate role between the above two categories of joint markings (see further discussion in Sect. 4.2.1.7).

Hackles. Hackles are common on surfaces of fractured glass (Figs. 2.23, 2.30), 2.31 and 2.33) but are very rare on natural joint surface. Even when present (Figs. 3.6a, 3.42c and 4.20a), they do not quite resemble hackles in glass, and their identification depends to some extent on observer's bias. The question whether a particular marking is to be classified as hackle or en echelon crack may arise in some cases, and although criteria for distinction have been suggested (Table 3.3), the determination is not always easy.

Cuspate Hackles. Cuspate hackles have sharp ridges of approximately constant width the radiate in a fringe around the main fracture (Figs. 3.41a and 4.6). They are characteristic fractures that develop in quarries by blasting (Fig. 3.60a), implying high energy conditions (supercritical crack velocities, Sect. 1.4.3). Their sharp peaks are distinct from the flat surfaces of en echelon cracks. Crack overlapping, which is a typical feature of en echelons, does not

Table 3.3. Main distinguishing features of en echelon, hackle and cuspate hackle cracks along joint fringes in rocks

Feature no.	Feature	En echelon	Hackles	Cuspate hackles
1	General shape	Distinct planes overlapping like shingles	Cracks fan out from a common centre and combine flat planes sub-parallel to the parent fracture with striae and disoriented surfaces	Ridges with sharp summits fan out from a common centre
2	Crack orientation with respect to parent fracture	Uniform	Non-uniform	Approximately uniform
3	Crack orientation in a given outcrop	Uniform	Non-uniform	Non-uniform
4	Crack propagation	Sub-parallel to each other	Radial	Radial
5	Surface texture	Smooth	Rough	Rough
6	Neighbour steps or striae	Alternating with echelons	Not well defined	Not developed
7	Plume markings on crack surface	Occasionally present	Occasionally present	?
8	Length/width ratios	Approximately systematic	?	?
9	Fringe orientation with respect to parent fracture	Coplanar	Inclined	Approximately coplanar

Fig. 3.41. a Cuspate hackles. A plume developed into right-stepping and left-stepping echelons at the *upper* and *lower parts* of the marking, respectively, but transformed laterally towards the left into cuspate hackles, suggesting a lateral cracking under conditions of greater stress intensity. From an Upper Cretaceous chalk in south England (see drawing in Fig. 4.6). **b** "Eye" joints in Frauenbach Quarzite of the Thuringian Schiefergebirge, representing the morphologic type c in Fig. 3.45. (Bankwitz and Bankwitz 1984)

occur in cuspate hackles. They have not been identified on fracture surfaces of technical glasses and they are uncommon on natural fractures. When they occur on joint surfaces they seem to serve the role of hackles on fracture surfaces of glass. There is a strong resemblance between cuspate hackles on joint surfaces and those on the fracture surfaces of some ceramics (Fig. 2.32), even if size scales are an order of magnitude different.

A fracture marking from the Charing Chalks in Kent, England, displays a curious type of fringe (Fig. 4.6). The echelon cracks occur above and below the plume, whereas cuspate hackles continue the trend of the plumes along the distal direction (left side of joint). This is interpreted as greater stress intensity along the lateral fracture propagation as compared to the upward and downward directions.

Both hackles and cuspate hackles should be distinguished form bifurcating plumes (Fig. 3.2g) and barb branching (Fig. 3.7). The former two morphologies occur in the fringe, whereas the latter two morphologies belong to the parent fracture.

3.1.7 Discoid Radial and Ring Joints

3.1.7.1 General Observations

Woodworth (1896) observed the tendency of joints in some argillaceous rocks to develop into warping or curving discoid shapes (Fig. 3.1e). Bankwitz (1966) and Bankwitz and Bankwitz (1984) describe similar features, such as warped joints in slates, which they call radial fractures, and ring structures (Fig. 3.41b). In these structures radial en echelon and step fractures develop around a circular parent joint, which is commonly convex or concave, and which may be either smooth or decorated by fracture markings. The parent joint is surrounded by a root zone, which may be prominent or occasionally very limited. The radial markings may be considerably longer than the circular parent joint (Fig. 3.29) and en echelons that leave the root zone form characteristic angles with the parent joint. Discoids often occur in clusters. In such cases radial fractures commonly end as downward-inclined fans below the adjacent radial fracture, or occasionally they fold upward. Bankwitz observes that, generally, an increase in the size of a radial en echelon is associated with a commensurate increase of the adjacent step.

3.1.7.2 Discoids of the Ramon Flint Clay

Discoids are very common in the lower Jurassic flint clay (Mishhor Formation) which occurs in the Ramon anticline, southern Israel. The discoids appear in groups, and adjacent individuals may vary considerably in size and shape, ranging in diameter from about 10 to 400 mm. They vary in shape from completely flat to conic end members. Most discoids are not flat but warped, with matching concave and convex sides.

a

b

Embryonic discoids, no larger than 10 mm in diameter, appear as concentric circular ridges that often surround a discontinuity (inclusion) at the centre (Fig. 3.42 a). Bigger discoids commonly deviate from circularity. Radial striae occur already in the "embryonal" stages, but stepping and fringes seem to develop with size.

Discoid joints are generally vertical, with deviations up to 15°. In adjacent discoids, plume axes may point in different directions although their joint surfaces usually maintain the same azimuth (Fig. 3.42 b). Rib markings may vary considerably. In flat fractures rib markings are symmetric and rounded, whereas in warped discoids rib markings are asymmetric (Fig. 3.26 e) and have sharp kinks (see Sect. 3.1.3.2) which are often associated with concentric fringes cracked by en echelons.

One conic discoid from the Ramon Flint Clay consists of a flat parent joint decorated by radial striae and two successive conic fringes (Fig. 3.43). In the first cone (closer to the parent joint) the crack angle $\theta = 76°$ (the crack angle is produced between the conic fracture and conic axis, Fig. 1.14 e). In the second cone $\theta = 72°$. Right-stepping en echelon cracks occur on both conic sur-

Fig. 3.42 a – c. Discoid joints in the Ramon flint clays. **a** A group of small discoids. At the *centre* of picture is an embryonal discoid with concentric rib markings (*arrowed*), diameter about 2 cm. **b** Three neighbouring discoids of different size and shape arranged, but of similar orientation. They include a circular flat parent joint with a sharp transition to a fringe of radial en echelon cracks, a flat parent joint with sub-concentric ridges, and a large discoid with a delicate plume on the parent joint, combined with sub-elliptic (partial) rib markings and prominent echelons on the fringe. *Bar* = 15 cm. **c** A group of discoids in Cenomanian chalks from Mt. Carmel, formed on joints striking 310°. The discoids display alternations of segmented rib marks and fringes of echelon cracks or hackles around the flat parent joint. *Bar* = 17 cm

Fig. 3.43. Fragment of a conic discoid from the Ramon flint clay. A curved shoulder separates between the flat parent joint (*lower left*) and the fringe of first-stage en echelon cracks (*upper right*). The second stage en echelon fringe is just visible at the *upper right corner* of the discoid. *Scale* 15 cm

faces, but the echelons on the first cone are much wider than the echelons on the second one, suggesting a diminishing of stress intensity.

The conic characteristics of some discoid joints resemble Hertzian fracture (Sect. 1.1.7.3). This resemblance supports the hypothesis that discoid fractures are a product of pore pressure that was locally transmuted into Hertzian stresses (Fig. 1.14).

Plotting of the crack angles θ of the flint clay discoids against the Poisson ratio υ of various materials (Fig. 3.44) suggests two alternative interpretations. An assumption of slow fracture propagation would result in $\upsilon \cong 0.31$ and $\upsilon \cong 0.27$ for earlier $\theta = 76°$ and later $\theta = 72°$, respectively, both implying a rock of considerable rigidity (Poisson ratios of granite and chalk are 0.25 and 0.30, respectively). An assumption of rapid fracture, on the other hand, would require υ values around 0.40 of an essentially unconsolidated material.

The fracture markings on the examined material (Fig. 3.43) are sharp, rather favouring the interpretation of slow fracture in a fairly consolidated material. The two fringes probably developed at different times (see also Sect. 5.3.8).

3.1.7.3 Discoids of the Carmel Chalks

Discoids are uncommon in chalk. A group of warped discoids occurs on near-vertical joints in Cenomanian chalks (Fig. 3.42c) of Mount Carmel, striking

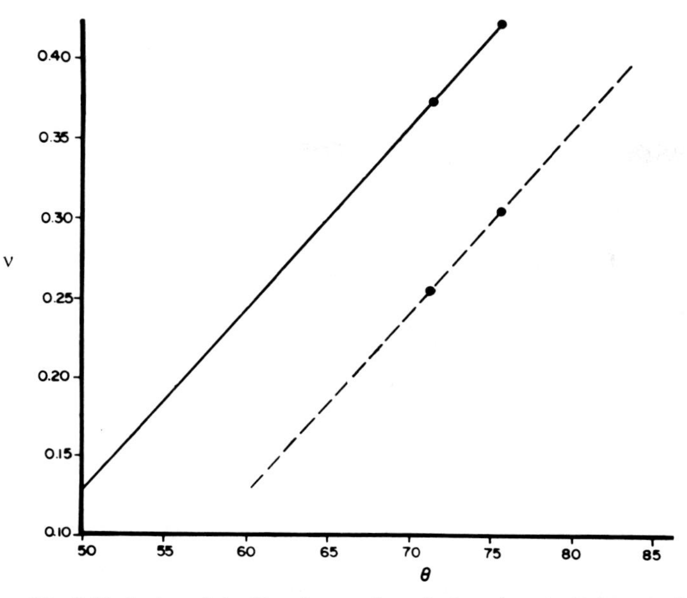

Fig. 3.44. A plot of the Hertzian crack angle θ against the Poisson ratio ν for various materials. The *solid line* represents the calculated principal stress trajectory predicted by Warren (1978) for dynamic conditions. The *dashed line* represents static experimental results. (After Bahat and Sharpe 1982). The *points* are the two new results showing what should be the alternative υ values on the *solid* or *dashed lines*

310°. Most discoids are circular, with diameters ranging from 20 to 40 cm. They display concentric rib markings which generally have sharp profiles. Unlike the ribs of the flint clay discoids, some of the ribs are segmented and they are bent somewhat upward. The concentric ribs alternate with fringes of echelons (hackles ?).

3.1.7.4 The Symmetry of Discoids

Bankwitz and Bankwitz (1984) divide the discoid joint surface into two or more segments which are interrelated by the symmetry of their fracture features, and superpose the symmetry elements of the joint on the stress ellipsoid. Accordingly, they view the joint surface as a two- or three-dimensional projection of the paleostress ellipsoid, and this enables them to determine the directions of the three paleo-principal stresses and maximum shear stress directions.

The novelty of their method lies in their sophisticated symmetry division of the joint surface. A simplified discoid appears flat or almost flat without segmentation (Fig. 3.45 a, b), whereas a complex discoid may be characterized by several symmetry elements and divided into two or four segments (Fig. 3.45 c, d). The model primarily assumes the coincidence of a principal stress direction with a plane of symmetry and the coincidence of the line of

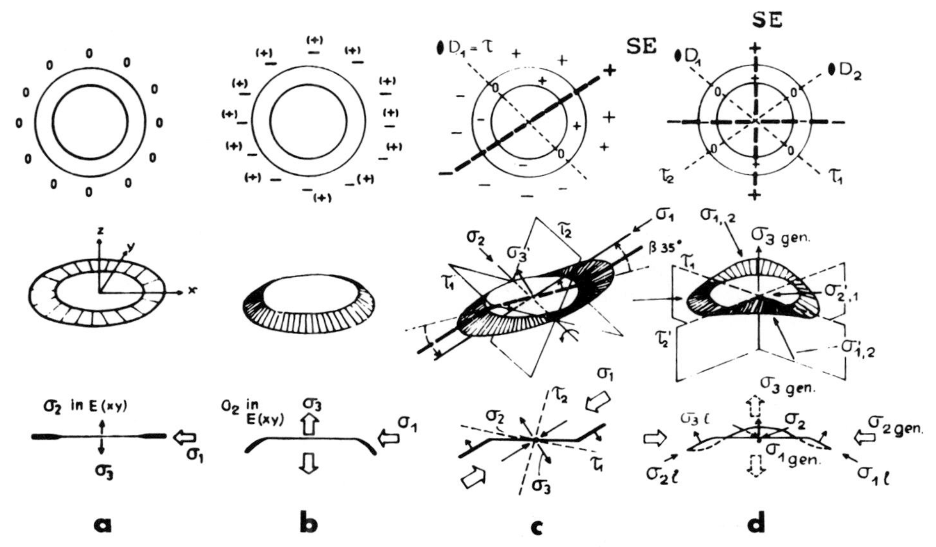

Fig. 3.45 a – d. Morphological classification scheme of fracture surfaces according to basic forms of the fringe as observed in geological fractures. (After Bankwitz and Bankwitz 1984). **a** Nearly planar (two-dimensional) joint; theoretically possible, though hardly ever observed in rock fractures. Joints are largely flat even though of slight curvature. **b** Shape of the fringe in a typical tensile fracture. **c** Characteristic shape of joint formed under compressive stress with superimposition of a tectonic compressive stress; the fringe lies in the plane of the direction of pressure. **d** σ_3 replaces σ_2 in the conventional Andersonian model (see Fig. 1.20 a, b). **a** and **b** non-segmental, **c** two-segmental, **d** four-segmental fracture surface, +, −, upward or downward bending of the fringe in relation to the joint surface. *SE* plane of twofold symmetry. D_1, D_2, 180° symmetry axes; $\sigma_1 > \sigma_2 > \sigma_3$ principal stresses; D_1 D_2 planes of shear stress = tau (τ)

intersection between the planes of maximum stress and the joint surface with the twofold rotation axis. Further assumptions are: (1) the parent joint and the fringe are the product of approximately the same stress field, therefore in some joints the principal stresses are inclined to the parent joints (Fig. 3.45 c) (2) The line of intersection of the planes of maximum shear stress is not unique, it can be either σ_2 or σ_3 and the two largest principal stresses σ_1 and σ_2 may alternate. The spatial representation in Fig. 3.45 c corresponds to a situation where the planes of maximum shear stress intersect along σ_2 and the shear stress attains the maximum level ($\tau = \pm\sigma_1 - \sigma_3/2$), whereas in Fig. 3.45 d these planes intersect along σ_3 corresponding to the lowest shear stress ($\tau = \pm\sigma_1 - \sigma_2/2$), implying that even if the joints develop at the theoretically lowest shear stresses these stresses have a significant influence on their morphology. (3) The determination of the symmetry elements in complex discoids is based on the geometry of crack features on the parent joint and the fringe that deviate from the 0 (zero) level of the flat parent joint, either as positive and negative or as left and right bends and steps. The plane of maximum shear stress intersects the flat joint along the 0 (zero) axis (Fig. 3.45 c, d). In general, Bankwitz and Bankwitz (1984) consider that the progressive morphological changes in

Fig. 3.45 reflect a corresponding change in ratio of shear strength to tensile strength in the rock. (4) The model may even be applied in the analysis of partial fractures (half and quarter joint surfaces) when the alignment of the partial fracture to the proper symmetry pattern is made. (5) This model is applicable in the investigation of large joints (mainly in magmatic rocks) and their stress histories, as well as in determining regional patterns of paleostress.

The analytical model of Bankwitz and Bankwitz (1984) is commendable. Nevertheless, a few shortcomings have to be taken into account: (1) The influence of pore pressure on the effective stresses (Secor 1965) and the contribution of the angle on internal friction (Jaeger and Cook 1979, p. 17) to symmetry directions are not included in the model. (2) Many discoids have circular, flat parent fractures and fringes which are "curled up behind and pulled down in front" (Woodworth 1896) (see Fig. 3.1 f). It is unlikely that σ_1 and σ_3 are inclined to the parent fracture and σ_2 is lying on this surface (Fig. 3.45 c), and to expect on the other hand that the circular fracture should reflect a condition of $\sigma_1 = \sigma_2$, as suggested by the authors. (3) There is no certainty that the discoids in these figures are really the product of a single stress field. A comparison with the experiment by Brace and Bombolakis (1963) (see Bankwitz 1978 and present Sect. 1.1.6) requires that the discoids in Fig. 3.45 b, c are the products of two stress fields, which produced the flat fracture first and then the fringe, following a rotation of the stress directions. An interpretation of a single stress field would probably imply that the discoids developed in a material that deformed plastically. (4) Figures 8, 13, 14 and 15 in Bankwitz and Bankwitz (1984) raise some misgivings – the application of a model derived from the warped discoids to flat large joints with partial fracture markings appears to be over-demanding. Are there sufficient assurances (probabilities or otherwise) that the symmetries interpreted for the flat large joints are justified? Quite often, expected bilateral fracture markings turn out to be unilateral (Bahat 1988 b). Further research on joint symmetry should nevertheless benefit from this pioneering model.

3.1.8 Fracture Mechanisms

On the basis of fracture surface morphologies two end members of fracture mechanisms in sedimentary rocks are suggested: (1) the ○ shape process manifested by concentric circular rib markings, and (2) the < shape process manifested by straight horizontal plumes.

3.1.8.1 The ○ Shape Fracture

Initiation of the ○ shape fracture (Type 3, Sect. 3.1.3.1) occurs remote from layer boundaries and it is not constrained by them. Fracture initiation occurs following a process of water percolation and the fracture propagates radially in a rhythmic pattern (Secor 1969). The result is a morphology dominated by concentric rib markings, which may develop in several styles. The more common pattern is that of smooth concentric undulations (Fig. 3.22) within the mirror plane (Fig. 2.31) which imply a slow sub-critical tensional process (Fig. 1.26). More rarely, the undulations may be embellished with hackles or en echelon cracks (Fig. 3.42c) which suggest an increase in stress intensity (Sect. 4.2.1.7).

Occasionally, fan-like plumes which consist of rectilinear barbs are superposed on smooth concentric undulations (Fig. 4.2a). The rectilinear radial barbs are analogous to striae (Sect. 2.1.2.2), a typical feature of low stress intensities as are the smooth undulations.

3.1.8.2 The < Shape Fracture

The < shape fracture is most typically represented by the S-type straight plume (Figs. 3.10, 3.11 and 3.18). Straight plumes probably develop by more than one mechanism, expressing alternative possibilities of palaeostress. Possibilities to be examined include: (1) remote palaeostresses, (2) geometric control by layer thickness, and (3) local stress gradients between axial zone and fringes.

Remote Palaeostresses or Geometric Control Leading to Plume Development. The S-type straight plume (Fig. 3.2a) characteristically occurs in stratified rocks, with its axis parallel or sub-parallel to the layer boundaries. In this case it is of significance for the geologist to know whether the plume axis develops in parallelism to the intermediate principal stress σ_2, while the maximum principal stress σ_1 is vertical, as suggested by Gramberg (1965, Fig. 1), or whether the plume axis parallels σ_1 while σ_2 is vertical (Kulander et al. 1979, Fig. 37).

A third possibility is that the orientation of the plume axis is mainly controlled by the layer boundaries (and layer thickness) and to a lesser extent, if at all, by the orientation of σ_1 and σ_2. Roberts (1961) observed that flaggy beds normally have well-developed plumes. Syme Gash (1971) suggests that a running fracture is dependent on the energy-to-thickness ratio: if the stratum or test plate is within the critical thickness, a running fracture with straight plume will result. But if the stratum is too thick, fan-like plumes will develop.

The concept of plume dependence on layer thickness is quite intriguing. Roberts (1961) shows the resemblance between the straight plumes on fracture surfaces of mild steel and of rock, yet this does not necessarily suggests a similarity in fracture mechanisms, and may merely demonstrate the nature of the geometric control of plume development. In detail, this means that the geometric constraints imposed on fracture propagation in thin plates may control the ultimate fracture pattern even more than the stress conditions. Let us consider the feasibility of this concept.

The sub-horizontal streamline curves in Fig. 3.46, which bear some resemblance to plumes on a fractured surface (disregarding the sub-radial lines in the figure), actually represent three different physical phenomena (Stevens 1974): (1) a low-speed liquid flow around a cylinder normal to the page; (2)

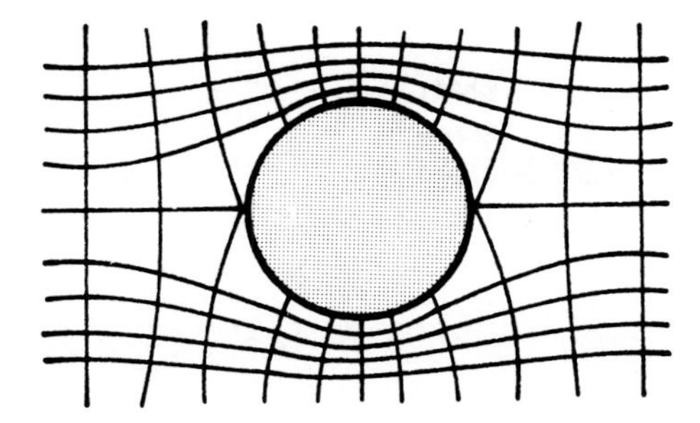

Fig. 3.46. Stream lines at low-speed flow around a cylinder. (Stevens 1974)

lines of electrical force in a uniform electric field and (3) lines of tension around a circular hole in a stretched plate (see for comparison Means 1979, Fig. 12.7). Those three phenomena are basically different but they all display the same type of spatial configuration. In analogy, it is possible to obtain look-alike (but not quite identical) patterns of straight plumes between closely parallel boundaries even by different mechanisms. Figure 3.47 a is a plume pattern which results from polycrystalline growth under constrained thermal and spatial conditions (Tomer 1988). The similarity between the plumes in Figs. 3.10 and 3.47 a supports the above concept regarding the importance of geometric constraints (note their scale differences). Orthogonality between concentric markings and radial crystal growth is maintained by the thermal process, resembling common relations between plumes and ribs (Sect. 3.1.4).

Geometric analogy of the plume pattern to the fracture morphology of a glacier surface is quite striking. Figure 3.47 b shows a giant straight plumes on a glacier surface, formed by numerous vertical crevasses. The glacier flows more rapidly at the centre; close to the banks friction slows it down and this results in lag plumes which curve towards the bank. Here, compliance is high at the centre and low along the banks. Nye (1952) shows that under such physical conditions the fracture pattern can develop into three styles determined by the curvature of the angle that the crevasses make with the banks (Fig. 3.48). When under longitudinal compression the ice expands sideways, the crevasses curve upstream and meet the glacier banks at angles smaller than 45° (Fig. 3.48 a). Simple shear drag along the banks produces crevasses at 45° to the glacier side (Fig. 3.48 b). Longitudinal extension along the glacier tends to open crevasses at 90° to that direction at the centre, crosswise to the flow axis, and to slightly curve downstream at the banks (Fig. 3.48 c). The wider the ice flow, the smaller the influence of geometric constraints on the plume pattern of the surface features.

Local Palaeostress Gradients from the Mid-Layer Towards the Fringes. Two interpretations can be proposed for the stress gradients across a plate or a rock layer, both based on the morphology of a straight horizontal plume on the face of a vertical joint, and yet both interpretations are essentially opposed. According to the first interpretation, fracture occurs due to stress which is maximal along the layer's median zone, decreasing towards the layer boundaries

Fig. 3.47. a Copper welding by electronic ray. (Courtesy of Avinoam Tomer, Nuclear Research Centre, Negev). The contact area is locally melted by the electronic ray and the plume pattern reflects the formed crystal texture. Note orthogonality between crystals and concentric undulations. Spacing between successive undulations is reduced from the centre to the boundaries of the contact area. The distance between upper and lower boundaries of plume (*between arrows*) is ~2 mm. **b** Tikke Glacier, northern British Columbia, after it surged. This glacier is 16 miles long. Low compliance along the banks results in lag plumes which curve towards the edges of the flow. (Schultz 1974, Plate 28)

(e.g. Tipper 1957; Kulander et al. 1979). The second interpretation regards the plume marking as evidence for an increase in the fracture driving force from the median midline zone towards the boundaries.

Tipper (1957), working on fractures in mild steel plates, maintains that the curvature of the barbs outward towards the plate surfaces seems to be accounted for by plastic flow which increases towards those surfaces with a corresponding retardation of velocity propagation. Also, arrested fractures show the centre of the plate fractures ahead of the surfaces. The fracture patterns on a glacier surface which reflect various degrees of retardation along the glacier banks (see preceding section) somewhat resemble these experimental observations.

Kulander et al. (1979) present hypothetical fracture surface illustrating the use of plume geometry to approximate past fracture fronts, fracturing stress distributions and relative fracture propagation velocities (Fig. 3.49). In their diagram, parabolic fracture fronts drawn at constant time intervals (and crossed orthogonally by plume barbs) indicate that actual fracture velocity and the

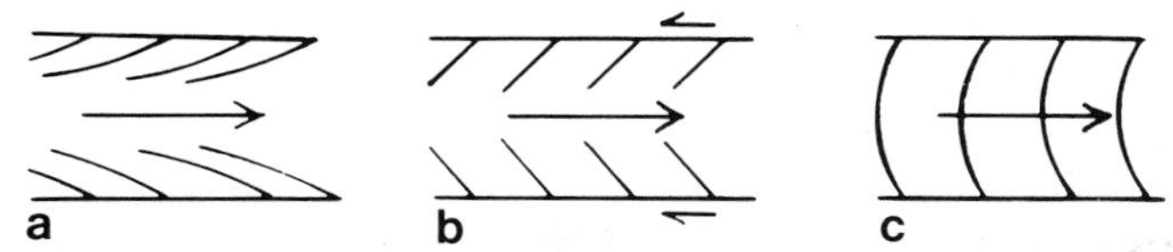

Fig. 3.48 a – c. Schematic fracture patterns on a glacier surface caused by the drag along the glacier banks. **a** A plume formed under longitudinal compression. **b** A plume formed due to shear along the banks. **c** Transverse arced crevasses formed due to longitudinal extension. (After Nye 1952 from Sugden and John 1976)

corresponding effective fracture stress (Sect. 1.6.1) increase to a maximum along the plume axis, and decrease towards the joint fringe (see AV length changes in Fig. 3.49). The diagram shows that the overall rate of fracture propagation parallel to the linear plume axis is constant between adjacent fracture front lines throughout the stratum (OV, Fig. 3.49). Constant velocity is necessary to maintain an equal radius of curvature at intersection points on successive fracture fronts in the overall fracture propagation direction (along OV axes). Kulander et al. further suggest that en echelon cracks would form at the top of the stratum where fracture velocities are lowest, due to low effective fracturing stresses.

However, there are observations and arguments which support the opposite view, that the breakdown of the main fracture into an array on en echelon cracks involves an increase, not a decrease, in effective fracturing stresses. These include:

1. Both continuous (Fig. 3.34) and discontinuous breakdowns (Fig. 3.35) generally show increases in size of step (the plumes become coarser in continuous breakdowns), in extent of echelon overlapping, and in overall fracture area as cracking proceeds further away from the mid-layer. These increases im-

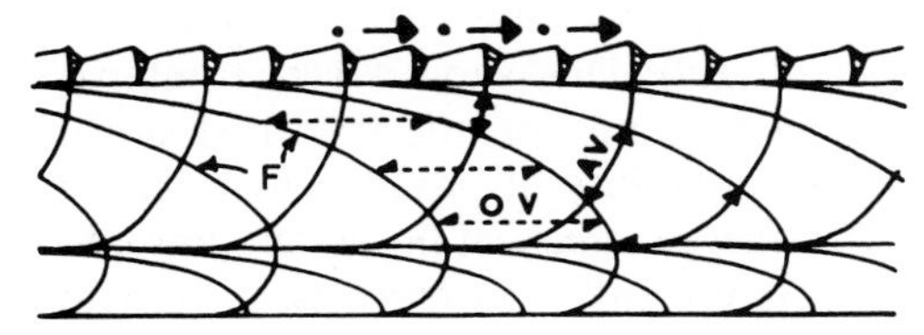

Fig. 3.49. Hypothetical fracture surface illustrating the use of plume geometry to approximate fracture parameters. Fracture propagates from left to right (*arrows* and *dots*). The varying barb lengths, crossing the propagation fronts *F* of past fractures are directly proportional to the average fracture velocity along these plume component lengths (*AV*). It follows that fracture propagation is fastest parallel to the plume axis, presents constant velocity (*OV*) along the stratum. The greatest effective fracturing stresses act in the lower stratum section. En echelon cracks and steps form in the top stratum section where, fracture velocities are lowest in response to low effective fracturing stresses. (After Kulander et al. 1979)

ply a proportional increase in the energy associated with fracturing (assuming that rock toughness is not significantly reduced during cracking).
2. The discontinuous transition from the main joint to the fringe zone is expressed by two features. The first is a root zone of small en echelon cracks. The second is the bulk of the fringe which consists of larger en echelon cracks further away from the shoulder which completes the development of the fringe. Although the sequence of formation is not established, it is assumed that the small en echelons develop either before or simultaneously with the large ones, but not after them. There is an analogy between these two features and the mist and hackle zones of a fractured glass (Figs. 2.31), with the implication of increasing stress intensities from the root zone to the larger en echelon cracks.
3. Plumes that initiate at the shoulder and propagate on the surface of en echelon cracks of the fringe (Woodworth 1896; Hodgson 1961 a) or initiate on the main joint surface and continue on the en echelon surfaces (Bankwitz 1966) have been known for quite a while. More recently, radial plumes were described on en echelon surfaces, but their origins were at some distance from the shoulder of the main joint surface (Fig. 3.35), indicating that fracturing of the fringe propagated both towards and away from the shoulder. A resemblant relationship between an adventitious Griffith crack ahead of the main crack was used by Congleton and Petch (1967) in their model of crack-branching. They suggest initiation of these surfaces at some distance from the main fracture due to a stress intensity increase at that distance (see concepts of nucleation by Preston 1926 and others, end of Sect. 1.4.4.1 and Sect. 2.1.2.4).
4. In the fringe zone the grooves (Fig. 3.12) are about three times greater than on the parent joint (Sect. 3.1.2.4);
5. Repetitions of second- and third-order step cracking (Fig. 3.38) attest to a great amount of stored energy in the fringe zone;
6. In discoid structures the fringe may be much larger than the parent joint (Fig. 3.29). The considerable area increase of the echelon planes and connecting steps hardly fits a trend of stress decrease at the transition from the parent joint to the fringe.
7. A compliance gradient opposite to that which occurs in the glacier (Fig. 3.47 b) is observed by Quackenbush and Frechette (1978) in glass fractured under controlled conditions (Sect. 2.1.2.5). They show a wedge (intersection scarp) which resembles the < straight plumes on the fractured surface that separates outside areas with higher availability of liquid, where fracture propagates faster than along the middle, where scarcity of liquid and low compliance delay the advancement of the fracture front. In remains to be determined to what extent the resemblance between the geometries of the < straight plumes and the intersection scarp (Figs. 2.28 a, b) reflects similarity in fracture conditions, and can be related to the effects of pore pressure on the development of plumes on joint surfaces. Possibly, water migrating laterally along the layer boundaries starts at some stage to percolate into the layer and facilitates a slight jogging ahead of the fracture front along these boundaries. The plumes in Figs. 3.10 and 3.11 indicate that fracture started within the rock layer at considerable distances from the layer boundaries.

Careful weighing seems to provide stronger arguments in favour of stress intensity increase from the median parts towards the layer boundary. However, there is a difference across a curved plume. The velocity of fracture propagation and the rock compliance are probably greater on the convex side of the curved barbs than on their concave one. It remains open to what extent these differences along the layer boundary influence the stress intensity conditions.

3.1.9 Fracture Markings in Thermally Deformed Rocks and in Granite

Fracture surface morphology has been mostly investigated in soft rocks, mainly in carbonate and slate lithologies subject primarily to tectonic deformation. Fewer studies have been carried out on fracture markings in magmatic rocks like basalts and granites, or on sedimentary rocks that were fractured by thermal stresses. Several fractographical observations on thermally deformed rocks are presented below.

3.1.9.1. Basalts

Fracture markings are common in volcanic and hypabyssal rocks, particularly on surfaces of columnar joints (Woodworth 1896). Yet, not many systematic studies have been made of these phenomena. Bankwitz (1978, Figs. 9 and 12) and Bankwitz and Bankwitz (1984, Figs. 10 and 16) adopt two approaches: (1) an analysis of the plumes and related markings, and (2) a perspective examination of the volcanic body. Their observations lead them to the opinion that macrojointing is from inside the body to the outside. On the plume scale, Bankwitz visualizes intermittent episodes of cracking, growing upward by regular increments, thus developing individual columnar joints growing up from the flow base. The intermittent growth, however, seems to follow variable patterns (Bankwitz 1978, Fig. 7).

DeGraff and Aydin (1987) follow the above two approaches and present convincing evidence (Fig. 3.50a), based on fracture marking analysis, that columnar joints grow incrementally from exterior to interior regions of solidifying basalt strata. They found that columnar joints display conspicuous bands (Fig. 3.50b) normal to the column axes. Each band contains a single plume and thus represents an individual crack, or joint segment, formed during a discrete fracturing event. Columnar joints form by initiation and growth of new cracks on the edges of older cracks. Each new crack begins at a point on the leading edge of the previous one, and propagates horizontally along a leading edge that is normal to the column axis in the zone where the thermal gradient results in stress concentration. Inward propagation of cracks toward hotter regions is limited by a decrease of stress associated with the brittle-ductile transition of lava (Ryan and Sammis 1978); outward and lateral propagation is limited by intersection with previous cracks and by low thermal stress in already fractured lava. Mechanical interaction causes a diverging crack to overlap, curve toward, and usually intersect the previous crack behind its edge, leaving a blind tip that points in the overall growth direction of the columnar joints (Fig. 3.50c).

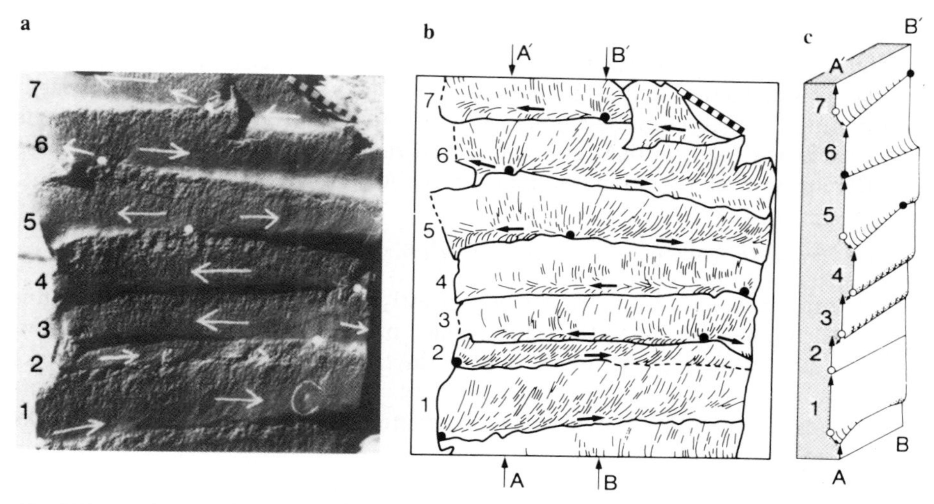

Fig. 3.50 a – c. Joint surface morphology in basalt. Markings on a column face near a flow base. *Bar* is in inches. **a** Horizontal bands of the column face, numbered upward according to sequence of formation. Each band has a single plume and represents an individual crack that formed during a discrete fracturing event. **b** Sketch of the column face, showing crack origins (*dots*), barbs (*short thin lines*) and joint boundaries (*continuous thick lines*). *Arrows* on horizontal plume axes show local propagation directions. **c** Oblique sketch of cracks on cutout section *A-A′-B′-B*. Profile of cracks along the cut *A-A′* shows plume axes (*open circles*) and vertical components of crack propagation (*arrows*)

Tier is the term used by DeGraff and Aydin to describe a set of regularly spaced, generally vertical to steeply inclined columnar joints that occur between two horizontal levels in a flow. Columnar fans (Fig. 3.51 a) commonly occur in upper tiers of two-tiered and multi-tiered flows (DeGraff and Aydin 1987). Bankwitz (1978) considers that the radial columnar joints grow and multiply by a branching mechanism. DeGraff and Aydin suggest that columns in fans are anomalous and result from change in the thermal stress regime. The early regime had nearly horizontal isothermal and isotensile surfaces, such that vertical joints grew down from the surface. A column fan would start forming when a master joint surface is suddenly chilled by water ingress, so that isothermal and isotensile surfaces are bowed downward near the joint and its tip (Fig. 3.51 b). Bankwitz and Bankwitz (1984, Figs. 10 and 16) present diagrams to illustrate concentric fracture evolution in the upper parts of columnar joint tiers.

3.1.9.2 Columnar Joints in Sandstone

Columnar sandstone, produced by the thermal metamorphism of an Upper Old Red quartz arenite is described from South Bute, Scotland, by Buist (1980). The columnar structure is thought to have developed as a result of the

a

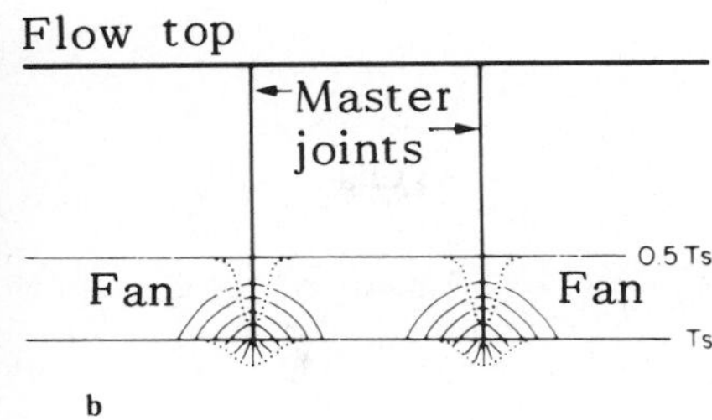

b

Fig. 3.51 a, b. Columnar fans. **a** Fanlike arrangement of columns around a vertical mastert joint (*arrow*) near the upper surface of a flow. **b** Inferred isotherms of cooling regimes that produce column fans. Isotherms are normalized to the solidus temperature, *Ts*. Early conductive regime (*solid* isotherms) produces sub-vertical master joints; later regime (*dotted* isotherms) caused by water convection in master joints produces columnar joints that grow radial to the master joints (Figs. 3.50 and 3.51 from DeGraff and Aydin 1987)

emplacement of a basic dyke via a fissure, accompanied by steam and volatile constituents of the magma. Buist mentions a related work by Macculoch (1829).

Columnar quartzite occurs in many outcrops in the Ramon anticline, Israel (Sect. 3.1.7.2) where Early Cretaceous dykes cut Jurassic sandstones (Inmar Formation). The columns appear in various styles, including vertical prisms which are separated by vertical fractures that appear to have started at the bottom and arrested after some vertical propagation upward, as well as inclined columns that occur in virtually all orientations. Columns showing transitions from incipient fractures (Fig. 3.52 a) to well-defined fractures are not rare. Often in good outcrops it can be seen that the prisms "choose" to be as perpendicular as possible to a close contact of the sandstone with the dyke.

Contrary to the abundance of plumes on surfaces of columnar basalts, plumes are very rare in the Ramon columnar quartzite. A petrographic examination of sections from columnar joints shows that the neighbour quartz grains contact each other along interlocking sutures (Fig. 3.52 b) which prevent surface microcracks from penetrating into the solid. The lack of microcrack penetrations (across the rock surface) which are typical of ridges that form plumes (Sect. 3.1.2.4 and Fig. 3.12) explains the rarity of this fracture marking in the columnar joints.

On rare occasions fracture markings appear on columnar joints (Fig. 3.52 c). A most intriguing occurrence is a series of cuspate hackles (resembling those in Figs. 2.32 and 3.41 a) on the surface of joints which are inclined at a dip of some 30° close to the top of a very thick tier. These cuspate hackles suggest violent fracture propagation (Sect. 3.1.6.6), which seems to fit Buist's explanation of fracture forced by steam and other volatiles.

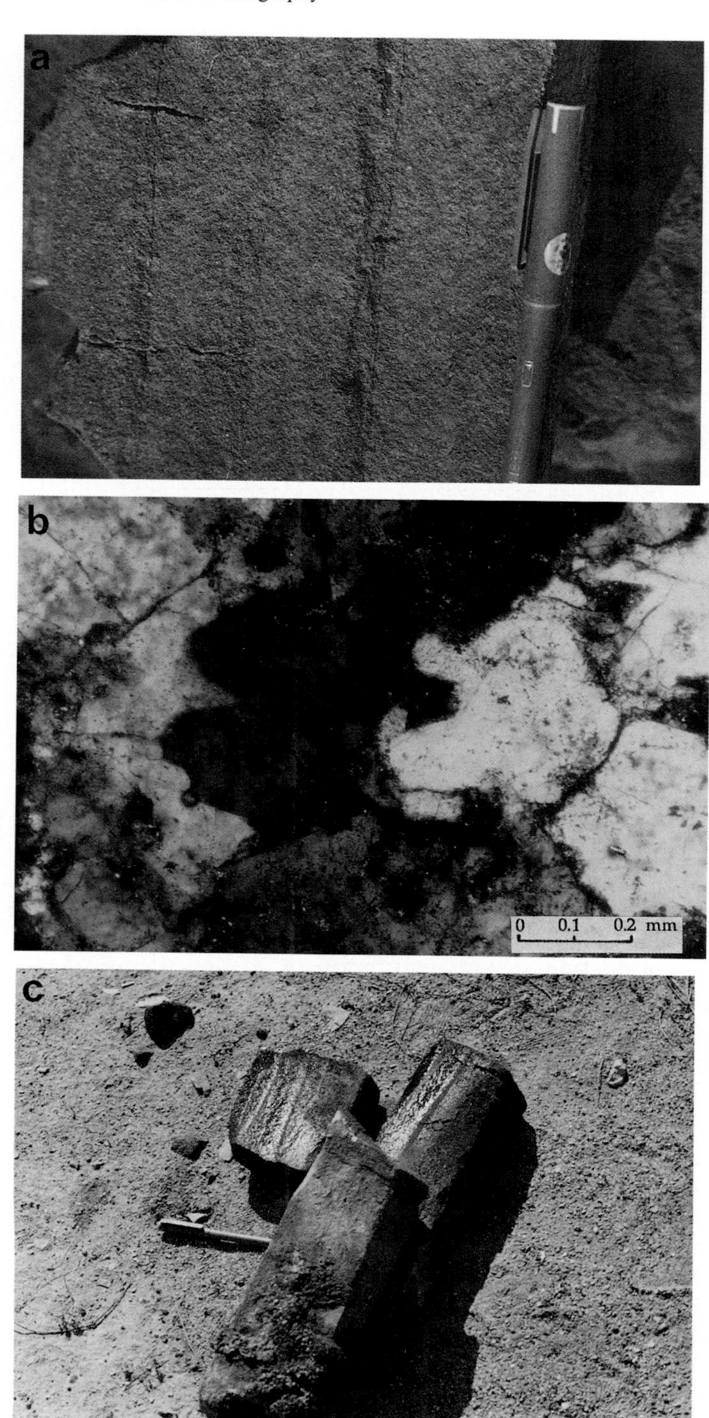
a
b
0 0.1 0.2 mm
c

3.1.9.3 Limestone Concretions

The Campanian in southern Israeli (Bentor and Vroman 1954) is represented by the Mishash Formation, which consists of a massive autobrecciated chert in its lower part and a chalky phosphorite series in its upper part. The latter include phosphate layers (of economic quality in synclinal basins) alternating with porcellanites, silicified phosphorites and limestone concretions of various size. These concretions, being more resistant than their chalk matrix, are usually weathered out, forming extensive fields of free-lying spheroids (e.g. in the Tsin Valley), ranging from about 20 cm to over 150 cm in diameter. The concretions have ellipsoid or lenticular shapes. Many concretions are cracked, and their fracture surfaces have some characteristic features:

1. The fractures are approximately planar, they cut through the concretions and often appear along the long or short diameters of the ellipsoid. As the concretions are randomly oriented, the planar fractures are also non-oriented.
2. The fractures have typical surface morphologies. The markings are mostly plumes that parallel the long axis of the elliptical section. Occasionally the markings display en echelon cracks, and undulations. In some cases the fracture surface displays markings that are in disharmony with each other, indicating several locations of fracture initiation (Fig. 3.53 a).

The above characteristics may imply fracturing by thermal stress. The ubiquity of such fractures suggests a common prevalent cause. The random orientation of the fractures, on the other hand, implies a mechanism that is independent of horizontality. The flat diametrical surfaces resemble the cracks derived from thermal stress by Preston (Fig. 2.1 a). The plumes and undulation markings suggest slow processes (Sect. 4.2.1.7). The rare concretions that were cracked by impact (Fig. 3.53 b) seem to emphasize the exception. Thermal stress is induced in the spheroids by intense day-time insolation and night-time cooling. The repeated setting-up of thermal gradients within the small and discrete volume of a spheroid appears to induce fatigue fracturing (Hodgson 1961 b) much as in the case of periodic pore pressure jointing (Secor 1965).

3.1.9.4 Exfoliation in Chalk

Chalk fragments often fracture in a concentric manner, resembling exfoliation in basalts. Occasionally these fractures occur as basin-like (Fig. 3.53 c) diametrical irregular surfaces, and very rarely they are flat. Characteristically, the fracture surfaces are marked by coarse annular undulations and delicate

Fig. 3.52 a – c. Columnar joints in sandstone. **a** Incipience of columnar joints. **b** An interlocking texture of quartz grains in a column. **c** Plumes on columnar joints

a
b
c

Fig. 3.54. Granite stone quarry near Mákrotin, West of Telc, Czechoslovakia. The central joint is approximately 40 m in height. Note sub-horizontal "giant" plume divided by two schematic symmetry elements. *M* change in joint morphology due to different lithology. (Bankwitz and Bankwitz 1984)

radial plumes, which coincide in a single origin. Specimens between 5 and 100 cm long exhibiting the above fracture style are found along Wadi Na'im (Sect. 5.2.2.1). This exfoliation seems to result from thermal stresses similar to those described in Section 3.1.9.3.

3.1.9.5 Granite

Bankwitz and Bankwitz (1984, Figs. 8 and 13) describe uplift joints (see Chap. 5) about 40 m high in granite quarries near Mákrotin, west of Telč, and about 80 long from Prosečnice, Sázava Valley, in Czechoslovakia. The large partial plume on the fracture surface indicates sub-horizontal propagation of the fracture (Fig. 3.54). Bahat and Rabinovitz (1988) present a quantitative analysis of a fracture marking in granite from Sinai (Sect. 4.2.1.5). The presence of fracture markings in granites of relatively coarse texture can be explained by the rapidity of fracture propagation in these rocks during the uplift

Fig. 3.53. **a** A limestone concretion, fractured by presumably thermal stresses, from the Tsin Valley, southern Israel. There appear to be two or more locations of crack initiation on the fracture surface. A delicate plume in the *upper part* suggests fracture propagation from left to right. The subparallel ribs in the *lower right* may fit the pattern of the delicate plume. Concentric rib markings with several echelons in the *lower left* imply a separate crack initiation there. *Bar* = 15 cm. **b** A bilateral plume *left of the pen* indicates fracture initiation at about the centre of the concretion. Fracture surface derived by impact ia shown *right of the pen.* Fracture initiation occurs at the lower part of the concretion (*at small shadowy dent*) and radial cuspate hackles are seen at the *right side* of concretion. **c** Exfoliation in chalk

process, which does not allow the graininess to obscure the fracture surface morphology.

3.2 Induced Fracturing in Rocks

This section deals with induced fracturing in rocks only (see also Sect. 2.3). Three aspects are presented, according to the technique applied: (1) controlled (laboratory) conditions; (2) coring, and (3) quarrying.

3.2.1 Controlled Laboratory Conditions

The results of four different experimental techniques are analysed below. Plumes are induced in the first three experiments, and Lüders' bands by the fourth.

3.2.1.1 Plumes Formed Under Axial Compression and Diametrical Compression

Gramberg (1965) induced fracture in cylindrical specimens of fine-grained brittle rock (one example is specified as lithographic limestone) by uniaxial compression and diametral compression techniques. In the former, compression is applied axially and in the latter (also known as Brazilian test), compression is applied at two opposite narrow zones parallel to the cylinder axis (for details, see Jaeger and Cook 1979, pp. 144 and 169). The fracture markings obtained by the two techniques are quite similar and have several characteristics features: (1) fracture origin is at the edge of the fracture plane, in both cases at the edge that parallels the compression direction; (2) the joint plane is divided into a central ragged wedge, with wavy plumes (Fig. 3.55) on both sides. In these plumes the barbs close to the edges which are normal to the compression direction approach the edges orthogonally. Gramberg suggests that in both cases fracturing followed principal planes, and may be caused in the same way by tension.

3.2.1.2 Plumes Formed by Bending and by Non-Orthogonal Extension of Artificial Rock Slabs

Cegla and Dzulynski (1967) investigated experimental plume development in weakly consolidated slabs of silt and silty sands of uniform grain distributions which settled from aqueous suspensions in cardboard boxes. Excess water was removed through the permeable cardboard walls, to ensure dilatant behaviour of the rock. Cegla and Dzulynski produced typical straight plumes (1967, Fig. 3) including two parallel opposing plumes on a single fracture surface (like the markings described by Woodworth in Sect. 3.1.1.1) by bending the slabs in various manners. Deviations from symmetry of plumes (Sect. 4.1.1.2) reflected oblique loadings in their experiment. They observed that in their samples frac-

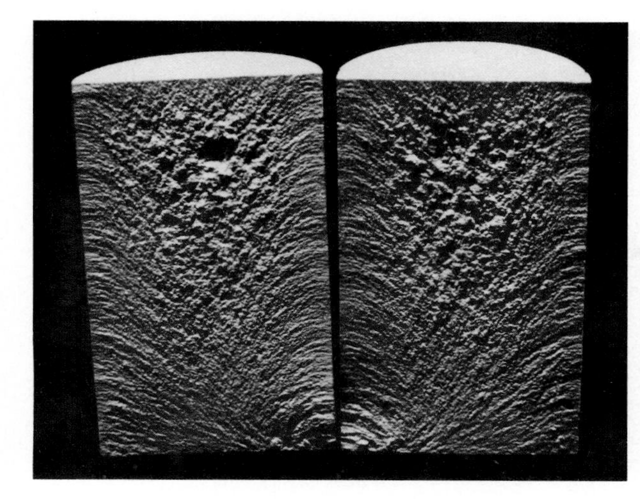

Fig. 3.55. Axial fracturing in the cylinder has occurred by the Brazilian tensile test. The fracture plane displays the fracture origin at bottom and divergent plumes propagating along the axial plane upward. Diameter of cylinder is 30 mm. (Gramberg 1965)

turing occurred in the transitional regime from ductile to brittle. Evidently, delayed fracture propagation favoured the development of plumes and rapid fracturing inhibited it. Generally, plumes initiated at upper or lower boundaries of the slabs, presumably at some imperfection, and propagated vertically inward, but then they changed their trend to parallel horizontally the slab boundaries.

Stoyanov and Dabovski (1986) applied oblique extension on soft wet clays (as well as on gelatine and acetylcellulose) by pulling basal plates below the investigated material. They induced upward and downward crack propagation. In one result they obtained second-order steps and plumes on the surfaces of large en echelon segments that developed in downwards fracturing. Stoyanov and Dabovski also induced three styles of fracture interaction and en echelon linkages (Sect. 3.1.6.2 and 5.9).

3.2.1.3 Lüders' Bands by Triaxial Compression

Friedman and Logan (1973) studied Lüders' bands in experimentally deformed sandstone and limestone. Although this fracture is marginal to the present subject, it is brought here for consideration due to its intriguing fractographic qualities. The following description is cited from their work. Lüders' bands are planar deformation features inclined along planes of high shear stress and along which cataclasis is concentrated. In Coconino Sandstone the bands begin to develop in the transitional regime(brittle-ductile) and are the major deformation feature in the macroscopically ductile regime. They are markedly different from shear fractures (or faults) of similar size that typically contain quartz gouge. The average angle between conjugate sets of bands bisected by the greatest principal compressive stress σ_1 increases from 75 to 109° as the effective confining pressure is increased. The corresponding angles between conjugate macroscopic shear fractures average 60°. The angle between the

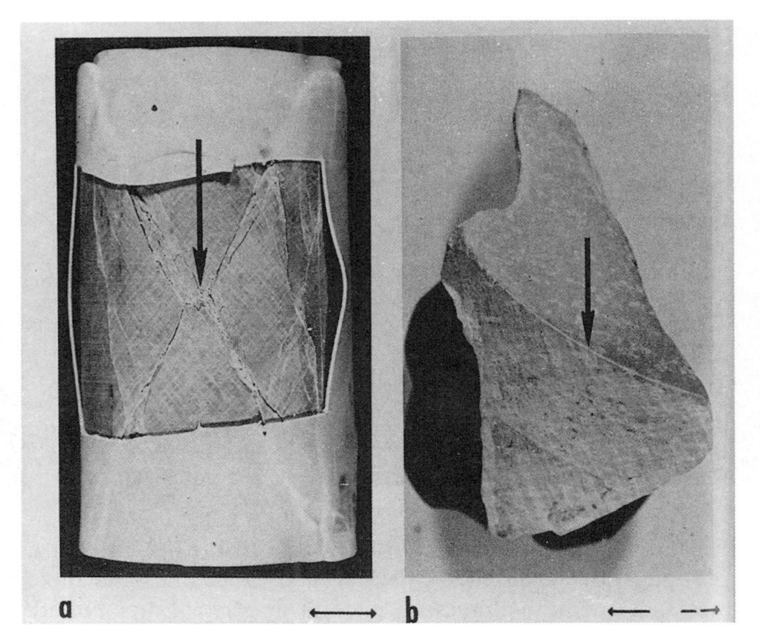

Fig. 3.56 a, b. Laboratory simulation of Lüders' bands in Solenhofen Limestone. **a** Conjugate Lüders' bands (*arrow*) and sub-parallel set of conjugate shear fractures. Average angle between conjugate bands is 77 and 52° between the shear fractures. **b** Fragment from same specimen shows edge (*arrow*) between cylindrical surface containing Lüders' bands and a polished portion of the original circular section. Note that Lüders' bands are developed only in the outer "skin" of the specimen. *Scale lines* below each photograph 1.0 cm. (Friedman and Logan 1973)

Lüders' bands is essentially independent of strain at fixed effective confining pressure. In Solenhofen Limestone, too, the bands are best developed in the transitional regime, as noted previously by Heard (1960). They do not form in the ductile regime. In the limestone the average angle between conjugate sets is independent of strain and strain rate; but, as for the sandstone, it increases from 75 to 103° with confining pressure (Fig. 3.56), and it is at least 20° larger than the corresponding angle between conjugate shear fractures which form after the Lüders' bands. Microscopic and SEM studies indicate that the features in both rocks are zones of intragranular and intergranular cataclasis, with negligible shear displacements.

Because the angle between conjugate Lüders' bands is a function of the effective pressure, the bands might be used to derive depth of burial at the time of deformation, provided that (1) the pore fluid pressure is assumed to be hydrostatic and (2) the orientation of σ_1, which can be the acute or obtuse bisector, is independently known. If (1) is unwarranted, then the angle between the bands could be used to infer pore fluid pressure, provided the depth of burial is known. This approach is discussed with reference to Lüders' bands occurring in naturally deformed Entrada Sandstone, Trachyte Mesa, Utah (Fig. 3.57).

Fig. 3.57. Entrada Sandstone showing development of possibly natural Lüders' bands from a drape fold at Trachyte Mesa, Garfield County, Utah. (Johnson 1970, Fig. 2.6, p. 40). *Notebook at lower right* is 12×18 cm. (Friedman and Logan 1973)

Friedman and Logan point out the similarities between stress bands in rock, flow or strain figures or Lüders' lines in metals as described by Nadai (1950) (Sect. 2.1.2.6).

3.2.2 Fractography Induced by Coring

Kulander et al. (1979) studied the fractography that develops in the course of coring operations, largely based on their experience with Devonian shales from the Appalachian basin, eastern U. S. They analyse the fractographic characteristics and formational modes of natural, coring-induced and handling-induced fractures. Their work presents possible criteria for distinction between natural and induced fractures in core samples. Excerpts from their work are cited below.

3.2.2.1 Coring-Induced Fractures

Coring-induced fractures are those that develop during the actual process of coring. The drilling process leaves, besides drilling marks on the core, characteristic fracture surfaces which enable three distinctions: (1) the time sequence of the core-induced fractures; (2) the origin of a particular set of core-induced fractures in relation to the drill bit; (3) direction of the principal tension that causes a particular set of core fractures. These three features display a unique symmetric geometry with relation to the cylindrical core section.

Disc Fractures. Horizontal and sub-horizontal disc fractures form in direct response to vertically acting principal tensile stress. These fractures are invariably inclined less than 15° to the horizontal. Horizontal coring-induced fractures cut entirely across the core or abut against earlier-formed natural or coring-induced fractures. Some disc surfaces consist of several coalescing cracks, each with its own individual origin (Fig. 3.58). The plumes generally first propagate toward the central core region.

The primary stress responsible for disc fracturing is attributed to unloading and bit pressure assisted by vibrations and torques that accompany the drilling process. Vertical tension which results from unloading causes horizontal disc fractures, whereas bit pressure produces inclined chipping circumferential to the disc.

Kulander et al. believe that most disc fractures form in the space between the drill bit and the scriber blades located within the core barrel and used to incise orientation groves. However, occasional peripheral chipping or scalloping of the disc fracture may indicate its formation immediately before the bit.

Petal Centreline Fractures. Vertical and inclined sections of petal centreline fractures, generally smooth curvi-planar to planar, and of a constant strike, form in response to a core-induced principal tension that rotates downward in a vertical plane from an inclined orientation to horizontal. The petal section

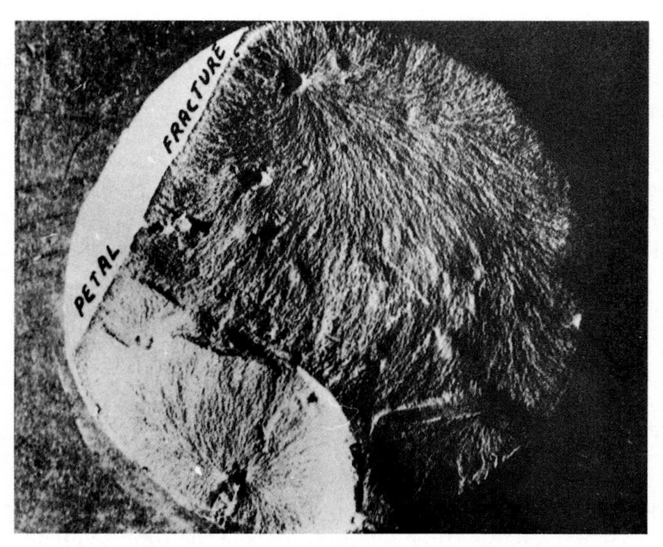

Fig. 3.58. Two distinct pyrite nodule origins on separate fracture chips forming a disc fracture surface approximately 7 cm diameter. Independent two plumes radiate from both origins and curve to meet the core boundary and earlier-formed petal fracture orthogonally. Note the discontinuity of plume across the fracture intersection separating the two origins. The intersection line is concave toward the second-formed fracture origin

Fig. 3.59. Parallel circular ribs (approximately four per cm) convex downcore (Figs. 3.58 and 3.59 from Kulander et al. 1979)

of these fractures is that portion that curves in a down-core direction from a dip angle generally between 30 and 75° at the core boundary to a vertical inclination within the core and parallel to the core axis. A continuity of both positions is attributed to a single fracture event (Fig. 3.59).

The exact origin is rarely discernible on petal-centreline fractures. However, the origin area is always located near the uppermost (upcore) inclined section of the fracture surface in proximity to the core boundary. It is likely that those origins formed below the cutting edge of the drill bit and have been subsequently drilled away. The origin may in some cases be associated with a large pre-existing flaw on a subsequently drilled core face.

Closely spaced arrest lines (approximately four per centimetre) are common and are arranged in a symmetrical fashion about the initiation site (Fig. 3.59). The closely spaced arrest lines indicate a slow propagation rate of a fracture that developed during drilling. The arrest lines are convex in a downcore direction, and symmetrical about an imaginary vertical line bisecting the fracture face.

Torsional Fractures. Torsional fractures in the cylindrical core are generated behind the bit but generally before the scribing blades. After initiation, they generally spread up and down the core at a 45° helix. Kulander et al. showed that fracture progress can be traced in shale cores by well-developed plumes and steps. Arrest lines are also common on these fracture surfaces.

Knife Edge Spalls. The knife edge spalls are coring-induced fractures directly related to tensile stresses caused by the scriber blades along the core barrel. The spalls appear as series of scalloped scars chipped along the orientation lines.

3.2.2.2 Handling-Induced Fractures

Handling-induced fractures are those induced during or after removal of the core barrel. These fractures are almost always caused by impact or bending. Fractures developing at this stage may by recognized by characteristic impact features (Sect. 1.1.7.3) or by fracture branching (suggesting a supercritical propagation, see Sect. 1.4.3 and 1.4.4).

3.2.3 Markings Induced by Quarrying

The most characteristic fracture markings caused by conventional blasting techniques as practiced in quarries, road cutting and other construction works are the multi-cuspate hackles MCH. The cuspate hackles CH (Sect. 3.1.6.6) have sharp ridges of approximately constant width that radiate in a fringe around a single parent fracture which is actually a deformed mirror plane. Multi-cuspate hackles are a CH series closely associated with the blasting

Fig. 3.60 a–c. Characteristic fracture markings caused by rock blasting. **a** A ruptured limestone block in a quarry. Note five separate mirrors *around pen.* Each mirror is surrounded by a curved fringe of cuspate hackles. **b** A mirror and long radial hackles in a ruptured chalk along a road quarry (Cenomanian, Upper Galilee, near Ma'alote). Typical length of hackles in this outcrop approximates the radius length of mirror plane. A 15-cm *bar* marks the point of fracture initiation. **c** A mirror surface that propagated from blasthole at left towards a joint normal to picture at right, along a road quarry on the Appalachian Plateau, U.S.A. Note clear curved mirror boundary, and ragged hackles above pen

origin, and forming together a common ragged surface (Fig. 3.60e). The various CH in these series may indicate propagations in different directions. In Fig. 3.60a for instance, the three CH above the pen and on its left side propagated from right to left, whereas the two CH below the pen propagated from left to right.

Lutton (1971) investigated the fracturing caused by pre-split blasting (a technique in which simultaneous or intermittent light charges are detonated in a line). He made a few intriguing observations, one of which was: "Hackle marks are intersected orthogonally by concentric steps which in effect connect en echelon set of radial fractures". On the basis of his illustrations (his Figs. 5 and 6) it appears that he described MCH which all propagated in the same direction in a periodic manner, and the concentric steps were in fact the mirror boundaries where the cuspate hackles started. Periodic growth and diminution of hackles along a propagating fracture under dynamic conditions is a phenomenon well known since De Freminville (Sect. 2.1.1.1). Lutton also noted that these markings are restricted above and below by clayey partings. In one blast he found concentric rib marks about a point of fracture origin.

Although the mirror sizes of the various CH in Fig. 3.60a are similar, the poor definition of the mirror boundaries renders it quite difficult to apply Eq. (2.7) to it. Occasionally however blasting may produce well-defined mirror boundaries (Fig. 3.60b).

The implication of Eq. (2.7) is that in natural fractures (joints) caused by low stresses (Bahat and Engelder 1984) mirror planes are expected to be large relative to the size of hackles, whereas in artificial cracks caused by quarrying techniques mirror planes are small relative to hackle lengths because of fracturing under high stresses. In natural fractures, short hackles radiate from a circular perimeter of relatively large radius (Fig. 4.20a). In artificial fractures, on the other hand, a small mirror plane is surrounded by long and narrow radial hackles (Fig. 3.60b). In a recent study (Bahat and Rabinovitch 1988), the fracturing stress of a natural joint was determined on the basis of mirror dimensions. Future studies may be able to determine the local fracture stress of mirrors in blasted rocks (as in Fig. 3.60c).

Chapter 4 Characterization and Classification of Fracture Surface Morphology in Geologic Formations

Fracture surface markings (FSM) in rocks are often overlooked due to their faint morphologies, and their exact features are revealed only to the careful observer. Their classification involves meticulous examination and comprehensive description. A qualitatitve characterization is presented in this chapter, using a set of complementary categories. Certain FSM features are also characterized quantitatively.

Some of the accompanying illustrations are drawings made from photographs, which serve to mark out fine details.

4.1 Qualitative Characterization of Fracture Surface Markings

4.1.1 Descriptive Parameters of Fracture Surface Markings

Following are 14 qualitative parameters, which can be used singly or in combination, to characterize FSM. They include geometry, symmetry, inclination, confinement to layer boundaries and maximum size, fringes, initiation, termination, relief, joint sets, orthogonality, periodicity, sequence, FSM overprinting and FSM on joint fillings.

4.1.1.1 Geometry

All FSM fall into two broad groups: (a) plumes and (b) rib markings or undulations. Distinction between these groups is based on the geometric differences between various types of plume and rib marking, as described in Chapter 3 (Figs. 3.2 and 3.19 to 3.25).

4.1.1.2 Symmetry

Symmetry is examined both parallel and perpendicular to the plume axis. Plumes may be quasi-symmetric when they are bilateral (Fig. 4.1 a, b, c), but when unilateral they are usually asymmetric. They can also be bilateral asymmetric (Fig. 3.10). Quasi-symmetry perpendicular to the axis is usually displayed by barbs on both sides, but in some types of joint, such as mud cracks (Fig. 3.17 d) and columnar joints (Fig. 3.50), barbs occur only on one side of the plume axis.

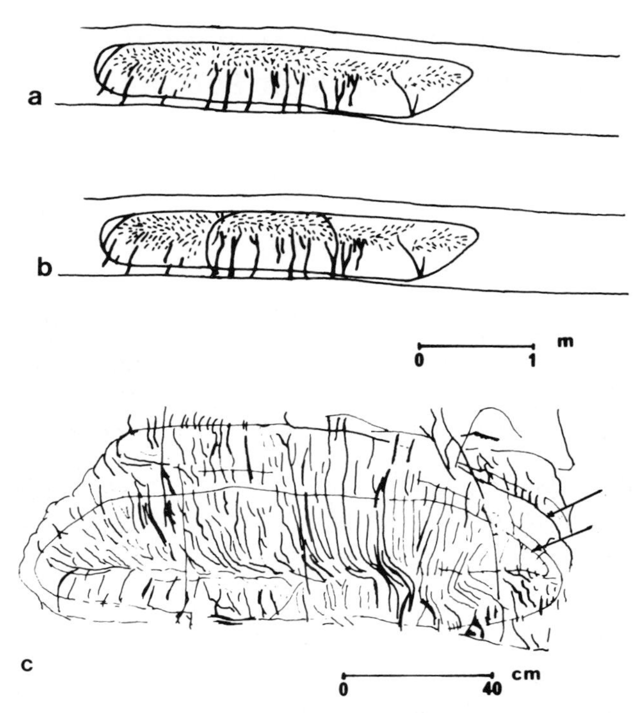

Fig. 4.1a–c. A bilateral plume with a gradual (continuous) breakdown of a fringe below the plume. Two distinct FSM are revealed by sun illumination at two different hours of the day. **a** A bilateral plume. **b** A distinction between an early semi-elliptical joint at the centre and the final bilateral plume (see photographs in Fig. 3.17a, b). **c** A coarse straight plume on a 345° joint in Eocene chalks near Beer Sheva showing horizontal bilateral propagation. Barbs propagate horizontally, upward and downward. The propagating plume lags behind along the median and is more advanced along the layer boundaries. Stages of fracture propagation are discerned as curved arrest lines at two different parallel levels on the fracture surface shown by *arrows*. This plume has no fringe (see photograph in Fig. 3.10)

Symmetric concentric rib markings (Fig. 4.2a) are distinguishable from compound circular rib markings which are asymmetric (Fig. 4.2b). Symmetric undulations may be circular (Fig. 4.2a) or elliptical (Fig. 4.3) (see Sect. 3.1.7.4 on symmetry of discoids, cited from Bankwitz and Bankwitz 1984).

4.1.1.3 Inclination

Axes of most plumes on fracture surfaces in sedimentary rocks are parallel or sub-parallel to layer boundaries. Deviations occur in thick layers or massive rock units, and on surfaces of en echelon cracks (Fig. 3.35). Vertical plumes are rare (Fig. 3.3).

When circular undulations on vertical fractures are elongated toward the elliptical, their longer axes are usually horizontal (Fig. 3.19). In some cases, ellipse or quasi-ellipse axes are sub-vertical (Fig. 5.23) or inclined (Fig. 5.19).

Fracture markings also occur on horizontal joints (commonly called sheets), and on horizontal bedding-parallel surfaces (Fig. 4.4).

4.1.1.4 Confinement to Boundaries and Maximum FSM Size

Plumes are mostly confined between layer boundaries, whose width (height) seldom exceeds 60 cm (Fig. 4.5). On the other hand, plume lengths may be great. A plume of 48 m length was observed in siltstone on the Appalachian Plateau (Fig. 3.18).

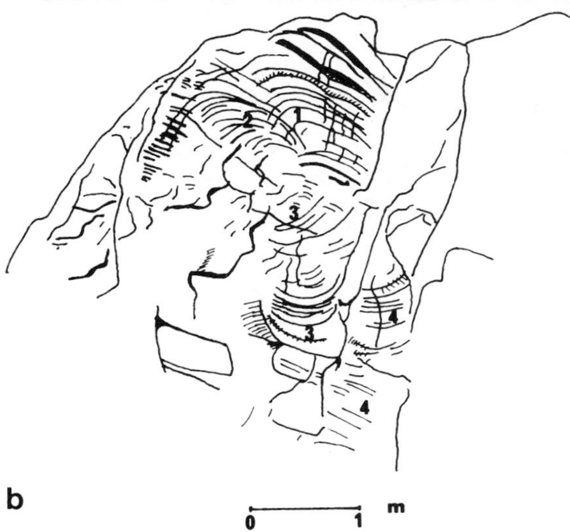

Fig. 4.2. a A joint surface with annular concentric rib markings and radial plumes in Senonian chalks (Menuha Formation) near the town of Arad. Part of the feature is eroded. *Key* provides scale. **b** A joint surface in Palaeozoic sandstones in Wadi Tweibeh, northern Sinai, showing asymmetric and compound circular rib markings. Four numbers represent propagation events from different initiation points (see type 4 rib markings, Fig. 3.23)

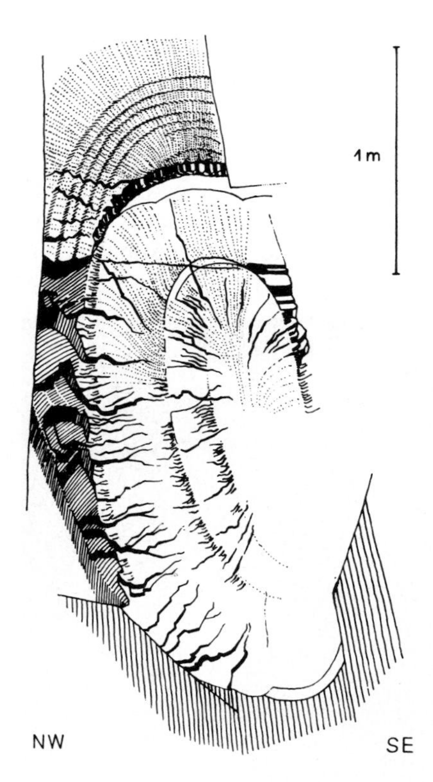

Fig. 4.3. Symmetric elliptical rib markings, plumes and a large fringe in Ordovician pencil slate, Thuringia, Germany. (Bankwitz 1965)

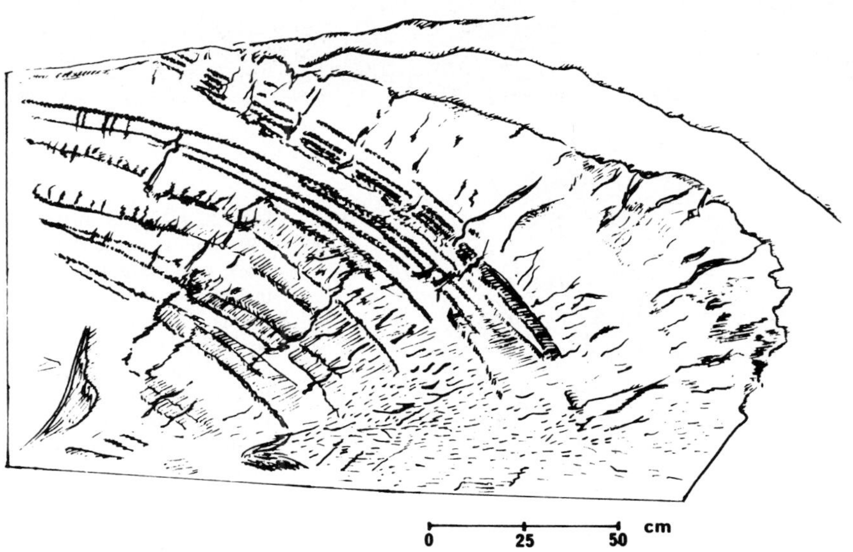

Fig. 4.4. Surface markings on a horizontal joint on the ceiling of a bell-shaped cavern in the Middle Eocene chalks near Beit Govrin. Note the concentric rib markings at *left* and radial hackles at the *upper right side.* Plume is seen at the *lower side* of drawing (see Fig. 3.6a)

Undulations are mostly confined to layer boundaries in carbonate rocks but may cut across thin shale beds. Very large undulations may occur in thick uniform formations (Figs. 5.23 and 5.19).

4.1.1.5 Fringes (Joint Peripheries)

Peripheries of FSM on planar joints are either smooth (Fig. 4.2a), or they appear with fringes of en echelon cracks (Fig. 3.41a) or hackles (Fig. 4.4). Fringes may result from either continuous or discontinuous breakdowns (Sect. 3.1.6.1). They are very common on discoid surfaces (Fig. 3.29).

Plumes (striae), rib markings, hackles, cuspate hackles (Fig. 4.6) and en echelons are various manifestations of crack morphologies which result from superposed tensile and shear stress conditions (Sect. 1.1.7.2). Whereas plumes and rib markings mostly occur on a joint's main surface, hackles, cuspate hackles and en echelons develop at the fringe. Fringes often line the upper or/and lower sides, and less frequently the distal sides of horizontal plumes (Fig. 4.6). En echelon cracks around large circular undulations often show preference for the lower part of the marking (Fig. 5.20a). Hackles, en echelons and cuspate hackles resemble each other (Sect. 3.1.6.6). However, criteria for their distinction are suggested in Table 3.3.

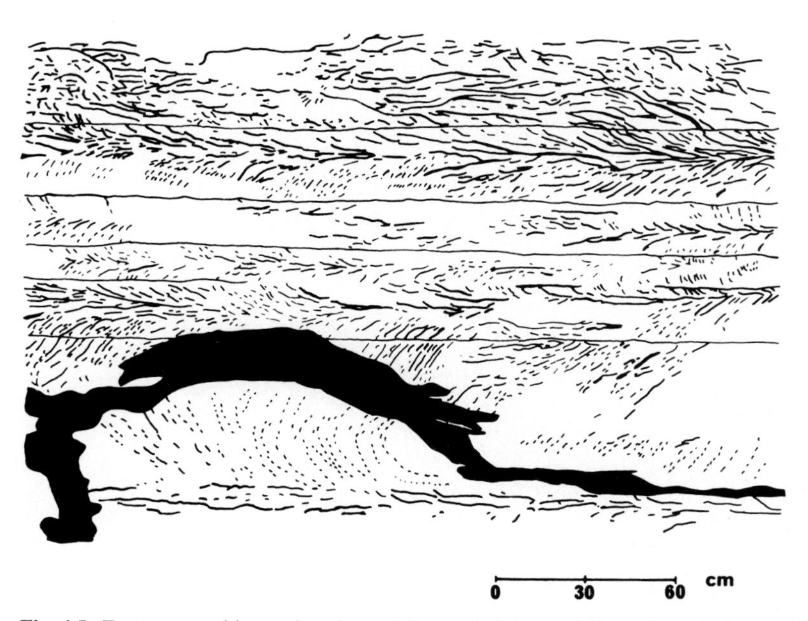

Fig. 4.5. Fracture markings of various styles, including straight and wavy plumes, in siltstones from the Devonian Appalachian Plateau at Watkins Glen, N.Y., showing some plumes that are confined between layer boundaries, whereas others cross-layer boundaries (see photograph in Fig. 3.8)

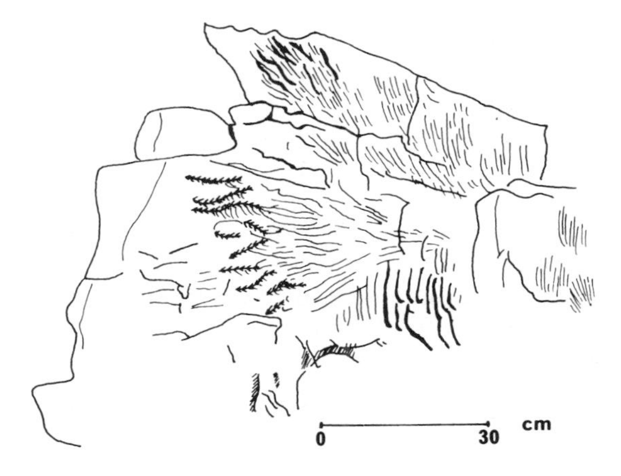

Fig. 4.6. Cuspate hackles at the distal fringe of a natural fracture marking from the Lower Chalks, Kent, England. The hackle surfaces are not flat in plan (Fig. 2.3), but have cuspate surfaces (marked here as *arrowed bars*) (see Fig. 3.41 a)

4.1.1.6 Initiation

Plumes tend to initiate either from the lower boundary (Fig. 4.7) or from the upper boundary of the layer or close to it (Fig. 3.18). They may initiate also from within the layer (Fig. 4.1 a) or from an earlier fracture (Fig. 3.17 c). Circular markings generally initiate within the layer (Fig. 4.2 a) but they also start at layer boundaries or from a previous fracture surface (Fig. 4.8).

Fig. 4.7. Plumes in Devonian siltstones from the Appalachian Plateau, Watkins Glen, N.Y., indicating independent propagation of the two adjacent joints, both initiating from the lower layer boundaries. Thickness of lower layer is 18 cm

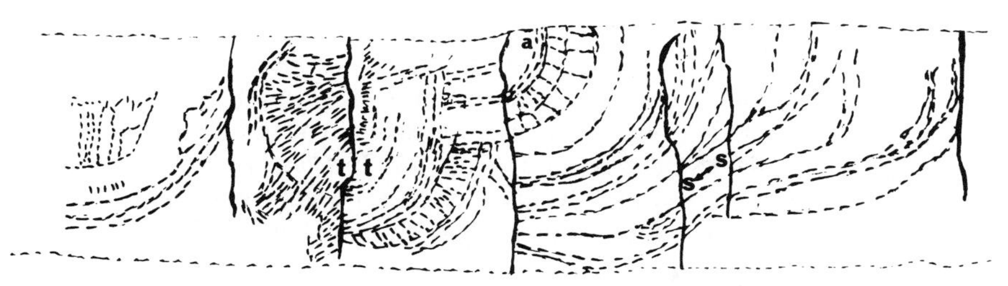

Fig. 4.8. Series of rib markings in a Lower Eocene chalk layer about 0.5 m thick, enclosed between chert beds about 0.07 m thick near Beer Sheva (Bahat 1988a, station 20). The *a* rib initiated at or close to the layer boundary. Fractures marked by *s* are single-layer strike joints (sub-normal to picture)) that cut earlier cross-fold joints. Fractures marked by *t* are semi-circular markings on surfaces of cross-fold joints (sub-parallel to picture) that initiated from an earlier strike joint (see Fig. 3.21)

Certain joint sets are "liberal" and allow fracture initiation at different points. On the Appalachian Plateau, for example, wavy plumes start from the lower or upper layer boundaries or even from within the thick siltstone layer. Other lithologies seem to impose strict limitations. Straight plumes in thin siltstone intercalations in shales from the Appalachian Plateau strongly prefer fracture initiation at the upper layer boundary (Fig. 3.18). On basal columnar joints formed in basalt flows, plumes initiate at the lower boundaries of their bands (Fig. 3.50). On the other hand, all examined plumes from the Eocene chalks around Beer Sheva start from within the layer (Fig. 3.10).

The different locations of joint initiation express various mechanisms of crack nucleation. In heterogeneous lithologies like siltstone intercalations interbedded in shales, or in rocks that are affected by strong thermal gradients like basalt, maximal stress seems to be concentrated along lines/bands of discontinuity. In layered sequences of uniform sedimentary lithologies, maximal stress concentration can be anywhere within the layer due to inclusion discontinuity, or less frequently along boundaries.

4.1.1.7 Termination

Hodgson (1961b) describes several styles of joint termination. Plumes may terminate laterally before reaching the distal edge of the fracture. This termination is often (but not always) marked by an arrest line (Fig. 3.10), implying that a joint may simply die out when the driving force for fracture propagation subsides. Abutting against a free surface (previous joint) at a high angle is another reason for a fracture to arrest. When a joint approaches a free surface at a low angle it often is influenced by it, mostly in the form of bending, before termination (Sect. 5.9.1). Joint termination on a free surface at a low angle is less frequent. Both bending termination (sub-orthogonal style) and low angle termination (asymptotic style) have been observed in the Appalachian Plateau in cross-fold joints of different sets.

Vertical fracture interaction between single-layer joints on overlying sedimentary layers is rare. These joints seldom cross boundaries, and when they do, they maintain coplanar relationships (stay on the same surface) (Fig. 4.5). Quite different is the rhythmic vertical fracture interaction on columnar joints in basalt (DeGraff and Aydin 1987; Aydin and DeGraff 1988) where horizontal fractures follow each other in an upward sequence so that the lower edges are curved and the upper edges straight behind (Fig. 3.50).

4.1.1.8 Relief

Plumes generally occur on planar surfaces and most of their relief is not deeper than 1 mm. However, on approaching layer boundaries in sedimentary rocks (Fig. 3.5) or band boundaries on columnar joints (Fig. 3.50) roughness often increases considerably (see also Sect. 4.1.1.12). Undulations, on the other hand, being wavy, may reach relief of several mm amplitude and in extreme cases even several cm (Figs. 4.9 and 5.23). Conic discoids (Fig. 3.43) are examples of high relief FSM.

Joints are often quite flat, but wavy joints are also common. Markings on such fractures will naturally assume wavy shapes. Curved plumes along the strike are fairly common (Fig. 3.17 a, b), as are to a smaller extent curved rib mark-

Fig. 4.9. A rib marking about 2 m long convexing to the right is shown *right* of Terry Engelder and scale. The rib cuts across many thin shale beds just below a 21-cm-thick siltstone bed marked by a straight plume (*upper left*) on the Appalachian Plateau, Watkins Glen, N.Y.

ings (Fig. 4.10) along the strike and along the dip (Fig. 5.19). Joint curving also occurs on horizontal (sheet) joints (Fig. 3.24).

4.1.1.9 Joint Sets

Rock fracturing typically results in several systematic sets of vertical joints (Hodgson 1961b). Regional data show that certain styles of FSM have affinities to specific joint sets (Sect. 4.2.1.3).

4.1.1.10 Orthogonality

Quite often two different sets occur at 80 to 90° to each other, a relationship known as orthogonal. Such sets may bear undulations FSM (Fig. 4.11). Orthogonal sets marked by plumes are rare.

4.1.1.11 Periodicity

FSM on surfaces of adjacent single-layer joints often display periodicity. Horizontal periodicity in plumes is an example. Plumes become incrementally longer along horizontal axes, attain a certain length, terminate, and then restart the cycle. Periodic plumes generally maintain the same width, but not the same length in all cycles. Lengths may vary considerably along a given layer (Fig. 3.18). Furthermore, adjacent plumes may propagate in opposite direc-

Fig. 4.10. Circular rib markings at a rounded corner formed by an intersection between two orthogonal single-layer joints in Lower Eocene chalks near Beer Sheva (see *pen* for scale)

tions (Figs. 3.34 and 4.12). Incremental plume growth also occurs on surfaces of columnar joints (DeGraff and Aydin 1987).

Periodicity along single-layer joints appears also in undulatory FSM. Generally, neighbouring rib markings maintain similar sizes (Fig. 3.21).

Cyclic undulations occur also on joint surfaces which cut thick rock layers (60 m or more). However, in these cycles size increments are not strictly regular, and small FSM may be observed adjacent to large ones (Fig. 5.19b).

4.1.1.12 Sequence

Many FSM on single-layer joints are formed during the burial stage of the rock history although such single-layer jointing may also occur during later stages. In Lower Eocene chalks of the Beer Sheva area, plumes which occur on joints that are thought to belong to an early diagenetic stage are coarse whereas plumes on a later set are delicate (Figs. 3.10 and 3.11, respectively). However, other factors that determine plume coarseness in these sets should also be considered (Bahat 1989).

Fractography is a useful tool in determining the fracture sequence in rocks. Otherwise, the determination of the sequence of joints cutting each other is most difficult. A young fracture cuts fracture markings of a previous joint (Bankwitz 1966, Fig. 9). Sequence of propagation of neighbouring cross-fold and strike joints is revealed by the cutting relations of rib markings (Fig. 4.8).

Fig. 4.11. Rib markings on orthogonal two joint sets in Senonian chalk, Lower Galilee, near Nazareth. The markings on the two surfaces approximately mirror each other. Bar = 15 cm

Fig. 4.12. Two joints that constitute a fracture plane intersect each other (near the scale contact with the rock). The plumes indicate that the fractures along the 21-cm-thick siltstone intercalation grew toward each other. The Appalachian Plateau, Watkins Glen, N.Y.

Bankwitz (1966, Fig. 23) shows that when a propagating joint produces en echelon cracks, the cracks that are formed close to a previous fracture are smaller than those farther away from it. The phenomenon is also known in fractures produced by blasting (Fig. 3.60c).

Kulander et al. (1979) present a scheme of various crossing relationships of joints which exemplify various fracture sequences (their Figs. 45–53). Outer undulations that deviate from concentricity with respect to the inner undulations are shown on the joint that parallels the page in Fig. 4.13, abutting another joint (oblique to page). This deviation implies that the oblique joint is the earlier, and that it influenced the development of undulations on the later joint that abuts it.

The sequence of fracturing as determined by fractographic means may be applied in conjunction with fracture-system architecture (Hancock 1985), which describes the various styles of joint contact and interaction by other criteria.

4.1.1.13 Overprinting of Fracture Markings

FSM overprinting is the term used to characterize two (or more) FSM that occur on the same plane of fracture. The various FSM have distinctive morphologies and they represent different events (Fig. 4.14), whose sequence may be determined by careful morphological analysis (see also Fig. 5.18).

Fig. 4.13. Ribs at *arrow* attempt to swing parallel to pre-existing fracture oblique to the plane of the photograph. (Kulander et al. 1979)

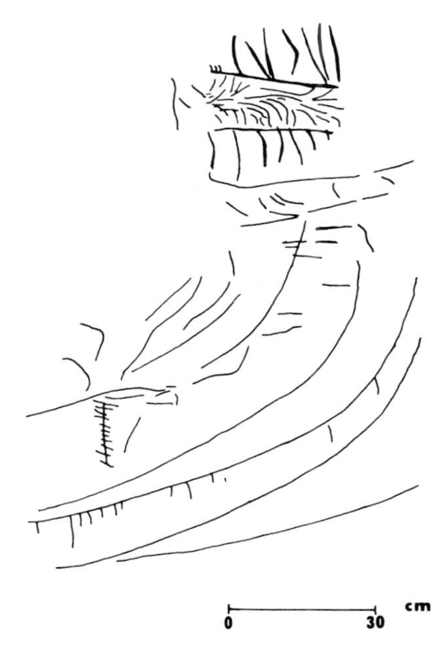

Fig. 4.14. A horizontal plume in the *upper part* of the drawing is cut by a series of undulations which convex downward in the *lower part* of the picture (see photograph Fig. 5.18a)

4.1.1.14 Fracture Markings on Mineral Fillings

Occasionally, fracture markings occur on minerals that precipitate along the joints (Fig. 3.7). These usually form fine-grained or cryptocrystalline films in single-layer joints (attributed to the burial stage). Coarse-grained fillings seem to be more associated with syntectonic and uplift events (Chapt. 5).

4.2 Quantitative Characterization of Fracture Surface Markings

4.2.1 Measureable Parameters of Fracture Surface Markings

In spite of the increasing body of information on FSM in rocks, and their correlations with FSM in technical materials (e.g. Kulander et al. 1979; Bahat 1979a), the methods for quantitative description of FSM in rocks are still embryonic. They tend to coalesce in practice into the following categories:

1. Studies on plume length in connection with:
a) plume waviness (Bahat and Engelder 1984);
b) plume periodicity and periodicity versus layer thickness;
c) joint length distribution (Bahat 1988b);
d) factors determining minimum fracture length.
2. Angular relationships in plumes (Roberts 1961; Syme Gash 1971).
3. Aspect ratio of fracture markings and joint dimensions.
4. Rib-marking profiles (Bankwitz 1966).
5. Determination of palaeostress on the joint surface (Bahat and Rabinovitch 1988).
6. Dimensions of en echelon cracks.
7. Relations between stress intensity, fracture velocity and morphology of the joint surface.

4.2.1.1 Plume Length

Plume Waviness. Plume waviness W_v is expressed by:

$$W_v = (l_m - l_p)/l_p \ , \tag{4.1}$$

where l_p is the length of the plume parallel to bedding and l_m is the maximal length of the plume axis. l_m may be longer than l_p for C-type plumes (Fig. 4.7) and they may approximate each other in S-type plumes (Fig. 4.1c and Bahat and Engelder 1984, Table 1).

Plume waviness that occurs before the stable propagation sets in (Fig. 3.18) seems to delimit the extent of crack incubation. At the early stage of fracturing, fracture growth is controlled by mixed modes regardless of scale (see Fig. 2.30b). Full understanding of this stage in rock fracturing requires further study.

Plume Periodicity and the Periodicity in Relation to Layer Thickness. In Devonian siltstone layers, oriented 335 and 004° at Watkins Glen, N.Y., the three plume end members (Fig. 3.2a, b, c) appear in various adjacent combinations. Periodic distances between rhythmic barb maximum intensities in C-type plumes (fan perimeters in Fig. 3.2c) seem to be approximately constant in a given layer, averaging 23.8 cm in a layer 36 cm thick (Fig. 3.7a). On this joint, which is marked only by rhythmic plumes, the periodicity in relation to layer thickness is 0.66. Another joint surface on a 30-cm-thick layer, is marked by a rhythmic C-type plume which also has elements of a straight plume (Fig. 4.5). Here, the average plume periodicity is 25 cm, and the periodicity in relation to layer thickness is 0.83. The significance of these different periodicities is not yet understood. Future investigation may possibly link these parameters with the process of fracture by pore pressure (Bahat and Engelder 1984).

Joint Length Distribution. Bahat (1988b) studied the joint length and plume length distributions in Middle Eocene chalks near Beer Sheva. The results for 45 joint lengths and 44 plume lengths measured along a 60-cm-thick layer, are summarized in Table 4.1 and in Fig. 4.15. The plumes, which are straight and delicate, fall into four groups:

1. Eighteen plumes of unilateral propagation (Fig. 3.17c) from NE to SW.
2. Eleven plumes of unilateral propagation from SW to NE.
3. Ten plumes of bilateral propagation, from a point on the joint surface towards NE and SW (Fig. 4.1a).
4. Five circular to near-circular plumes (aspect ratio of the ellipses ranging between 1 and 3) which propagated radially in all directions (Fig. 4.16, plume a). They are termed radial plumes in Table 4.1.

Bilateral plumes tend to occur on longer joints than unilateral and circular plumes. The mean lengths of the unilateral joints is 66.5 cm, of the bilateral joints 155.6 cm, with standard deviations of 30 cm and 60.6 cm, respectively,

Table 4.1. Mean parameters and standard deviations of fracture and plume lengths in a single chalk layer

Type	Mean (cm)
Unilateral fracture	71.6 ± 42 (36)
Unilateral fractures (not including the 250 cm fracture in Fig. 4.15a)	66.5 ± 30 (35)
Bilateral fractures	155.6 ± 61 (9)
All fractures	88.6 ± 57 (45)
Unilateral plumes	66.5 ± 34 (29)
Unilateral plumes (not including the 190 cm plume in Fig. 4.15b)	62.5 ± 26 (28)
Bilateral plumes	138.9 ± 49 (10)
All plumes	82.4 ± 48 (44)

Number of parentheses indicates population size.

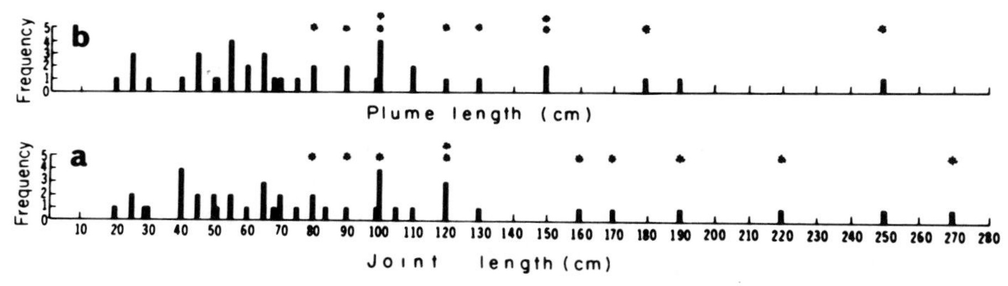

Fig. 4.15 a, b. Distribution of joint and plume lengths on a 60-cm-thick chalk layer from the Middle Eocene near Beer Sheva (outcrop Y, Bahat 1986a). **a** Joint lengths. **b** Plume lenghts. *Asterisks* in **a** designate joints with bilateral plumes; *asterisks* in **b** designate bilateral plumes

suggesting a bimodal length distribution. Neither joints nor plumes shorter than 20 cm were observed.

Factors Determining Minimum Fracture Length. The absence of joints shorter then 20 cm and the maximum of unilateral plumes within the 40 to 70 cm range (Fig. 4.15) rather than a power law distribution (Segall and Pollard 1983) raises the question of the reasons for lack of short fractures. The mechanism that causes the observed fracturing may be rationalized on the basis of Secor's

Fig. 4.16. A bed (*between vertical arrows*) with two adjacent delicate plumes. Plume *a* on the *right* is almost circular (group 4). Plume *b* is unilateral. It initiates close to plume *a* and propagates from right to left along a distance of 4.5 m. Parts of the plume are not seen in the picture, except for some of the coarser barbs (*inclined arrows*) which display uniform plume orientation. Figs. 4.15 and 4.16 are related to different layers

theory (1969) that the growth of a tension fracture at depth in the earth's crust is a slow process, consisting in detail of numerous short episodes of crack propagation interspersed with longer periods of quiescence during which pore fluids from the surrounding rock percolate into the crack and wedge it open. Field observations by Bahat and Engelder (1984) support this model. Accordingly, a quiescent period is required for a critical volume v_{crit} of water pore fluid [Eq. (1.60)] to percolate and build up pressure before failure criteria for fracture [Eq. (1.59)] can be reached to initiate fracturing. Commensurately, v_{crit} would require wedging fractures with *minimal lengths.* It appears that in the investigated layer such fractures have not been shorter than 20 cm.

4.2.1.2 Angular Relationships in Plumes

Two useful plume parameters, the median angle m and the interface angle i (Fig. 3.13) were suggested by Roberts (1961) and Syme Gash (1971). Otherwise plumes are designated by type (Fig. 3.2) or by joint set.

The median angle m which is formed during the early growth of the plume is small (generally below 35°). The initial part of the plume may display some waviness, before curvature is established. In such cases m is read at the approximate point where plume curvature becomes regular (Fig. 4.17 a). The interface angle i which is formed during the late growth of the plume is variable and may in some plumes exceed 90°. The characteristic herringbone profile of the plume (Fig. 3.2 a) illustrates how plume propagation lags behind along the median and is more advanced along the layer boundaries (see Sect. 3.1.8.2). In the case of bilateral plumes, barb angles undergo rapid change near the point of initiation, and measurements in this area are meaningless. The relation between m and i varies fron joint to joint implying different fracture processes.

A plot of m versus i for different plumes (Fig. 4.17 b) demonstrates several ratios which allow inferences as follows:

1. Syme Gash's observation (1971) that i is diagnostic and m is not (Sect. 3.1.2.5), is essentially supported by these data, although some variations of m appear to be consistent for certain plume specimens.
2. One would assume that in layered rocks the propagating plume would approach the layer boundary, which is a free surface, at right angles. Nevertheless, most of the measured plumes do not reach i values of 90°.
3. The two mud-crack samples (Nos. 1 and 2 in Fig. 4.17 b) display straight plumes either below or above the horizontal plume axis, but not on both sides of the plume axis, and they reach i values of 90 or even 100° (Fig. 4.18 a, b).
4. Straight plumes from different fracture provinces may differ significantly in their angular relationships. The straight plume 345° from the Appalachian Plateau (No. 5 in Fig. 4.17 b) has an exceptionally low i value. The coarse plume from the Lower Eocene Beer Sheva chalks (No. 6) differs considerably from the plumes of the Middle Eocene chalks (Nos. 7, 8).
5. The only samples which display fringes (of en echelon cracks) are those from the Middle Eocene chalks (Nos. 7, 8). These samples have i values that

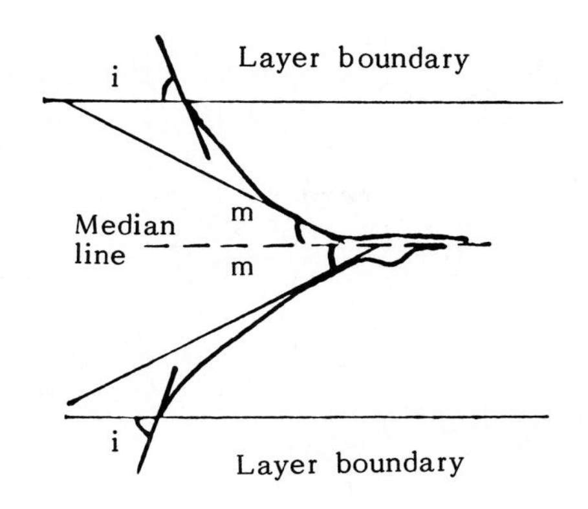

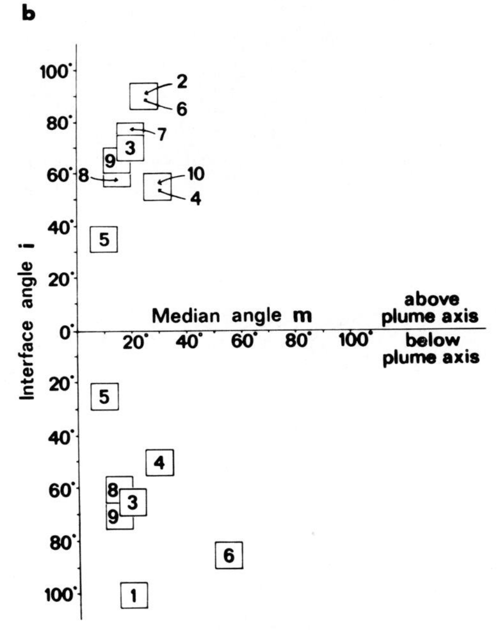

Fig. 4.17. a Schematic relationship of plume initial curvature, *m* is measured where plume waviness ends. **b** A plot of the median angle m versus the interface angle *i* above and below the median (see Fig. 3.13). Data from ten plumes: *1* mud crack; *2* mud crack; *3* 335° wavy plume, Devonian, N.Y.; *4* 335° wavy plume, Devonian, N.Y.; *5* 345° straight plume, Devonian, N.Y.; *6* straight coarse plume, Lower Eocene, Beer Sheva; *7–8* straight delicate plumes, Middle Eocene, Beer Sheva; *9* straight delicate plume, Middle Eocene, Shivta; *10* plume from a metamorphic Precambrian rock, Sinai

range from 55 to 80°. They correspond to group B of Syme Gash (Fig. 3.14). The other plumes have lower or higher i values. Future research may tell how useful these parameters can be.

b

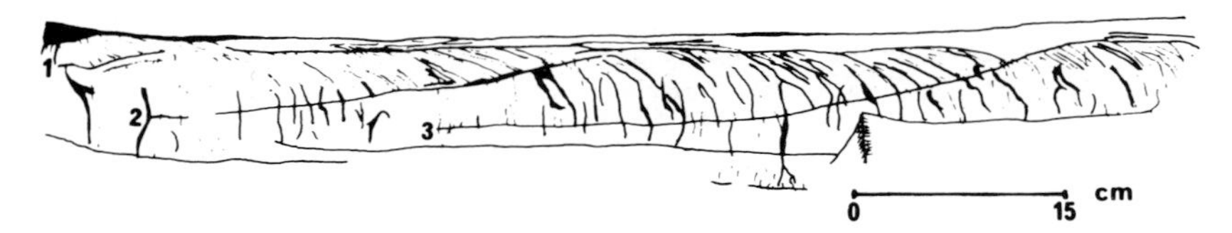

Fig. 4.18 a, b. Mud crack. Four stages of plume propagation are distinguished by three curved arrest lines. **a** Photograph. **b** Drawing

4.2.1.3 Aspect Ratio of Fracture Markings and Joint Orientation

The aspect ratio (the length/height ratio) of FSM on joint surfaces appears to differ according to joint set. Bahat and Engelder (1984) found wavy plumes with small aspect ratios (maximum 10) on 335° joints of the Appalachian Plateau, and straight plumes with large aspect ratios (maximum 200) on the 344° joints. In Lower Eocene chalks around Beer Sheva the plumes on 309 and 345° joints (Fig. 4.1 c) have aspect ratios up to about 5, and rib markings on 328° joints have aspect ratios of 1 to 2 (Fig. 4.8). In all the above cases the FSM sizes approximate the respective joint dimensions. The full implication of these relationships is not clear. It has been pointed out (Bahat 1987 a) that the aspect ratios may be partly dependent on the angular relationship with respect to the principal palaeostress direction. In the case of the Appalachian Plateau, the connection between aspect ratio and joint orientation corresponds to lithological differences; this, however, is not the case in the Beer Sheva area, where affinities of FSM to particular joint sets also appear in the same lithology.

4.2.1.4 Rib-Marking Profiles

Bankwitz (1966, Fig. 32) considers the angular relationship between fringe and parent fracture in discoids. He designates α, which is the deviation angle of various planes in the fringe from the parent joint, and β, which is the kink angle between adjacent planes in the fringe (Fig. 4.19). Both α and β may be larger than 90°. The angles can be measured quite accurately, because the fracture surfaces of different inclinations in discoids often intersect sharply (occasionally along cuspate boundaries). The angular determination is useful in characterizing the three-dimensional aspects of such markings (Sect. 3.1.7).

Discoids with $\alpha > 90°$ are found in fringes with strongly changing inclinations (Fig. 3.26e) which possibly reflect several fracture stages under different stress conditions. The angular determination shown in Fig. 4.19 is not applicable to these discoids since they consist of wavy rib markings which do not display sharp angular contacts. Another method has to be devised for expressing the surface geometry of these markings.

4.2.1.5 Determination of Palaeostress on the Joint Surface

Most studies of FSM on joint surfaces have been made on sedimentary rocks, limited generally to qualitative characterization. Bahat and Rabinovitch (1988) applied principles of quantitative fractography derived from ceramics in the analysis of a tectonic joint in granite. Their method enables calculation of the fracture-causing palaeostress.

The following general assumptions were applied:

1. Fracture mechanics of glass and ceramics may be adapted to brittle solid rocks.
2. Initial fracturing occurs under sub-critical stress intensity conditions ($K_I < K_{Ic}$, Sect. 1.4.1.3).
3. As the fracture length $2c$ reaches the critical size $2c_{cr}$ (Sect. 2.1.3.1), K_I attains the value K_{Ic} and spontaneous propagation starts.
4. Post-critical fracturing takes place under constant tensile stress with a negligible pore pressure influence.
5. Unstable fracture propagation and branching (Sect. 1.4.4) occur in feldspar which constitutes about 70% of the rock volume.

Bahat and Rabinovitch determine the fracture stress by two methods. The first is based on the ratio of the mist-hackle radius r to c_{cr} (Fig. 2.31). This ratio has been theoretically postulated (Anthony and Congleton 1968), and empirically shown (Bansal 1977; Mecholsky and Freiman 1979) to be a constant for various glass and ceramic materials. This method is not directly related to the fracture geometry, since c_{cr} must be determined empirically from the radius r which may cause an increased error. The second method, on the other hand, directly uses the radius r together with Griffith's equation

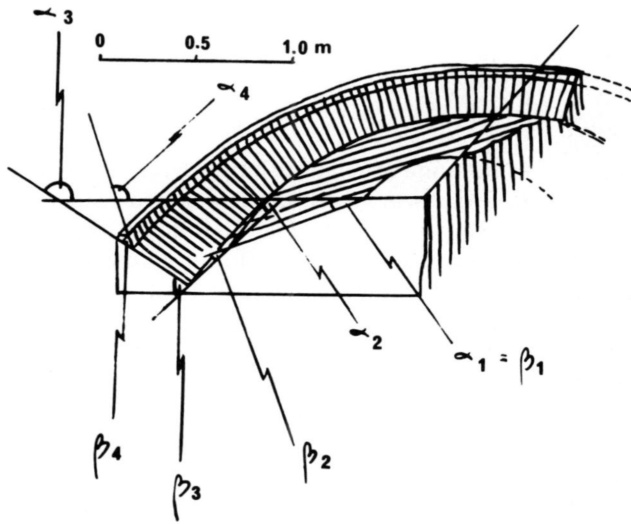

Fig. 4.19. A schematic discoid, showing the angles of deviation α between the parent fracture and various fracture planes of the fringe, and the kink angles β between these fringe planes. (After Bankwitz 1966)

modified for bifurcation by Congleton and Petch (1967). This straightforward method in which geometry and stress are directly related considerably reduces the error. While the first method only yields an order-of-magnitude estimate of the palaeostress, the second method determines it to within a reasonable range. Therefore only the latter method is elaborated below.

Hackling is the consequence of fracture branching. According to Congleton and Petch [1967, their Eq. (3) modified], this branching occurs when

$$\sigma_f = 2M(E\gamma/\pi)^{1/2} \cdot r^{-1/2} , \tag{4.2}$$

where E, γ and M are Young's modulus, the surface energy, and the enhancement factor for fast versus slow propagation (see below), respectively. Using relevant estimates for these constants, Eq. (4.2) yields the fracture stress σ_f directly.

Fracture Markings and Microcracks. Fracture markings commonly occur on joint surfaces of fine-grained rocks such as chalks, sandstones, slates and even lavas or hypabyssal rocks. They are rare in medium-grained rocks like granite due to the increased effect of grain boundaries in masking the delicate fracture morphology. Exceptionally well-developed fracture markings on a joint in granite (Fig. 4.20a) were found in Wadi Zrara in eastern Sinai at coordinates 28° 40′/34° 17.5′.

The granite jointing is ascribed to the great uplift that started in the late Neogene (Garfunkel and Bartov 1977; Bahat 1980a), proceeding at an average rate of 0.1 – 0.2 mm/a (Kohn and Eyal 1981).

The point of fracture initiation is at the extreme upper right side of the granite block (Fig. 4.20a, point A). The radial striae suggest that the fracture propagated in a hemi-circular front from right to left, forming the circular line

Fig. 4.20. a Fracture surface in granite from eastern Sinai. Fracture initiation at *A*. Curve *B* marks the circular perimeter of the hackle boundary (the mirror-mist boundary from Figs. 2.30 and 2.31 is obscured by rock graniness). Note radial striae between *A* and *B* and hackles fanning out beyond *B*. Distance from *A* to *B*, r = 360 mm. **b** Thin section of granite from eastern Sinai. Crossed nicols. A microcrack 4 mm long (see distance between x and y) runs along the boundary between two feldspars (*dark*). Two microcracks cross a quartz grain (*white*) near *c*

(curve B) which marked the boundary of the mirror plane. From this boundary and further to the left a region of hackles developed. A distinction can be made between the radial striae on the mirror plane which, although clearly visible, are rather superficial, and the radial marks in the hackle region which cut considerably deeper.

The granite is of equigranular, fine- to medium-grained texture (Fig. 4.20b). Microscopic examination reveals microcracks along the grain boundaries, typically between quartz and feldspar grains (Brace et al. 1972). Occasionally, microcracks cross quartz crystals, and may locally form anastomosing networks. Fractures can be traced along several adjacent biotite grains. Straight or curving traces that follow such microcracks (ignoring short curvatures) reach approximate lengths of 4 mm. Accordingly, initial microcrack length $2c_{ci}$ ranges from 4 mm to 5 mm (the latter is the maximum straight grain boundary of quartz).

The values used in the calculation by the second method are: $E = 5.0 \times 10^{10}$ Pa (Segall and Pollard 1983); $\gamma = 4.06$ J/m^2 for microcline (Atkinson and Avdis 1980); r = 360 mm measured on the fracture marking (Fig. 4.20a). Here it is assumed that branching occurred in the feldspars. M ranges between $2\sqrt{2}$ and $5\sqrt{2}$ (Congleton and Petch 1967). Equation (4.2) yields the range $2.4 < \sigma_f < 6.0$ MPa, with an average of 4.2 MPa. Bahat and Rabinovitch find by their first method (not elaborated here) that σ_f ranges between 2.9 and 10.1 MPa with an average of 6.5 MPa for a shallow flaw and between 4.6 and 15.8 MPa with an average of 10.2 MPa for a semi-circular flaw. They also determine independently (by a third, rather speculative method) the value 6.0 MPa for the assumed effective fracture stress σ as a function of c_{ci} for $c_{ci} = 1.9$ mm. The latter result agrees well with the σ_f values obtained by the first and second methods, and the observed grain boundary size ($2c_{ci}$ ranges from 4 to 5 mm).

4.2.1.6 *Dimensions of En Echelon Cracks*

Several en echelon dimensional parameters are useful for description.

Length/Width Ratios. Crack width W, and w (when w is the exposed W in the area where overlapping does not occur, Fig. 3.33) are measured on horizontal sections (normal to the length of echelons). In outcrops which provide only vertical sections (e.g. road quarries), measurements are limited to crack length l and w. Bahat (1986a) observed the relationship $l \cong 2.5$ w in the Middle Eocene chalks near Beer Sheva (Fig. 4.21).

Overlap of Straight Non-Coplanar Segments. The overlap of W increases percentagewise with the twist angle ω (Fig. 3.33). Therefore, variations in ω have to be considered in comparing overlaps of different joint sets. Caution has to be taken in comparing overlaps of en echelon cracks of various size on the same joint that may indicate different fracture conditions. There is particularly a difference between large echelons and small ones associated with the

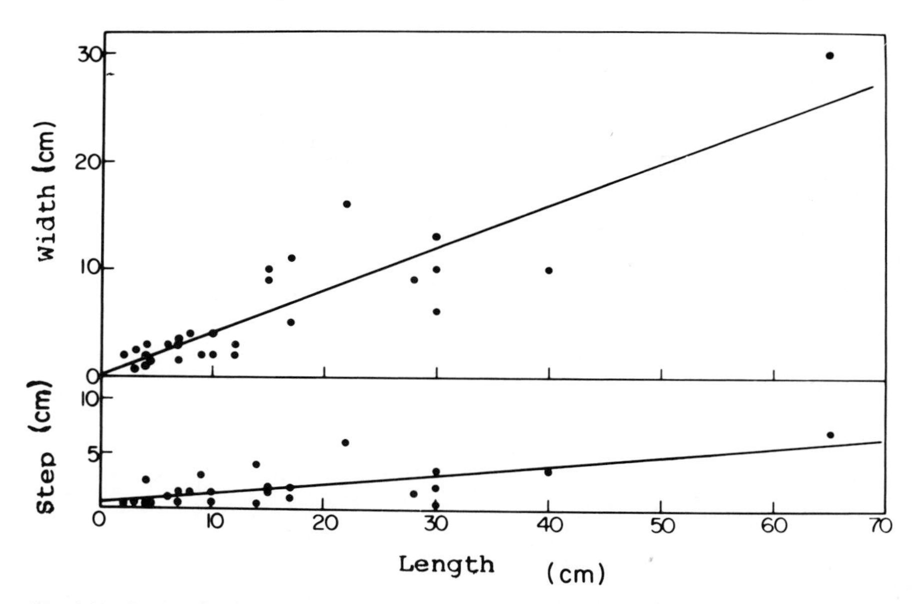

Fig. 4.21. Geometric dimensions of en echelon cracks. Length versus step (*below*) and length versus width w (*above*). (After Bahat 1986a)

root zone (Sect. 3.1.6.1). In set 357° from the Middle Eocene chalks near Beer Sheva (Bahat 1986a, outcrop X), $\omega = 20°$ and W overlaps for the large segments generally vary from 30% to 70%.

Overlap of Curved Non-Coplanar Segments. The theory and implications of this type of crack interaction are referred to in Sections 2.1.2.2 and 5.9 (Fig. 5.30).

Measurement of Area Produced by en Echelon Segmentation. The transition from a single joint surface to a series of overlapping en echelon cracks causes an increase of joint surface area. The change is gradual along continuous steps (Fig. 4.1 a) and abrupt in discontinuous fringes (Fig. 3.35). Noteworthy also is the contribution of the root zone (Fig. 3.35) to the area enlargement. Reasonable field estimates of the fracture area may be made only on outcrops where adjacent horizontal and vertical sections of the same layers are well exposed for the measurements of both W and l.

Joints 3(N) and 5(N) from Bahat (1986a, outcrop X, Table 3) approximately meet the requirements for joint area estimates. The mean lengths of small (6 cm) and large (14 cm) en echelons from the breakdowns above the 3(N) joint are multiplied by the respective width sums of the echelon segments (30 cm and 126 cm) in Fig. 4.22. The sums of en echelon areas obtain for 3(N) is 1944 cm^2 along a 50-cm joint length. When the latter length is multiplied by the length of the largest echelon an unsegmented area of 750 cm^2 is obtained.

Fig. 4.22. Fringes of fractures 3 N at *left* and 5 N *right*, from Middle Eocene chalks near Beer Sheva in horizontal section with small en echelon cracks along the root zone, and larger ones further away from the shoulder. The *portion of scale* in picture is 8 cm long. (Bahat 1986a)

The ratio of sums of en echelon areas to the unsegmented area is 2.59. It reflects only the added area due to breakdown above the joint, namely, only about half the increase in area (or less if the echelons below the joint are larger than those above it) due to en echelon cracking. The equivalent ratio for 5 (N) for half area increase is considerably lower: 1.4. The range between these two ratios gives an idea regarding the increase of area involved in the en echelon segmentation of 3 (N) and 5 (N), which represent a typical en echelon overlapping. Greater ratios for other joints are not rare. When echelon overlapping does not occur (e.g. Pollard and Aydin 1988, Fig. 22c), segmentation may involve an area reduction.

4.2.1.7 Relations Between Stress Intensity, Crack Velocity and Morphology of the Joint Surface

Previous studies on fracture propagation in glass (Kerkhoff 1975; Michalske 1984) and in polymers (Ravi-Chandar and Knauss 1984a) correlate the crack velocity V and the stress intensity factor K_I (Fig. 1.26) with fracture surface morphology in these materials (Sects. 2.1.2.2 and 2.1.2.5). Similarly, various fracture morphologies in rocks may be related to specific regions on the V versus K_I diagram. Here, the morphologic and mechanical characteristics of striae (Chap. 2) and ridges, barbs and plumes (Chap. 3) are assumed to be analogous (see also Sect. 2.2 on terminology).

Kerkhoff (1975, and Sect. 2.1.2.3) observes that an arrest line is induced in plate glass at propagation rates below $V = 4 \times 10^{-5}$ m/s and $K_I < 0.73$ MPa/m$^{1/2}$, which correspond to the range of regions I and II. Concentric rib markings on joint surfaces are occasionally interpreted as indicators of arrest lines, or loci of fracture hesitation, which would suggest low values of V and K_I. The fairly frequent occurrence of rib markings superposed by radial plumes on joint surfaces (Fig. 3.19 and 4.2a) suggest that these plumes are formed under fracture conditions similar to those that produce rib markings, that is, in the range of regions I and II.

The rhythmic plumes from the Appalachian Plateau (Fig. 3.2c) probably reflect a fracture process under pore-pressure conditions (Bahat and Engelder 1984; Engelder 1985) with a periodic increase of barb (striae) intensities in region II. Quite possibly the periodic smooth fracture that follows the rhythmic increase in barb intensity (fan perimeter in Fig. 3.2c) represents conditions at the lower limit of region I.

The striae shown in Fig. 2.28c developed on a smooth fracture surface of soda lime silica glass in water when the velocity of fracture propagation reached about 10^{-2} m/s at around $K_I = 0.7 - 0.8$ MPa/m$^{1/2}$ (Michalske 1984, Fig. 1), which was above the mid range between the stress corrosion limit, $K_I \cong 0.3$ MPa/m$^{1/2}$, and the fracture toughness $K_{Ic} = 0.9 \pm 0.1$, that is, in region III (Figs. 1.26 and 2.28a).

The lack of arrest lines along the very long straight S-plume from the Appalachian Plateau (Fig. 3.18) indicates that this plume possibly propagated faster than the wavy or rhythmic C-plumes (Fig. 3.2) (Engelder et al. 1987). Whereas propagation of the C-plumes periodically vacillates between the stress corrosion limit and region II (Fig. 1.26), the S-plume seems to resemble the behaviour of striae (Fig. 2.28c) and propagates mostly monotonously in region III (Sect. 5.2.3.1).

Ravi-Chandar and Knauss (1984a), on the other hand, simulate striae within the mirror plane in a (Homalite-100) polyester glass and assign to the mirror plane K_I values in the approximate 0.5 MPa/m$^{1/2}$ – 1.0 MPa/m$^{1/2}$ range. The K_{Ic} and K_b values for this material are 0.5 MPa/m$^{1/2}$ and 2.04 MPa/m$^{1/2}$, respectively (Kobayashi and Ramulu 1985), where K_b is the K_I at onset of branching. Hence, according to Ravi-Chandar and Knauss, striae may be formed in a polyester glass above K_{Ic} compared with conditions below K_{Ic} which are required for the development of striae in soda lime glass, as cited above from Michalske (1984). This descrepancy is surprising and may be interpreted either by differences in the properties of the two materials (soda lime is basically an elastic material, whereas the polyester glass yields more to ductile deformation), or by the experimental conditions (dry Homalite-100 versus soda lime glass in water). For the time, it is considered that jointing follows conditions more similar to those described by Michalske (1984) and Kerkhof (1975) than those implied by the experiment of Ravi-Chandar and Knauss (1984a).

The range of K_I conditions under which en echelon cracks develop is not known. En echelon cracks have been observed in borosilicate glass after ap-

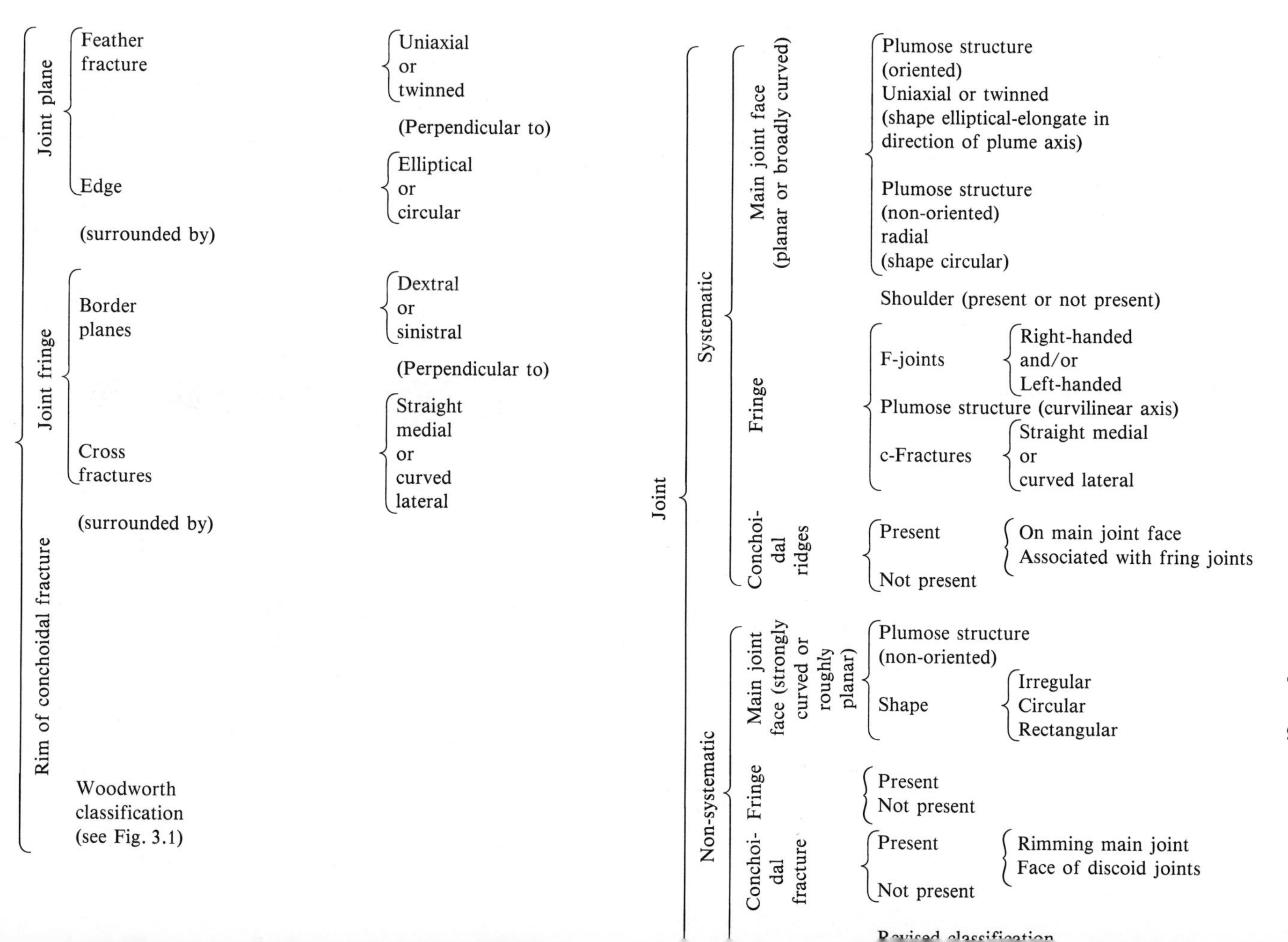
Joint
Joint plane
Feather fracture
Uniaxial or twinned
(Perpendicular to)
Edge
Elliptical or circular
(surrounded by)
Joint fringe
Border planes
Dextral or sinistral
(Perpendicular to)
Cross fractures
Straight medial or curved lateral
(surrounded by)
Rim of conchoidal fracture
Woodworth classification (see Fig. 3.1)
Joint
Systematic
Main joint face (planar or broadly curved)
Plumose structure (oriented)
Uniaxial or twinned (shape elliptical-elongate in direction of plume axis)
Plumose structure (non-oriented) radial (shape circular)
Shoulder (present or not present)
Fringe
F-joints
Right-handed and/or Left-handed
Plumose structure (curvilinear axis)
c-Fractures
Straight medial or curved lateral
Conchoidal ridges
Present
On main joint face
Associated with fring joints
Not present
Non-systematic
Main joint face (strongly curved or roughly planar)
Plumose structure (non-oriented)
Shape
Irregular
Circular
Rectangular
Fringe
Present
Not present
Conchoidal fracture
Present
Rimming main joint
Face of discoid joints
Not present

plication of stress with an intensity factor less than the fatigue limit (Wiederhorn and Johnson 1973, Fig. 5). Figure 3.39 (Sect. 3.1.6.5) implies growth of echelons at low K_I values. The K_I range in which en echelon cracks occur probably varies considerably, as it seems to depend on the K_I/K_{III} ratio (Sommer 1969).

Hertzberg (1976, Fig. 14.9) analyses the fracture conditions of a turbine rotor made of steel. This analysis is cited in Sect. 2.3.2.1 (Fig. 2.49). The clear implication from this analysis is that the en echelon cracks developed at close to K_{Ic} conditions.

Cuspate hackles (Fig. 3.41 a and 4.6) and most probably hackles (Figs. 3.6 a and 4.20 a). indicate fracture conditions of $K_I > K_{Ic}$, perhaps even $K_I \cong K_b$. Cuspate hackles replace hackles in rock fractures, mostly due to graininess and porosity (Fig. 2.32).

4.3 Classification of Fracture Surface Markings

Fracture markings may appear in a multitude of shapes and combinations. Some 14 qualitative parameters of FMS have been presented above, and more are conceivable. Naturally, the student of FSM stands in need of a practical system of morphological classification, which may be based on alternative or complementary sets of criteria. Some classifications are briefly presented below.

1. Woodworth (1896) divided the joint surface into three main parts: (a) the joint plane, (b) the joint fringe, and (c) the rim of conchoidal fracture.
2. Hodgson (1961 a, b) classified the joints into two main groups: (a) systematic and (b) non-systematic. Systematic joints occur in sets within which the joints belonging to each set are parallel or subparallel in plan and which display FSM. In contrast, non-systematic joints display random, curvilinear traces in plan and section and lack oriented FSM. Each group was again subdivided into three, according to a modified Woodworth division (Table 4.2). Hodgson summarized his morphological concepts in two schematic block diagrams (Figs. 3.5 and 3.19), which present plumes and undulations as two major categories. This division has basically been adapted by various authors (e.g. Bankwitz 1966; Hancock 1985).
3. Bankwitz and Bankwitz (1984) divided the joint surface into segments interrelated by the symmetry of the FSM. They viewed this segmentation as a projection of the palaeostress ellipsoid from which they derived the directions of the palaeo-principal stresses (for details of their concept, with some critical remarks, see Sect. 3.1.7.4).
4. Kulander et al. (1979) and Bahat (1979 a) preferred to emphasize the systematic relationship between particular morphologic features on the joint surface and their analogies on fracture surfaces of glass, ceramics and other technical materials.

◄

Table 4.2. Classification by Woodworth (1896) and revised classification by Hodgson (1961 a) of joint surface markings

The first three classifications enable to identify and describe FMS and to characterize fracture parameters to a fairly advanced degree (Chap. 3). The fourth classification assumes the applicability of fracture mechanics to the analysis of FSM. However, none of the above classifications (perhaps with the exception of Bankwitz and Bankwitz) has really helped to promote FSM as an extensive tool in tectonic analyses of joints.

5. A different approach is the one presented in the present chapter: joints in sedimentary rocks are divided into three major categories which are associated with concepts of the Huttonian geological cycle, the burial, syntectonic and the uplifting stages. A fourth, the post-uplift stage, is added for a corresponding jointing category. This division is based both on distinct joint morphologies typical to each stage, and on characteristic fracture markings with affinities to specific jointing stages. Conversely, tectonic interpretation can then be based on combined classifications of fracture morphology and fracture surface morphology. The approach put forward in the present study offers some useful genetic and structural implications (see Chap. 5).

Chapter 5 Tectonofractography

5.1 Application of Joint Surface Morphology in Tectonics

A major obstacle in joint analysis has been a lack of criteria for assignation to tectonic categories. Although concepts have been defined regarding burial and uplift jointing (Price 1966, 1974; Engelder 1985), these were not linked in a straightforward manner to field data.

In the present chapter three classes of joints are recognized, in concordance with the three major tectonic phases of the Huttonian cycle: burial, syntectony and uplift (Engelder 1987). A fourth class of lesser importance is related to the post-uplift phase. This classification is based on field patterns of joint morphology and joint surface-morphology, and is mainly applicable to stratified rocks.

Both the classification of joint marking styles (Chap. 4) and their linking with particular tectonic processes (present chapter) involve a number of ambiguous and sometimes unknown parameters. Several assumptions need to be made, and certain problematic angles have to be considered. They are addressed in the next sections. All new geological measurements in this book were made by the author.

5.1.1 Assumptions

1. The ubiquity of crack markings in technical materials (Chap. 2) and in rocks (Chap. 3) justifies the assumption that fracture surface markings (or morphology) FSM reflects various failure mechanisms.
2. Investigations of the growth of fatigue cracks in glass (Sect. 1.4.1) and in rocks (Scholz 1972; Atkinson 1984) show the great reduction in material strength at low rates of crack propagation. Most jointing in natural environments is assumed to reflect fatigue conditions (Kendall and Briggs 1933; Hodgson 1961 b).
3. Rhythmic FSM, including alternating concentric undulations (Fig. 3.22) as well as periodic plumes in rocks (Figs. 3.2c and 3.2e), support the model of jointing by pore pressure applied periodically to the rock (Secor 1969; Bahat and Engelder 1984). Pore pressure is assumed to be a dominant mechanism of jointing.
4. Experimental fracturing (Chap. 2) demonstrates that fracture markings also develop in dry environments under simple extensional conditions (when no pore pressure is involved). By analogy, it is assumed that FSM may form on joints in dry rocks also in nature.

5.1.2 Problems Involved in the Transition from Material Science to Tectonophysics

When adapting the fractography of technical materials (glass, ceramics, metallic substances, etc.) to the fractographic effects of tectonics, some basic differences must be taken into account:
1. Material scientists deal with the morphology of fractures that start at the surface and penetrate into the examined body. Joint surface markings on rock exposures, on the other hand, belong mostly to fractures that initiate at depth, and propagate to all directions (see exception in Sect. 5.4.5.2).
2. Material scientists generally prefer tensional conditions for experiments (with some noted exceptions, e.g. Figs. 1.11c and 1.11d). Therefore they consider tension positive. On the other hand, specialists in rock mechanics must consider wider stress ranges, from compressional conditions at depth to tensional conditions at the surface of the crust. They prefer to regard compression as positive. Hence, for material scientists σ_1 is the greatest tensile principal stress, σ_2 is the intermediate, and σ_3 is the least tensile principal stress. Earth scientists often take σ_1 as the greatest compressional principal stress and σ_3 to be the least compressional principal stress.
3. Whereas material scientists generally use for the three principal stresses $\sigma_1 > \sigma_2 > \sigma_3$, earth scientists rather consider the effective principal stresses $\bar{\sigma}_1 > \bar{\sigma}_2 > \bar{\sigma}_3$ which represent the differences between the respective principal stresses and the pore liquid pressure $\bar{\sigma}_j = \sigma_j - P$.
4. Material scientists monitor the stresses prior to, during and after failure. Earth scientists measure the strain and calculate the causative stresses.
5. Size scales are different. Material scientists usually deal with features of experimental samples, which range from several mm to several tens of cm. Fracture markings in rock exposures vary from cm scale to many tens of metres.
6. Material scientists perfer samples of uniform chemical, physical and textural properties in which residual stresses are largely eliminated or controlled. Earth scientists have to work with rocks that are inhomogeneous in these respects.
7. Material properties in experimental conditions are usually well controlled, unlike under natural conditions.

5.2 Burial Jointing

The burial phase is the historical stage that includes sedimentation, downwarping and diagenesis, all preceding the phases of syntectonic deformation and uplift. The surface morphology of burial joints may preserve affinities to the burial phase, which are distinguishable from those of the subsequent phases. The morphologic character of burial joints were studied in rocks from two provinces: Eocene chalks near Beer Sheva, Israel, and Devonian siltstones of the Appalachian Plateau, eastern U.S.A. On outcrops of the Appalachian Plateau, burial and syntectonic joints are closely associated in many outcrops, and they are described together in Section 5.2.3.

5.2.1 Early Burial Joints in Lower Eocene Chalks Near Beer Sheva

5.2.1.1 Geology and Sequence of Events

Events during the Early Eocene (represented by the Mor Formation) include consolidation of the chalk, single-layer jointing (i.e. joints that are confined to single beds, and which do not cross layer boundaries) both normal and parallel to the subsequent fold axes (cross-fold and strike sets, respectively), and the consolidation of nodular chert. Many of the single-layer joints in the Mor Formation are crossed by penecontemporaneous faults (Fig. 5.1 a), but their surfaces never curve toward or into the fault planes. Also, the pattern of joint spacing in the Lower Eocene chalks is nowhere affected by adjacent faults.

Thus single-layer cross-fold jointing evidently took place before faulting in the Lower Eocene, during the early burial stage. There is, however, some evidence suggesting that although the cross-fold joints existed at the time of faulting, some of the strike joints might have formed after development of these faults (Bahat 1988 a). Jointing in the Mor Formation seems to have been a response to low horizontal stresses during the burial phase, as is indicated by slight bending of the chalk layers (Bahat 1985) and by the regional pattern of these cross-fold joint sets (Bahat and Grossmann 1988). The magnitudes of these stresses, however, were much smaller than those of the syntectonic phase.

5.2.1.2 Fracture Surface Markings of Joints of Different Sets

Three types of surface markings occur on the single-layer, cross-fold joint surfaces in the Lower Eocene chalks (Bahat 1987 a): (1) Coarse horizontal axial plumes, with barb widths ranging from 0.5 cm to several centimetres (Fig. 3.10) and quite large median m and interface i angles (Fig. 4.17); (2) circular rib markings (Fig. 5.1 b); and (3) transitional variations between the two (Fig. 3.6 b). The three types may occur together in the same outcrop.

About half (56%) of the 225 FSM examined on joints in four outcrops around Beer Sheva bear circular markings, and these occur on joints of a set of average strike 328° (N 32° W). Sixty five samples (29%) are axial plumes, and these occur on joints belonging to sets of average 309 and 344° strike. Transitional markings make up 12%, and about 3% are other combinations. It seems that circular markings develop when the minimum principal stress σ_x is horizontal and normal to the joint surface, and horizontal σ_y equals vertical σ_z. The axial plumes represent the same conditions with respect to σ_x but with horizontal principal stress different from the vertical, i.e. $\sigma_y \neq \sigma_z$. Furthermore, the circular markings on the 328° joints appear to have been formed along the direction of the horizontal principal stress, and the axial plumes on sets 309 and 344° were formed on joints at low angles to that principal stress, probably in a conjugate pattern (Stearns 1968). Surface markings of the three joint sets usually have no fringe zones (see Sect. 3.1.6).

a
b
c
0
30
cm

Joint sets with circular markings and joint sets with plumes occur in separate rock layers, even in the same outcrop (Fig. 3.28). The reason for this clustering is not clear, and may be related to the lithomechanical properties of the rock (Sect. 5.2.2.3).

The different FSM of the various joint sets is strikingly displayed at station 20 (Figs. 5.1 b, c), where two sets almost touch each other. The question arises whether they developed sequentially or simultaneously. They could develop sequentially only if the circular joint on set 328° developed first, since it propagated radially and could maintain a limited approximate square surface (aspect ratio 1), leaving space for the plumed joint of set 344° to develop later. On the other hand, if the plumed joint of set 344° developed first, with its tendency to develop a large length-to-height ratio (aspect ratio >3 see Fig. 3.10), it would transect a large area and probably would prevent the development of the neighbouring 328° joint. Simultaneous formation, though not impossible, is improbable, even though adjacent coaxial and inclined fractures have been experimentally obtained by a single loading (e.g. Hobbs et al. 1976, Fig. 7.31).

5.2.1.3 Determination of Fracture Stresses

Bahat (1989) calculated the approximate pore pressures and fracture conditions of cross-fold, single-layer burial joints at shallow depths in the Lower Eocene chalks of Beer Sheva, Israel.

For calculation, the gradient of overburden pressure σ_z of 25 MPa/km is adopted from Voight and St. Pierre (1974) and from Mimran (1977), the tensile strength S of the water-soaked chalk is taken as 0.5 MPa (Koifman and Flexer 1975), and the Poisson ratio is taken as $\upsilon = 0.29$ (Hayati 1975). The relationship between S, σ_x, the horizontal minimum effective principal stress $\bar{\sigma}_x$ and the pore pressure P is (Jaeger and Cook 1979, p. 225):

$$\bar{\sigma}_x = \sigma_x - P = -S \ . \qquad (5.1)$$

Also

$$\bar{\sigma}_z = \sigma_z(1-\lambda) \ , \qquad (5.2)$$

where λ is the ratio of P to σ_z, and (Johnson 1970, p. 213):

$$\sigma_x = \sigma_z(\upsilon/1-\upsilon) \ . \qquad (5.3)$$

Fig. 5.1. a A normal fault crossing vertical single-layer joints in Lower Eocene chalks near Beer Sheva (Fig. 5.29, Station 20). Dark chert nodules occur as horizontal beds and along the two dipping fault planes. The *horizontal rod* measure at centre is 90 cm long. **b** Two fracture markings on adjacent single-layer joint surfaces in the same rock layer from the outcrop shown in **a**. The undulatory marking on the *right* is oriented 328° and the plume marking on the *left* is oriented 342°. **b** Photograph. **c** Drawing

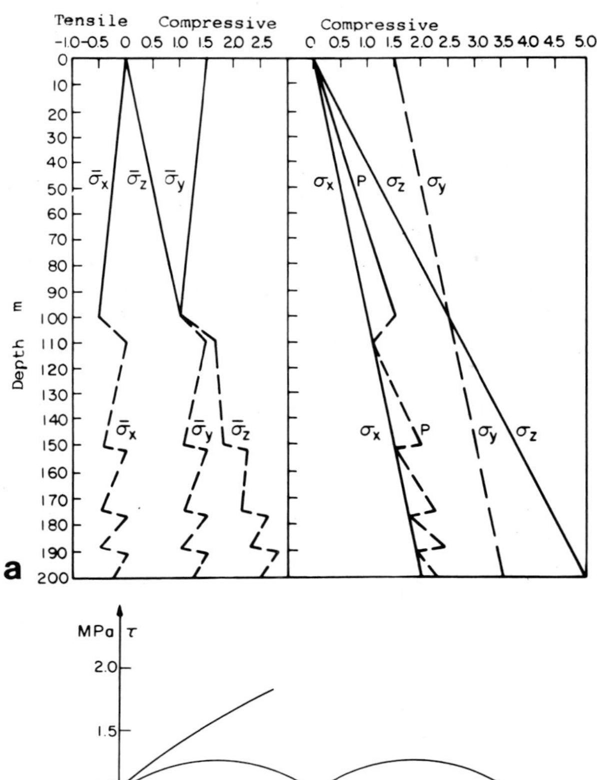

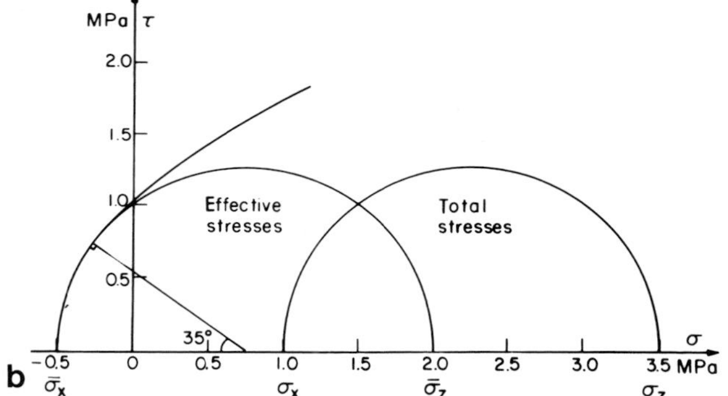

Fig. 5.2. a. Change of stress with depth. σ_x, σ_y and σ_z are the two horizontal and the vertical principal stresses, respectively, *P* is the pore pressure, and $\bar{\sigma}_x$, $\bar{\sigma}_y$ and $\bar{\sigma}_z$ are the respective effective stresses. Reductions of effective stresses occur periodically at pre-jointing pressures, when pore pressures increase to certain peaks. The reverse occurs on jointing. (After Price 1974, Fig. 7). **b** Mohr diagram showing the relationships between the effective stresses $\bar{\sigma}_x$ and $\bar{\sigma}_z$ and between the total stresses σ_x and σ_z, at joint-producing intensities

Although there are strong theoretical arguments against Eq. (5.3) in favor of a lithostatic stress state ($\sigma_x = \sigma_y = \sigma_z$) during burial (McGarr 1988), the ubiquity of burial joints seems to require extensional conditions. The latter are supported by the orthogonality and spacing relationships of the cross-fold and strike sets (Sects. 5.6.1.1, 5.4.6 and 5.6.1.7).

Fracturing in the Lower Eocene chalks occurred in two stages. Initially, extensional joints oriented 328° with circular rib markings developed at about 100 m depth, where pore pressure P = 1.5 MPa, the horizontal maximum effective stress $\bar{\sigma}_y$ and the vertical effective stress $\bar{\sigma}_z$ both equal 1.0 MPa, and $\bar{\sigma}_x = -0.5$ MPa (Fig. 5.2a). By comparison, Fraissinet et al. (1988) estimate that jointing in chalks in the Paris Basin occurred at depths of about 60 m, with $\bar{\sigma}_x$ ranging between -0.5 and about -0.9 MPa.

The axial plumes which characterize the joints of sets 309 and 344° suggest that these joints developed at depths greater than 100 m (Fig. 5.2 a), in a direction normal to σ_x where $\sigma_z > \sigma_y$ (Gramberg 1965). These conditions appear to initiate the onset of jointing in the examined chalks, although $\sigma_y > \sigma_z$ conditions may also be considered (see Sect. 3.1.8.2).

For further calculations, the following assumptions are made:

1. The Mohr diagram (Fig. 5.2b) shows that if the angular difference of 35° (344 − 309°) represents a dihedral angle of conjugate hybrid sets (Hancock and Al-Kadhi 1978), an effective stress difference ($\bar{\sigma}_z - \bar{\sigma}_x$) of 5 S during fracturing is implied. The application of Griffith's parabolic envelope (left side of Fig. 5.2b) for the 35° dihedral angle is justified, as representing the average homogeneous field of major tension and minor shear (Bahat 1987 a). On a smaller scale, however, jointing is a result of stress gradients (Sect. 3.1.8.2).
2. If at $h = 0$, $S = 0$ (initially the rock has no strength), Eq. (5.1) would require $\bar{\sigma}_x = 0$ at that depth.
3. Lines σ_x and σ_y, according to Eq. (5.3), have the same slope (Fig. 5.2 a). Lines σ_x, σ_y and σ_z are approximated to have constant slopes (Means 1979, Fig. 12.5).

Accordingly, the hybrid (extensional-shear) plume-marked joints of sets 309 and 344° were formed at a depth of about 150 m when $P \cong 2.0$ MPa, $\bar{\sigma}_x \cong -0.5$ MPa, $\bar{\sigma}_y \cong 1.0$ MPa and $\bar{\sigma}_z = 1.8$ MPa. Fracturing then continued as the sedimentary sequence continued to accumulate, possibly down to 200 m or 400 m. The formation of the fracture sequence of set 328° before sets 309 and 344° is supported by the observations presented in Section 5.2.1.2 and Fig. 5.1 b.

5.2.2 Jointing in Middle Eocene Chalks South of Beer Sheva

5.2.2.1 Late Burial Jointing

Two prominent, single-layer joint sets, 031 and 357°, occur in the Middle Eocene chalks (the Horsha Formation) in Wadi Na'im, 9 km south of Beer Sheva (in outcrops X and Y from Bahat 1986a, which correspond to stations 37 and 40 in Fig. 5.29). These joints typically have straight, delicate plumes, with barb widths smaller than 0.5 cm, and with fringe zones along one or both layer boundaries (Fig. 3.11). The m and i angles (Fig. 4.17) are generally smaller than those of the coarse horizontal axial plumes. No circular fracture markings occur in these sets.

Fracture interaction in two successive chalk layers of joints oriented 028° (Bahat 1987 b) is quite intensive (Fig. 5.30). Fractures of set 028° also interact with previous joints from set 344° (which is unrelated to set 344° in Sect. 5.2.1) and this serves as evidence that the 028° fractures did not develop at an early stage but at some time in the late history of the rock.

The two joint sets 357 and 031°occur in two separate outcrops, 400 m apart. Both display fringes with en echelon cracks, oriented 337 and 053°, respectively. A gradual rotation of the azimuths of the joints and en echelon cracks is observed between the two outcrops via five chalk outcrops along the 400 m strip (Fig. 5.3 a). This rotation implies NNE-SSW compression and en echelon cracking due to unloading of this compression (Fig. 3.40). It also suggests local block tectonism (Bahat 1986 a).

Joints in Middle Eocene chalks were also examined some 32 km south of Beer Sheva. Two joint sets with mean 340 and 328° azimuths occur in this outcrop. Two en echelon sets with mean 324 and 005° azimuths form the fringes of set 340° and two en echelon sets with mean 324 and 338° azimuths are associated with set 328°. Both the joints and the fracture markings are very similar to those of outcrops X and Y near Beer Sheva, suggesting that they were formed by similar effective stresses. The differences in azimuth, however, indicate that the stresses at these two locations (23 km apart) acted in different directions, supporting other evidence for block tectonics.

In the Beer Sheva area the fracture markings on the Lower Eocene joints (Mor Formation) are quite different from those of the Middle Eocene (Horsha Formation) (Figs. 3.10 and 3.11), even though the chalks are lithologically similar and outcrops are only several km apart. They appear to have been produced by different tectonics events: the Lower Eocene burial joints are the product of regional compression from NNW (Bahat and Grossmann 1988) and the Middle Eocene joints (outcrops X and Y) are the result of local compression from NNE (Bahat 1986 a). Although jointing probably took place at shallow depths (several tens to several hundred meters), the joints in the younger chalks (Horsha Formation) were most likely formed at shallower depths than those in the older chalks (Mor Formation) (Bahat 1989).

5.2.2.2 Plane Stress Fracture of en Echelon Cracks at Shallow Depths

Joints in layered Middle Eocene chalks around Beer Sheva almost invariably have straight fringes of en echelon cracks along their boundaries (Figs. 3.11, 3.17 a). The transition from a flat to a slanted fracture, also termed shear lip, is well known to metallurgists (Sect. 1.2.4). In Fig. 1.17 c, a flat fatigue crack at the left is shown at an early stage propagating at right angles to the maximum tensile stress under plane strain conditions. Later, depending on the stress level, crack length and ratio of plastic zone size to plate thickness, the crack slants and completes its propagation at 45°, reflecting tensile plane stress conditions. Slight deviations from pure tension would impose mode III shear on the tested material. The mixed modes I and III will result in en echelon segmentation (Sect. 1.1.7.2) of the slanted crack.

When drawing an analogy between the laboratory and natural jointing in rocks, two allowances have to be made. First, the actual slant would be smaller than 45°, and would be determined by the relationships $\omega = 45° - \beta/2$ when ω is the twist angle between the initial flat crack and the en echelon crack, and

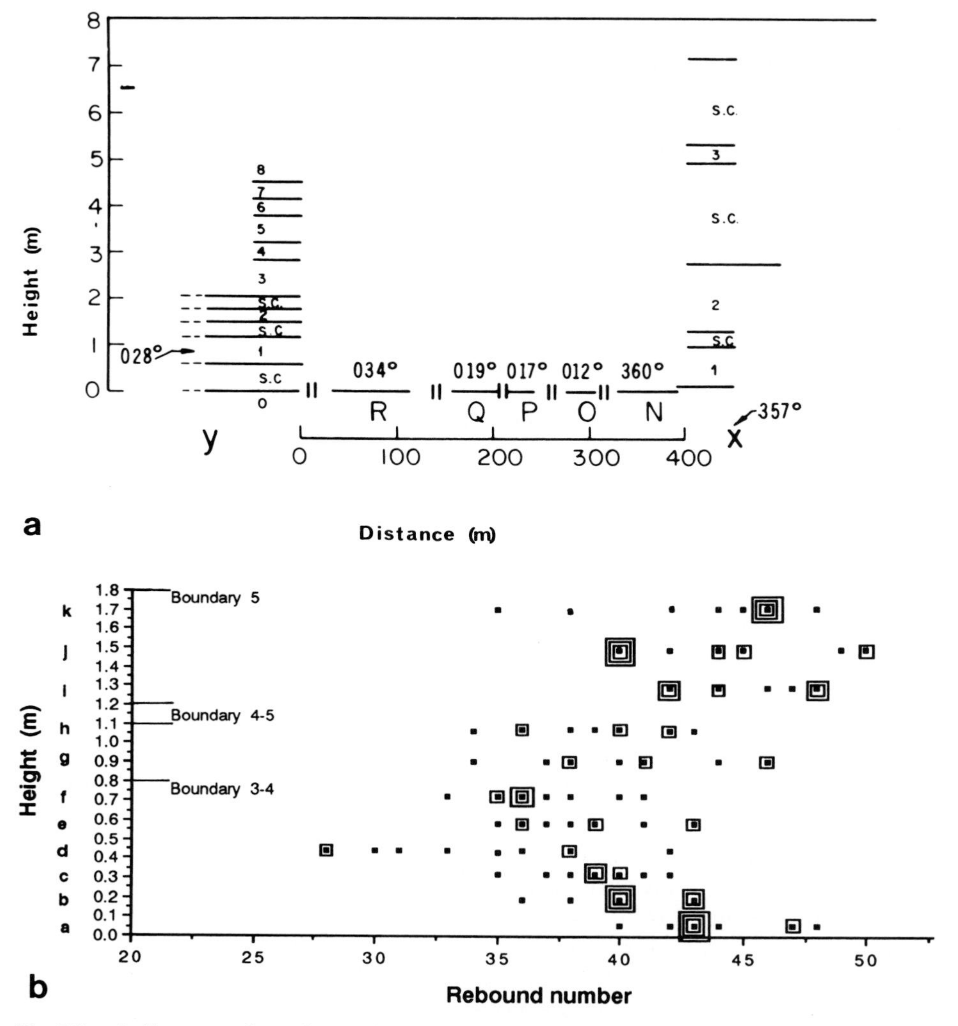

Fig. 5.3. a A diagrammatic section of the outcrop from Wadi Na'im (Fig. 5.14). Outcrops *X* and *Y* (corresponding to stations 37 and 40 in Fig. 5.29) including floors *N* to *R* between them showing rotation in mean azimuths of joints from 357 to 028°. Numbers *1–3* in *X* and *0–8* in *Y* designate hard chalk layers in vertical sections. Beds of soft chalk *s. c.* alternate with some of the hard chalks. In *Y*, beds *0–2* and alternating *s. c.* occur also as platforms in horizontal sections. The *two close small vertical lines* between floors *N* and *R* represent talus. **b** Rebound numbers obtained by Schmidt Hammer tests across three consecutive chalk layers from the Middle Eocene near Beer Sheva (station 40, Fig. 5.29). *Letters* designate different heights. *Each square* represents a single measurement. Results are given from one location (out of four locations examined)

β is the angle of internal friction. In the Middle Eocene chalks around Beer Sheva $\omega = 17°$ (Bahat 1986a).

Secondly, no fringes developed along boundaries of joints cutting the Lower Eocene layers, which were formed at greater depths than joints in the

Middle Eocene chalks (Sect. 5.2.2.1). Neither did fringes develop along the burial joints that were formed at considerable depths in siltstones of the Appalachian Plateau (Sect. 5.2.3.2). It appears that en echelon cracking is inhibited beyond certain magnitudes of confining pressure. Similarly, splitting in extension which occurs parallel to the direction of loading due to uniaxial compression is inhibited under moderate confining pressures (Holzhausen and Johnson 1979). Tschegg (1983) considers retardation of mode III cracking and calls attention to the Mode III crack closure effect which results from interference between mating crack surfaces. This effect is dependent on crack length, crack depth and specimen diameter. Tschegg also observes that in steel and aluminium, fatigue crack propagation by Mode III is 10 to 50 times slower than Mode I crack growth in the same nominal stress intensity range. Hence, there appear to be various mechanisms, not all fully known, that inhibit en echelon cracking, and it is conceivable that moderate overburden pressure is one of them.

5.2.2.3 Fracture Markings and Lithomechanical Properties

Figure 5.3a shows layers 3, 4, and 5 in a Middle Eocene chalk cliff (outcrop Y). Although of different thickness, they appear to be lithologically similar. Yet layer 5 is profusely marked with plumes on its joint surfaces whereas they are rare in layers 3 and 4. Two reasons are possible:

1. Jointing was more or less contemporaneous in the three layers, but due to inherent lithological differences each layer responded differently to the joint-producing stresses.
2. Jointing occurred at different times, under different stress conditions.

Alternative 1 was partly checked by a series of elasticity tests, using the Schmidt Hammer Concrete Test (Arkin and Bahat 1989 unpubl.). Altogether, 535 hammer rebounds were measured across the three layers (180 cm) at four locations. Results (Fig. 5.3b) indicate somewhat lower rebound numbers near layer boundaries, and maxima occur in the upper third of layers 4 and 5. A minimum occurs within layer 3, suggesting a “hidden” layer boundary. The highest rebound number occurs along the plume-marked level of layer 5. This level coincides with uniformity and the zone of maximum carbonate density. Rock alterations are minimal along this zone, which is less weathered than layer boundaries and is less impregnated by clay. Hence, it appears that plume-marked joints are preferentially formed along more indurated chalk zones.

5.2.3 Burial Joints and Syntectonic Joints in the Appalachian Plateau, U.S.A.

5.2.3.1 Two Jointing Phases

Early burial joints and late syntectonic joints are closely associated in many outcrops of the Devonian Catskill Delta of the Appalachian Plateau, particularly in the Fingers Lakes District, New York.

Engelder (1985) classifies the joints that develop during burial into tectonic and hydraulic categories. These joints are formed at depth, prior to uplift, in response to abnormal fluid pressures. Tectonic compaction may cause abnormal pore pressures which result in tectonic joints at depths between 0.5 to 2.5 km. Compaction by overburden may cause abnormal pressures which result in hydraulic joints at depths greater than 5 km. During the Alleghanian Orogeny, at least 50 m.a. after the beginning of the deposition of the Catskill Delta (Engelder and Oertel 1985), one set or more of cross-fold joints (termed Ib) was formed in the deeper portion of the Delta. Another set (termed Ia) was formed during the main phase of the orogeny (Permian or later, Engelder and Geiser 1984). Ib joints, striking 335°, cut siltstones and stop at siltstone-shale interfaces, whereas Ia joints, striking 345°, cut the shales. The two sets were formed within the depth range of tectonic joints.

By analog with industrial hydraulic fracturing (Engelder 1985), the least principal stress in sandstone during burial may be smaller than the least principal stress in shales. Thus it may be assumed that the abnormal pressures that produced the Ib joints in the siltstone – whose lithological properties somewhat resemble those of sandstone – were smaller than those that caused jointing in the shales. The Ia joints within the shales appear to have been formed as massive ruptures. This is indicated by the S-type plumes (Fig. 3.2a) of the 345° set in the thin siltstone beds that alternate with the thick shales (Fig. 4.9). In one outcrop, S-type plumes show no arrest lines in more than 40 m outcrop length (Fig. 3.18). The joints in the shales must have propagated through larger hydraulic fluid reservoirs compared with the restricted reservoirs through which the short, single fractures of the siltstone beds propagated (Fig. 3.7). Engelder et al. (1987) raise the possibility that the long S-type plume (Fig. 3.18) is typical of swifter propagation as compared to the C-type plumes, which are short and which occur at frequent intervals, thus suggesting a slower process of jointing (see Sect. 4.2.1.7).

A large band of subvertical arrest lines (convex towards the right in Fig. 4.9) in the shales occurs just below a node in the siltsone bed. The shape of the arrest lines indicates that this long joint propagated horizontally from left to right. The node in the intercalated siltstone (Fig. 3.18) and the arrest lines suggest that the fracture which forms the large joint in the shales simultaneously induced fracture in the siltstone. The fractures arrested in the two rocks as they interacted with fractures approaching from the opposite direction.

The two sets show a consistent difference in their patterns of joint spacing. In joints of the Ib set, spacing is regular (without great differences between minimum and maximum) and rather wide. A typical spacing is about 1 m in a layer about 0.5 m thick (Fig. 5.4a). In the Ia joints, on the other hand, spacing is irregular and often very narrow, amounting to a few cm in a layer several metres thick (Fig. 5.4b).

Thus, two jointing stages, one formed during the burial phase and one during the later syntectonic phase (both "tectonic joints" according to Engelder) are distinguished in the New York outcrops. An early stage, that of Ib jointing in siltstone, was induced by a low, abnormal fluid pressures at a slow rate, and

a

b

c

a late stage, that of Ia jointing, occurred during the main phase of the orogeny, induced by higher pore pressures, which possibly propagated at higher velocities. The two fracturing stages occurred at considerable depths.

5.2.3.2 Multiple Sets of Cross-Fold Joints in the Appalachian Plateau

The distinctions between systematic and non-systematic fractures which show considerable morphological difference (Sect. 4.3), and between orthogonal or sub-orthogonal sets which often belong to intersecting sets of cross-fold and strike joints, are relatively simple. Much more difficult can be the discrimination between cross-fold sets which occur at small angles to each other. According to various tectonic models, cross-fold joints are parallel, sub-parallel, or at an acute angle to a palaeostress direction (Stearns 1968; Engelder and Geiser 1980). The possibility that a cross-fold joint set represents an actual palaeostress direction makes correct discrimination among these sets particularly important (Bahat 1987a).

Three main joint directions (010, 327 and 350°) were distinguished by Nickelson and Hough (1967, Plate 3) in the Canton area, Pennsylvania. In addition they identified a fourth direction, 303°, of apparently lesser importance.

In Upstate New York the resolution into cross-fold sets has been a more laborious process, and for a long time only two sets were identified. Sheldon (1912) established two main directions in the Ithaca region. She observed that the angle between these joints varied as a function of rock hardness. A difference of 10° represented best the angular relationship between the two main directions (Sheldon 1912, Fig. 5). Wedel (1932) also observed two major groups of cross-fold joints, about 15 to 20° apart. Parker (1942) found these joints to cluster around two average azimuths, about 19° apart, and later (Parker 1969), he corrected the angular difference between Sheldon's two directions to a range between 9 and 38°. The angular difference between the two main cross-fold sets reported by Engelder and Geiser (1980, Fig. 4) in most parts of New York State, and in the central area in particular, was around 30°.

It is fairly evident that the early investigators in New York did not consistently compare the same main directions. More recent data (Bahat 1982, unpubl.) show that Sheldon's two directions 001 and 351° correspond (after declination correction) to the Ic and Ia directions (Table 5.1, stations 1 and 2). There is no reference in Sheldon's work to a direction which corresponds to Ib, which clearly exists in the Ithaca area (mostly in siltstone). Most of

Fig. 5.4a–c. Cross-fold joints cutting Devonian siltstones and shales in the Appalachian Plateau. **a** Two cross-fold sets at Watkins Glen, New York. The joint in shales at the wall strikes at 348° (note the *arrowed* 21-cm-thick siltstone intercalation at the *top*), whereas the joints in siltstone cutting the platform strike 335°. **b** Cross-fold joints within the upper portion of the Genesse group in Tauhannock Falls State Park, New York. The early joint (set Ib) within the thin siltstone bed is counter-clockwise from the later joints (set Ia) cutting the underlying shales. The siltstone bed is 19 cm thick. (Engelder 1985). **c** Crescent-like fractures in shales are cut by the large 348° joint shown in **a** (scale = 50 cm)

Table 5.1. Orientation of joint sets (in degrees) in parts of southern New York and northern Pennsylvania

Station no.	Location	Sets					Angle produced by Ia and II	Stratigraphic group
		Ib	Ia	Ic	II	III		
New York								
1.	Ithaca, route 13 (low)	336 ± 3 (3)	353 ± 2 (6)	359 ± 1 (3)?	077 ± 3 (4)	063 ± 3 (3)	84	Genesse
2.	Ithaca, route 13 (high)		345 ± 2 (9)			066 ± 4 (5)		Genesse
			350 ± 2 (6)					
3.	Watkins Glen, routes 414 × 79	334 ± 3 (9)	341 ± 1 (3)	004 ± 2 (5)				Genesse
4.	Watkins Glen, routes 414 × 14	335 ± 1 (10)	345 ± 2 (10)					Genesse
5.	Montour Falls, route 14	334 ± 2 (5)				066 ± 2 (3)		Genesse
6.	Montour Falls, route 224	335 ± 2 (5)		008 ± 2 (3)				Genesse
7.	Montour Falls, hospital (low)		346 ± 1 (4)			066 ± 8 (3)		Genesse
8.	Montour Falls, hospital (high)		349 ± 1 (3)	012 ± 1 (3)	084 ± 1 (3)		95	Genesse
9.	Horseheads, route 14	330 ± 1 (4)			089 ± 1 (3)			
		325 ± 1 (3)						Lower West Falls
10.	Binghamton, route 17			360 ± 1 (3)		062 ± 2 (3)		Lower West Falls
11.	Owego, routes 17 × 96		353 ± 1 (3)		088 ± 1 (3)		95	Lower West Falls
12.	Owego, route 96		340 ± 2 (5)		073 ± 3 (3)		93	Lower West Falls
13.	Owego, routes 96 ± 38		343 ± 1 (3)		087 ± 3 (3)		105	Lower West Falls

Pennsylvania								
14.	Gillet, route 14	329 ± 1 (3)				065 ± 2 (3)		Middle West Falls
15.	Snederkerville, route 14			014 ± 2 (3)	089 ± 2 (3)			Middle West Falls
16.	Snederkerville, route 14	337 ± 1 (4)			072 (1)			Middle West Falls
17.	Troy, routes 14 × 514		343 ± 1 (2)					Middle West Falls
18.	Canton, Back Road			011 ± 3 (3)	085 ± 1 (3)			Middle West Falls
19.	LeRoy, route 414		343 ± 2 (3)	009 ± 1 (3)	090 (1)		107	Middle West Falls
20.	W. Franklin, route 414	332 ± 1 (2)	343 (1)		074 (1)		91	Middle West Falls
21.	Franklindale, route 414		354 ± 2 (4)					Middle West Falls
22.	Monroeton, route 414	335 ± 2 (5) 325 ± 3 (6)	343 ± 2 (6)	358 ± 1 (3)				Middle West Falls
23.	Towanda, route 220		348 ± 5 (11)					Middle West Falls
24.	Towanda, route 220	335 ± 1 (3)	344 ± 2 (4)	355 ± 2 (3)				Middle West Falls
25.	Ulster, route 220			359 ± 2 (3)				Middle West Falls
	Number of deviations	13	17	11	11	6	7	
	Standard deviation	4	4	6	7	2	7	
	Mean	333	346	004	082	065	96	

Data in this table are taken from unpublished report (Bahat 1982) submitted to ARCO Oil and Gas Company. Numbers in brackets are amounts of measurements.

Sheldon's measurements were taken in shales, where set II is dominant (hence her observation that "the strike set is the most important"). This set, which indeed consists of strike joints, is rare in the thick siltstones of the area. It is obvious, on the other hand, that the large angular difference reported by Engelder and Geiser (1980) corresponds to the angular difference between the Ib and Ic sets in most outcrops in the central and western parts of New York State (though possibly not in the eastern part).

A comparison of the above discrepancies with the two reports by Parker is illustrative. In an earlier study (1935, Plate II), Parker found the dominant joint direction in the Ithaca area to be about 347° (set Ia), though in a later study (1942, Plate 4) it was found (probably on different outcrops) that 335° (Set Ib) is by far the most dominant cross-fold direction. Each number in Table 5.2 is an already processed value for a 15-minute quadrangle that represents a mean result of various measurements (ten joints in Binghampton, Parker 1942).

The means in Table 5.2 for the Ib, Ia and Ic sets are 331, 346 and 005°, respectively. These are quite close to the means obtained in Table 5.1, thus supporting the postulation of three main directions in the region.

At the intersection of routes 14 and 414 at Watkins Glen a 345° joint cuts through a series of parallel joints of different strike (Fig. 5.4c). The traces of

Table 5.2. Orientation of joint sets (in degrees) and angular relationships in parts of southern New York and northern Pennsylvania. (After Parker 1935, 1942)

Quadrangle no.	Quadrangle	Sets					Angle produced by Ia and II
		Ib	Ia	Ic	II	III	
Parker (1935)							
1.	Ithaca		347	359	079		92
2.	Watkins Glen	331		002	075		
3.	Elmira	331	350		070		80
4.	Waverly	334					
5.	Owego		341	360	075		94
6.	Appalachian		350	018		067	
7.	Binghamton		354	013		068	
Parker (1942)							
1.	Ithaca	335				066	
2.	Dryden		345			067	
3.	Elmira	325	340		070		90
4.	Waverly	330			076		
5.	Owego		339			060	
6.	Appalachian		344	005		065	
7.	Binghamton		354	009		065	
8.	Sayre	330	348			067	
9.	Towanda		340	356	083		103
Number of quadrangles		7	12	8	7	8	5
Standard deviation		3	5	7	4	2	7
Mean		331	346	005	075	066	92

the latter (forming a series of sub-parallel crescents convex to the left) have smoother surfaces than the 345° joint and are marked by delicate undulations. It is difficult to measure the exact direction of the traces of the sub-parallel crescents. These are oriented clockwise from 345° (the Ia direction), and appear to pre-date it. The sub-parallel crescents seem to belong to the Ic direction. Joints oriented 335° (the Ib direction) are found adjacent to the other two.

Joints belonging to sets 346 and 004° occur in the lower part of a large outcrop at the intersection of routes 414 and 79 west of Watkins Glen. Joints of set 333° occur further up the road. To date, these two outcrops are the only ones known in Central New York where three cross-fold sets can be observed at close proximities.

Table 5.2 lists only two cross-fold directions for each quadrangle. With few exceptions, this is generally the case in Table 5.1.

5.2.4 Common Features of Burial Joints

In spite of differences in the histories of the Appalachian Plateau in the U.S.A. and the Beer Sheva area in Israel, the burial joints at both localities appear to have features in common. These features, summarized below, enable to define constraints of time and tectonic framework on the burial stage at each of the localities. Common features include (Table 5.3):

1. The development of vertical planar joints in approximately horizontal layers, only slightly deformed by horizontal stresses.
2. Horizontal propagation of the joint-forming fractures, which is generally confined to single beds by layer boundaries. This is indicated by the persistent horizontal direction in all styles of plumes, reflecting control by horizontal stresses.
3. Unless opened by subsequent tectonic or/and uplift processes, joints of the early burial phase generally maintain very slight openings (Fig. 5.5). Mineralization along these joints is slight (less than 1 mm) or absent.
4. Burial joints are generally closely and regularly spaced rarely more than several tens of cm apart. The spacing appears to be related to layer thickness (Ladeira and Price 1981). For a given layer thickness spacing is a little wider in the strike joints than in the cross-fold joints (Table 5.4 and Bahat 1988a).
5. Burial fractures that from at great depth generally are devoid of fringes (Figs. 5.4a, b). But when formed at shallow depths (several tens or hundreds of metres) fringes may develop (Fig. 3.11). En echelons occur along layer boundaries, more often in association with axial plumes than with circular rib markings.
6. Most if not all burial fractures in sedimentary rocks are probably caused by diagenetic processes that release fluid pressures. This is indicated by the periodic nature of the single-layer joint markings (Bahat and Engelder 1984). Periodicity is a fundamental concept in the Secor theory (1969) of fracturing by pore pressure, involving the rhythmic increase and decrease of liquid pressure in the rocks. These rhythms cause periodic episodes of fracture initia-

Table 5.3. Styles of fracture surface morphology on burial joints

No.	Style of fracture surface morphology	Rock description	Rock stratigraphy	Location	Joint inclination	Vertical size of marking (m)	Horizontal size of marking (m)	Special remarks	Source of figure number in text
1	Delicate rhythmic plume	Siltstone	Devon	Appalachian Plateau	Vertical	0.3	1.5		Fig. 3.2c
2	Delicate wavy plume	Siltstone	Devon	Appalachian Plateau	Vertical	0.6	1.5		Fig. 3.2b
3	Delicate axial plume	Siltstone	Devon	Appalachian Plateau	Vertical	0.25	1.5		Fig. 3.2a
4	Delicate axial plume	Indurated chalk	Middle Eocene	Northern Negev	Vertical	0.3	2		Fig. 3.11
5	Coarse axial plume	Indurated chalk	Lower Eocene	Northern Negev	Vertical	0.5	1.5		Fig. 3.10
6	Delicate radial plume	Indurated chalk	Middle Eocene	Northern Negev	Vertical	0.30	0.30	On echelon cracks	Fig. 3.35
7	Concentric circular undulations	Indurated chalk	Lower Eocene	Northern Negev	Vertical	0.50	0.50		Figs. 3.21, 5.1b
8	Coarse axial plume	Limestone	Miocene	Rhinegraben Budenheim quarry	Vertical	0.2	0.40		

Fig. 5.5 a, b. A single-layer joint in a prismatic block from the Mor Formation near Beer Sheva (scale is 30 cm). **a** The two parts of the block are intact along the hair-line joint. **b** The opened block reveals matching markings on the joint faces

Table 5.4. Joint spacing in single-layer cross-fold and strike sets

1 Station	2 Joint set (deg)	3 Mean spacing (m)	4 Layer thickness (m)	5 Normal to a length (m)
20	Cross-fold 328	0.17	0.53	4.0
20	Strike 051	0.26	0.53	5.0
20	Cross-fold 328	0.23	0.40	4.0
20	Strike 051	0.33	0.40	5.0
20	Cross-fold 328	0.21	0.42	7.0
20	Strike 051	0.27	0.42	8.0

Station 20 (Fig. 5.7 from Bahat and Grossmann 1988) is represented by three chalk layers (column 4). The length of exposure along which the mean joint spacing was determined is given in column 5.

tion, propagation and arrest, as is illustrated by axial plumes, and more obviously by rhythmic plumes and concentric rib markings (Figs. 3.2a and 3.34; 3.2c and 3.7a; 5.1b and 5.5b, respectively). Engelder and Oertel (1985) present a scheme of relationship between lateral compaction, abnormal pore pressure and jointing in the Appalachian Plateau.

Both the conjugate joint model (Stearns 1968; Bahat and Grossmann 1988) which follows orthorhombic symmetry (Wilson 1982 and Fig. 3.40c present volume) and the joint rotation concept (Engelder and Geiser 1980; Bahat 1986a) which conceives a pattern of rotated joints (Fig. 3.40a) bisected by the horizontal compression component may be applicable to burial joints (a third, alternative concept, regarding rotation of the horizontal compression while maintaining coaxiality with the joints, is not considered here). Accordingly, burial joints may be the products of both regional (tens km distances) and areal (hundreds m distances) stresses.

Burial joints and their fracture markings may develop over a fairly long time span, from shortly after initial burial (1 m.a. or less in chalks) up to a long time after that (50 m.a. according to Engelder and Oertel 1985). These joints may develop both at shallow depths (several tens or hundreds m in chalks) and at great depths (several km, Engelder 1985). The shallow joints probably form in response to small stress differences.

5.3 Syntectonic Jointing

Syntectonic joints include those fracturing phenomena which are associated with intense deformation. They often reflect regional stress, or the resolved or rebounded stresses that derive from it. Syntectonic joints are often different from those of the burial and uplift phases, and may display different FSM characters. Following is a description of syntectonic joints from eight localities, which include: (1) Santonian chalks in Israel; (2) Upper Cretaceous chalks in east France; (3) Upper Cretaceous chalks in south England; (4) Upper Palaeozoic shales in the Appalachian Plateau, New York; (5) the Entrada Sandstone, Utah; (6, 7) Middle Eocene chalks near Beer Sheva, Israel; (8) Lower Jurassic flint clays in Machtesh Ramon, southern Israel.

The first three provinces display syntectonic joints in Upper Cretaceous chalks. The American localities demonstrate jointing in shales and sandstone, respectively. The Eocene chalks near Beer Sheva display jointing that developed in specific tectonic settings. The flint clay of Ramon exhibits discoid jointing, linked to the syntectonic stage.

5.3.1 Syntectonic Jointing in Santonian Chalks, Israel

Chalk of the Santonian Menuha Formation (Bentor and Vroman 1954) in Israel, and its equivalent in neighbouring countries, are biomicrites mostly interpreted as of outer shelf (Flexer 1971). They are thickest in the shallow synclines of the Sigmoid Fold bundle known as the Syrian Arc, which has been developing intermittently since the Turonian and are often cut by transversal normal faults of Tertiary-Quaternary age (Picard 1943). The faults are characteristically associated with vein networks. In central Israel these veins consist of coarse clear calcite, occasionally mixed with chalk fragments, which in some cases appears deformed as if sheared and of a general syntectonic aspect (Mimran and Michaeli 1986). Stable isotope analysis of these vein fillings from the western flank of the Fari'a anticline, central Israel, has shown that they crystallized in an environment influenced by meteoric water at some post-Santonian stage (Mimran 1985). Calcite- and chalk-filled veins occur as systematic sets in many Santonian outcrops elsewhere in Israel. The veins often cut unstratified massive chalks (2 – 3 m thick or more). The Santonian chalks are moderately or occasionally strongly folded. The veins commonly are vertical but occasionally they deviate up to 20° from the vertical. A close examination of such veins reveals occasional shear displacements on some of them, but smooth joint surfaces on others, occasionally with perfectly developed fracture markings which rules out lateral displacements along these surfaces.

Jointing seems to have occurred in multiple events: some joints are tilted and others are vertical, spacing in different sets varies from about 1.5 cm to about 1 m (vertical joints with 1 – 2 m spacings quite likely belong to the uplift stage, Sect. 5.4).

5.3.1.1 Jointing in Santonian Outcrops from the Judean Desert

Three outcrops of Santonian chalks, all belonging to the widespread Menuha Formation, were investigated in the 'Arad area.

The Yatir Outcrop. (About 24 km west of 'Arad on the 'Arad-Tel Shoqet highway). Some fractures strike 010°, and others strike 317°. Fractures of the 010° set are somewhat wavy and deviate about 10° from verticality, those of the 317° set are straight and vertical. Calcite mineralization occurs in all fractures. The calcite veins of the 010° fractures (about 10 mm thick) are offset along the 317° veins. The calcite is not fragmented at the contact, which indicates that mineralization postdated displacement. Concentric rib markings occur on the calcite surface along a 294°-striking joint. Spacing of the 317° joints averages 10 cm. Sub-parallel joints do not seem to interact with each other (differing in this respect from some burial joints, Bahat 1987b). There are also joints which strike 280°, abutting on earlier fractures of strike direction 360°. Hence, there are at least two fracture generations. The early ones strike NNE and are somewhat tilted, while the later, NW-striking fractures are not. Fracture opening and mineralization are considerable in all the fractures.

The Nahal Ye'elim Outcrop. This outcrop occurs in Ye'elim Valley, some 4 km southeast of 'Arad, and consists of hard limestone of the (Turonian) Netser Formation, which dips 10° toward 050°. The limestone is overlain by moderately folded chalks of the (Santonian) Menuha Formation. Joints that deviate up to 10° from the vertical cut the chalks and reach 5 m or more in height. The dominant strike is NNW (333° ± 6.7°, in 21 random measurements). Spacing varies between 20 and 80 cm. Joints striking NNW also occur in the hard limestone. Some of these joints are confined to the limestone whereas others seem to have continuity in the chalk.

The Menuha chalks are overlain by endemically folded chert of the (Campanian) Mishash Formation. Spacing of the joints decreases from 20–40 cm to less than 1 cm in chalks that occur close to the folded chert. Joints 4–6 m high spaced about 1 cm apart (Fig. 5.6) are an impressive sight. A greater deviation from verticality occurs in the joints closer to the folded chert (up to about 20°). In so doing, the joints in this outcrop appear as a convergent fan with respect to the folded chert. Fracture convergence in foliated formations is associated with competent rocks (e.g. Park 1983, Fig. 10.15). If an analogy with the latter structure is permissible, it would suggest that fracture occurred in a cohesive rock during a late stage of rock consolidation. Fracture in a chalk of low cohesiveness would have developed a divergent fan of joints. An alternative explanation is that the deviation of the joints from verticality conventionally implies tilting of the joints from their original vertical orientation due to folding of the chalk.

The local average strike of the cross-fold joints which underlie the folded chert is 324° ± 4.7° (19 random measurements). Concentric undulations on a joint surface that bends along the strike suggest that the fracture initiated on a surface striking 295° and ended on a surface striking 324°. Fracture markings are very common on these joints, rarer in joints traversing the Santonian chalks that overlie the Netser Formation. A few NE-trending joints abut on the 324° striking joints, implying an early NW set and a later NE set. There is, however, also an indication of a reverse sequence.

The same minerals, mostly gypsum and calcite with traces of aragonite, celestite, quartz and montmorillonite, occur in fillings of the NW- and NE-striking joints. The filling consists of a 1-cm-thick dark, coarse-grained layer rich in gypsum, covered by a 1-mm layer of predominantly white fine-grained calcite.

Fig. 5.6 a, b. A decrease of joint spacing upon approaching a zone of deformation. **a** A deformed chert bed of the Mishash Formation (Campanian). Note the strong bending at *right*. **b** Close-up view of jointing in the chalks below the strongly folded chert bed. Spacing may be as narrow as 10 mm between joints that are more than 1 m long. Note rib markings on one joint surface near the pen

Steinitz (1974) regards the genesis of the Mishash chert as a diagenetic process, involving increase of volume and associated intraformational contortion and dislocation. Thus, if the 324° joints were caused by stresses associated with chert genesis and chert folding, they were formed during the Late Campanian-Maastrichtian-Palaeocene. They could not develop after beginning of the (non-deformed) Eocene.

The 'Arad Outcrop. (About 4.5 km SE of 'Arad, on the 'Arad-Dead Sea highway). This outcrop shows a prominent set of dominantly vertical joints, striking 317°. Many rib markings appear on the joint surfaces, including a few spectacular annular undulations (Fig. 3.22). The centres of the concentric fracture markings are generally well away from the lower or upper boundaries of the chalk layer, indicating that fracturing initiated in the interior. Plumes are relatively rare in Santonian chalks. On a 037° joint there occurs in plume which is cut by a later 317° joint. One 070° striking joint is cut by a later 319° striking joints. Thus, the fracturing sequence was: early jointing 037 and 070°, then joint 317°. A rib marking occurs next to a plume on a joint striking 070°. These two markings do not seem to fit a common rational geometry and may signify fracturing of a single joint in two stages, by different stress fields? (a similar problem is visible in Fig. 3.53).

Spacing in set 317° becomes gradually narrower along the outcrop in one direction (Fig. 5.7). No structural reason for this is evident in the area, but the

Fig. 5.7. Heavily jointed Santonian chalks near 'Arad, the Judean Desert. *Bar* = 1 m. Note a gradual decrease of joint spacing *from right to left*, and some transitions from vertical to subvertical joints at the *centre* of the outcrop

structural map (Aharoni 1976) shows a nearby normal fault of 40 m throw, striking E-W, a few hundred m north of the outcrop. The Ye'elim and 'Arad outcrops are both located in the convergence area of the Kidod and Qana'im lineaments (Aharoni 1976), i.e. in a zone of regional dislocation. The associated changes in joint spacing is similar to situations described elsewhere by Pohn (1981) and Shepherd et al. (1982).

The fracture sequence of the NE and NW joints in the 'Arad outcrop is generally the reverse of the Ye'elim outcrop. However, the frequency of the NW-trending joints is by far greater than that of the NE joints (Fig. 5.8). Fillings of joints striking 298 and 317° are mineralogically quite similar to those of the Ye'elim outcrop, suggesting that mineralization postdated the younger joints.

5.3.1.2 Distinctive Characters of Syntectonic Joints in the Santonian Chalks

The following characteristics distinguish the syntectonic joints in the Santonian chalks from those of the burial phase:

1. Joints predominantly trend NW-NNW. Less frequently they are oriented NE. Other directions generally represent local conditions.
2. Displacement and considerable opening of the fractures, which were later filled by calcite from predating meteoric waters (Mimran 1985).
3. Joints are mostly vertical or sub-vertical, occasionally tilted. Both vertical and tilted joints may occur in the same outcrop (Fig. 5.7).
4. Spacing is generally closer than 1 m, but near areas of dislocation the spacing may narrow considerably, and large joints (several meters high) may be 1 – 2 cm apart.
5. Joints in a given set do not interact with each other even at small spacings.
6. Joint orthogonality is unpronounced in syntectonic joints (compare Fig. 5.8 to Fig. 5.26, Sect. 5.6.1.1).

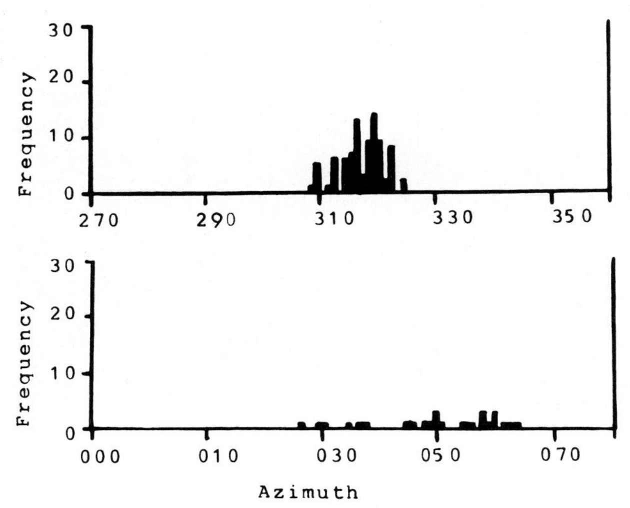

Fig. 5.8. Strike orientation histogram of syntectonic fractures in Santonian chalk, from the 'Arad outcrop in the Judean Desert

Fracture markings on the syntectonic joints are mostly concentric, and plumes are notably uncommon. Concentric annular markings, with centres well within the chalk beds, are very common, implying radial fracture propagation. Full circles are not rare. Fracture markings vary in shape and size even when occurring on neightbouring joint surfaces. Some markings are small and perfectly circular (Fig. 4.2a), others are large (Fig. 3.22). This is quite different from the general uniformity of FMS size displayed by the burial joints of Eocene chalks (Fig. 5.1b).

5.3.1.3 Contemporaneous Markings on Parallel Joints

It is usually assumed that a given set of joints is formed under the same set of conditions. Nevertheless, such conditions may operate over an extended period, and the joints are not necessarily the result of a single event. In not a few cases, evidence suggests that jointing was produced through cycles of increase and decrease of effective stresses (Secor 1969). Consequently, adjacent joints of the same set need not be a result of the same effective stresses except when bearing very similar fracture markings. Such repetitive fracture morphology is relatively rare, which supports the view that joints of a given set generally result from boundary conditions of unnegligible variations. Repetitions of similar fracture markings may be seen where spacing between adjacent joints is small (several cm). At the Israel localities such cases are rare in Eocene chalks, but common in Senonian chalks (e.g. Fig. 5.9).

Fig. 5.9. Two parallel joints with almost identical rib markings. Senonian chalks Upper Galilee. *Bar* = 0.25 m

5.3.2 Upper Cretaceous Chalks in East France

The Coniacian chalks in the eastern part of the Paris Basin appear in a quarry near Chepy, some 12 km southeast of Chalons-sur-Marne (Coulon and Frizon de Lamotte 1988a). The chalk is white and hygroscopic, due to high porosity (approximately 40%) and limited diagenesis. The cracks are divided by horizontal partings, some of which are layer boundaries (mostly along marly interlayers) and most of which are horizontal uplift fractures. The latter are recognizable where they cut fracture markings on vertical joints. No single-layer jointing is observed. The two dominant joint sets, visible in the lower part of the quarry, strike along 330 and 040°. Joints of the 330° set are never less than 6 m high, and their spacing varies irregularly between about 1 cm and 100 cm. Individual fractures from this set may reach 10 m or more in length. Occasionally there is a slight opening, lined by mineralization. The two sets abut on each other, indicating reverse sequences.

Plumes are very common on set 330°. They become coarser as they spread from their initiation points and transform into en echelons. This marking, the Chepy-type-plume, differs from the classical Hodgson pattern (Fig. 3.5) in two ways: (1) the plume may be oriented in any direction along the vertical surface, and correspondingly the en echelon fringe may be straight or curved, sub-horizontal, sub-vertical, or inclined (Fig. 5.10a). (2) The en echelon cracks appear somewhat disorganized, resembling continuous patterns (Sect. 3.1.6.1, Fig. 3.34), yet they initiate along well-defined perimeters as in discontinuous patterns (Fig. 3.35), and small cracks often give place to larger en echelons. Details of this style of fracturing are shown by a single joint marked by multiple elements (Fig. 5.10b). The plume pattern in Fig. 5.10c shows radial growth from one or more initiation sites. At some stage continuous breakdown of radial en echelon segments occurred. Rhythmic relationship seems to have developed between the echelons and the elliptical arrest lines (resembling Fig. 3.39). No plume periodicity (see Sect. 4.1.1.11) is observed at Chepy. Set 330° from the lower part of the quarry seems to be replaced by a 310° set in the upper part.

Set 040° occurs at all levels of the quarry. It resembles set 330° in its morphologic and fractographic characters although its joints are usually somewhat smaller, and undulatory markings that are rare on 330° joints are more common on the 040° set. An unusual marking was observed on a 040° joint (Fig. 5.10d): a sub-horizontal plume separates into two branches that then transform into en echelon fringes on two separate joints. Such unusual branching of a vertical joint may occur under conditions of intensive stress (Bahat 1979a).

The observed directions of the joints in the Chepy quarry, 040° and 330–310° are possibly related to two Oligo-Miocene extensional events (Coulon and Frizon de Lamotte 1988b, Fig. 2). Accordingly, the NW-SE direction (divisible into two discrete directions 290 and 310°) is orthogonal to the Rhine Graben, and the other, NE-SW direction is sub-orthogonal to it, being normal to the Dutch Graben. If these joints developed during the extensional

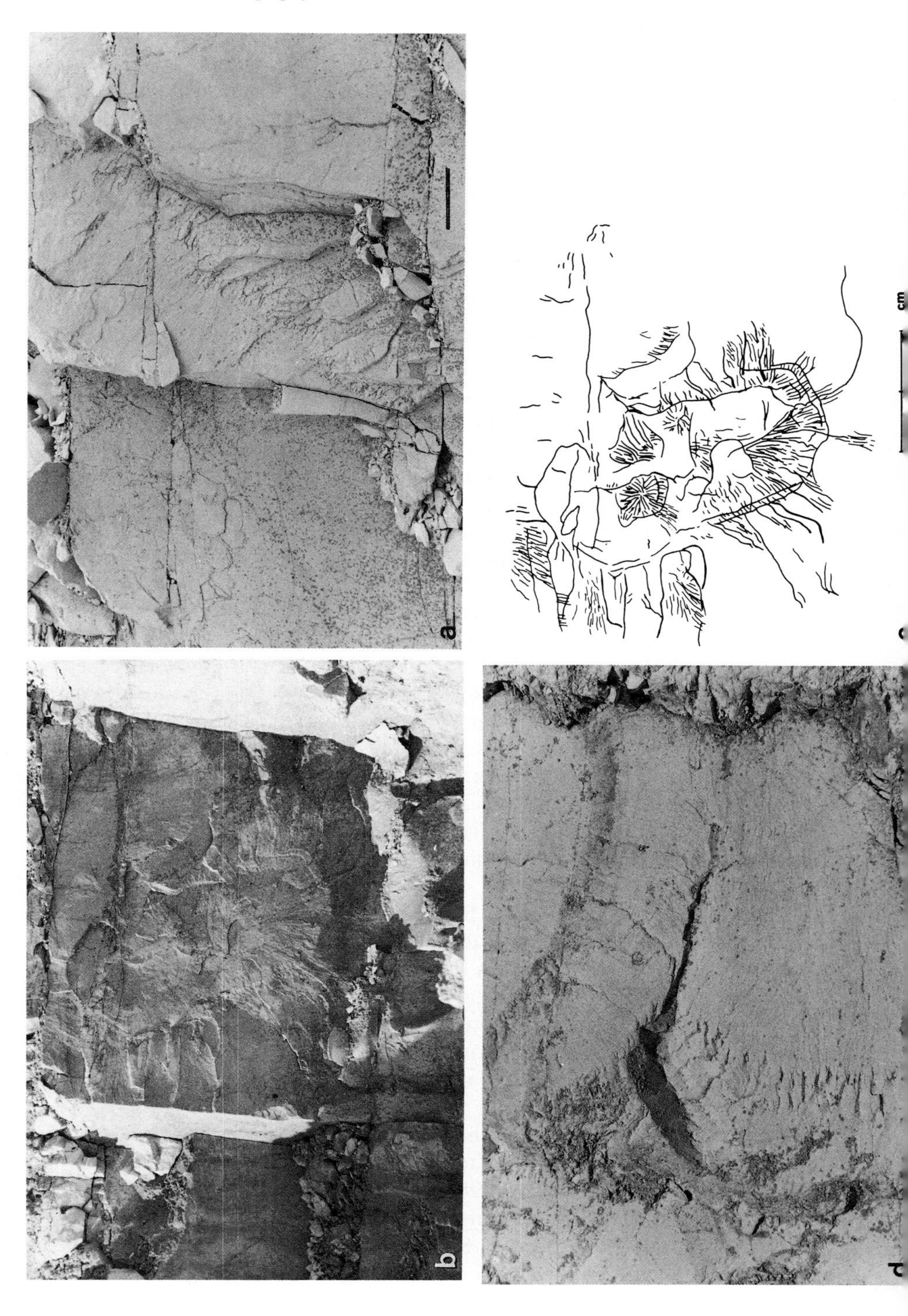
a
b
c
d
cm

period suggested by Coulon and Frizon de Lamotte, they would rate as syntectonic fractures in Coniacian chalks, formed some 60 m. a. after sedimentation.

The suggestion of syntectonic fracturing is supported by the irregular spacing and general high ratio of joint length/joint spacing in the Chepy joints. These properties and the lack of periodicity and the non-confinement of joints by layer boundaries testify against the possibility that these fractures are a product of the burial phase.

5.3.3 Upper Cretaceous Chalks in South England

The Wealden district in south England is a western extension of the Paris Basin. The structural history of the Wealden district can be divided into three phases (Gallois 1965): (1) Palaeozoic platform, (2) Mesozoic downwarping and sedimentation; (3) Tertiary uplift and folding. The uplift is a large, slightly domed structure, stretching from northern France through The Channel, Kent, Sussex and Surrey to the Hampshire Downs. In south England it is approximately 200 km (east to west) long and 80 km wide. Erosion of the domal structure exposed the thick chalks of the Upper Cretaceous at the periphery of the dome in a horseshoe shape open towards east. The chalks are subdivided into lower, middle and upper units (Fig. 5.11) (see also Kennedy and Garrison 1975).

5.3.3.1 Overall Jointing and Fracture Morphology

Some 24 outcrops were examined in southern England, and the observations are summarized in Table 5.5. Several joint features in the chalks are of special interest.

1. Single-layer SL vertical joints are very rare. Where present, they form irregular series of cracks in outcrops 7, 9, 11, 12, 13 and are questionable in outcrops 2, 3, 4 (Table 5.5). SL joints are practically absent in outcrops with flint layers.
2. Systematic multi-layer ML vertical joints are rare. They are, however, prominent in outcrops 9 (azimuth 315°) and 17 (azimuth 290°) and occur also in outcrops 5, 8 and 20.
3. Fracture markings are rare in the chalks of south England. They develop almost exclusively in medium-layered (50 – 100 cm) hygroscopic chalks of the Lower Grey Chalk (outcrops 2, 3, 4), in the Middle White Chalks (outcrops 7, 9) and in the upper Middle Chalk (outcrops 12, 13).

Fig. 5.10 a – d. Fracture markings on syntectonic joints in Coniacian chalks of the Paris Basin near Chepy, Champagne. **a** An inclined en echelon fringe is cut by late horizontal fracture. Note secondary steps on the lower large echelon crack. *Bar* = 20 cm. **b** Radial plumes, concentric undulations and a concentric en echelon fringe. **c** Drawing of **b**. **d** A wavy plume propagating downward and sidewise from right to left. A sub-horizontal step divides the plume into two separate surfaces on which two differently curved en echelon fringes have developed (see Fig. 3.2 g)

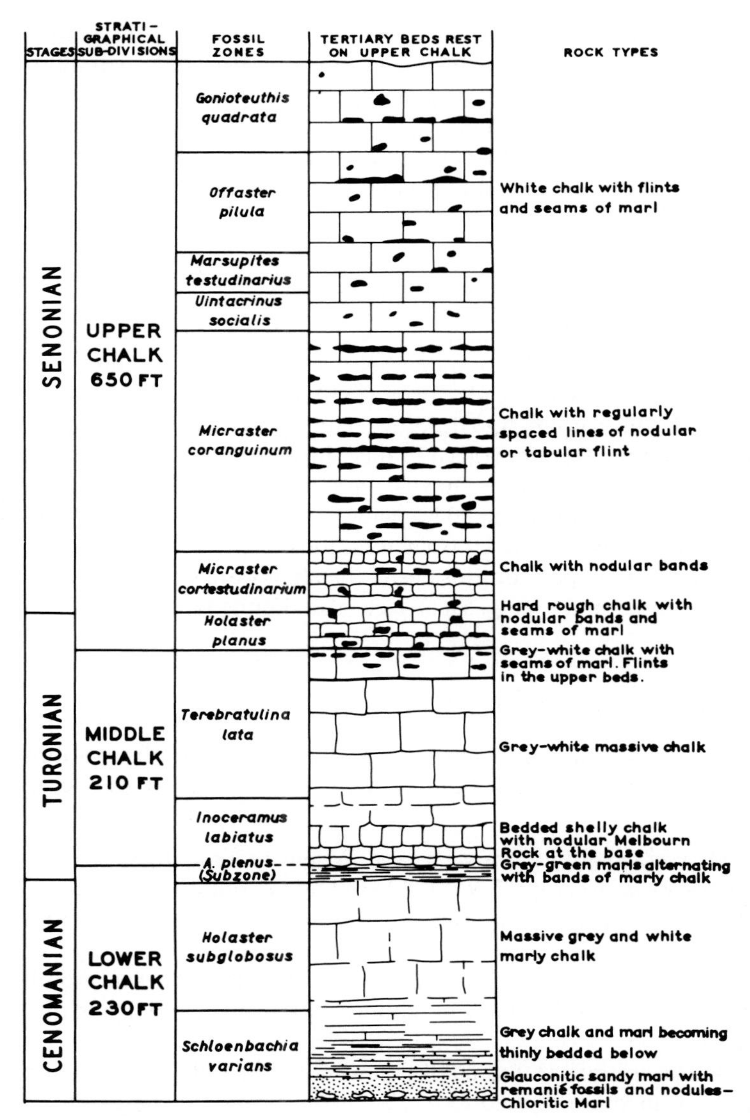

Fig. 5.11. Generalized section of the chalk of the Wealden District. (Courtesy of the Director, British Geological Survey). (Gallois 1965)

4. In the Grey Chalk (outcrops 2, 3, 4) fracture markings occur on surfaces of non-systematic irregular joints.
5. In outcrop 7 (Charing, Kent), fracture markings are well developed, though exhibiting uncommon morphologies. These include a composite pattern of vicinal plumes (resembling Fig. 3.4b), a feature not characteristic of systematic SL joints. Another fracture marking propagated concordantly with an earlier

Table 5.5. Systematic jointing and fracture markings in South England chalks

1 No.	2 Outcrop	3 Location	4 Stratigraphy	5 Rock properties	6 Systematic single-layer SL joints	7 Fracture markings	8 Systematic multi layer ML joints	9 Remarks
1	Folkstone quarry	Kent	Lower C	Marl and nodular chalk	–		–	
2	Betchworth Hills, quarry near railway station	Surrey	Upper Lower C	Grey hygroscopic chalk	Systematic jointing is questionable, subvertical early grey fractures and late white fractures	Abundant, and enriched in marking styles including secondary en echelon fringes on grey and white fractures	Scarce at the upper part of outcrop	Characteristic rock splinters caused by last year thermal stresses
3	Oxted (north) quarry	Surrey	See No. 2	See No. 2	Approximate No. 2	Abundant, in cluding SL markings	–	See No. 2
4	Glynde near Lewes quarry	Sussex	See No. 2	See No. 2	Approximate No. 2	Abundant	–	See No. 2
5	Eastbourne (west of Borough) beach cliff	Sussex	Lower Middle C	Stratified nodular at lower part and unstratified nodular at upper part of outcrop	–		Bended along the strikes $050 \pm 10°$ and $330 \pm 10°$	
6	Brook quarry	Isle of Wight	Lower Middle C	Nodular and fragmented chalk	–		–	Cracks are inter and intra nodular

Table 5.5 (continued)

1 No.	2 Outcrop	3 Location	4 Stratigraphy	5 Rock properties	6 Systematic single-layer SL joints	7 Fracture markings	8 Systematic multi layer ML joints	9 Remarks
7	Charing (Beacon Hill)	Kent	Middle Middle C chalk	Layers of white hygroscopic chalk	Dominant 315° also few 350°	Individual and composite markings are abundant on 315° joints, also constrained between inclined multilayer joints oriented 045°	Inclined striking 045°	
8	New Westgate Thanet beach-cliff	Kent	Middle C	Fragmented layered chalk	–		Few 325° and 360°	
9	Canterbury Bramling (Bridge)	Kent	Upper Middle C	Layered white hygroscopic chalk with flint beds	Irregular 030° see No. 7	–	Dominant 315°	
10	Dover beach-cliff	Kent	Middle C		–	–	Massive (2 m thick) chalk layers with some E-W and N-S ML joints in lower part, alternating chalk layers and flint beds in upper part	
11	Beer harbour (low) beach-cliff	Devon	Upper Middle C	Stratified nodular chalk and some alternating flint beds	Possibly limited to several chalk layers		Some inclined	

12	Dartford west quarry	Kent	Upper Middle C	Layered chalk and some flint beds	Set 315° in some layers with rare fractures subor-thogonal to 315°	Infrequent	–	
13	Dartford east quarry	Kent	Upper Middle C	Alternating chalk layers with flint beds	Few 040°	Markings on some 040° joints	Few 090°	
14	Beer harbour (high) beach-cliff	Devon	Upper Middle C	Nodular chalk scattered and in bands, flint nodules scattered and in a bed	–		–	–
15	New Haven (west of harbour) beach-cliff	Sussex	Upper Middle C Upper C	Alternating chalk layers and flint beds	–		–	
16	Seaford beach-cliff	Sussex	Upper Middle C Upper C	Occurrence of layered chalk and flint nodules	–		Possibly some	
17	Margate (Captain Digby beach-cliff)	Kent	Upper Middle C Upper C	Layered chalk with some chalk nodules and flint nodules	–		Common at 290° and some suborthogonal to it	
18	Lewes (south of the tunnel)	Kent	Upper Middle C Upper C	Layered chalk with flint nodules among them	–		–	
19	Ramsgate harbour	Kent	Upper C	Layered chalk with flint within the layers and among them	–		–	

Table 5.5 (continued)

1 No.	2 Outcrop	3 Location	4 Stratigraphy	5 Rock properties	6 Systematic single-layer SL joints	7 Fracture markings	8 Systematic multi layer ML joints	9 Remarks
20	Deal quarry	Kent	Upper C	Alternating chalk layers and flint beds	–		310 and 040° sets	
21	St. Margaret's Bay beach-cliff	Kent	Upper C	Alternating chalk layers and flint beds	–		?	
22	Calver Cliff beach-cliff	Isle of Wight (east)	UpperC	Strongly inclined (vertical) chalk layers with alternating flint beds	–		–	
23	Campton Bay (along the beach)	Isle of Wight (west)	Upper C	Strongly inclined chalk layers with alternating flint beds	–		?	Prominent sheeting (late horizontal fractures) and post sheeting some vertical fractures (Fig. 5.27b)
24	Campton Bay (at the upper cliff)	Isle of Wight	Upper C	Fragmented chalk with flint beds	–		–	

C In the stratigraphy column designates chalk units following Fig. 5.11. This is an approximate designation, and not claimed to be precise.

Fig. 5.12 a, b. Fracture markings in Cretaceous chalks from southern England. **a** Fragments of radial plumes initiating at the *upper part* of picture (see *vertical arrow*) propagating downward and left. Note a late undulation (*inclined arrow*) that developed concordant to an inclined older multi-layer fracture (roughly normal to picture) at the *left side* of the undulation. Note that late horizontal fractures cross the plumes. **b** Two adjacent spiral plumes evolving from wavy plumes that started between horizontal partings (in **a** and **b** horizontal fractures are 30 – 40 cm apart)

inclined fracture, producing an arrest line parallel to the inclined fracture (Fig. 5.12a). Rare spiral plumes (Fig. 5.12b) and cuspate hackles (Fig. 3.41a) also occur in the Charing outcrop. A spiral plume was previously found by Ernstson and Schinker (1986) in limestone (Fig. 3.2f). Cuspate hackles are typical of blast surface fractures (Fig. 3.60), but in Charing these markings occurs on a joint surface having the same orientation as other natural joints (see also Fig. 4.6).

5.3.3.2 Syntectonic Characteristics

The fractures at Charing, depicted in Fig. 5.12 and Table 5.5, are classified as syntectonic rather than burial joints because:

1. They developed late in the history of the rock, after previous jointing.
2. Jointing is only partly systematic and spacing is irregular.
3. Various types of fracture markings of various sizes occur in close proximity.

On the other hand, some features are untypical of uplifting jointing:

1. Fractures are generally small (less than 1 m), and so are the fracture markings.
2. Spacing is narrow (up to tens of cm).

5.3.4 Upper Palaeozoic Fractures in Shales of the Appalachian Plateau, New York

These fractures have been described above (Sect. 5.2.3). The Upper Palaeozoic joints strike 345° and form the Ia set.

5.3.5 Jointing in the Entrada Sandstone, Utah

Hodgson (1961b) describes joints and their fracture markings in sandstone formations from Arizona and Utah. From his descriptions, it appears that jointing in the various sandstones could have developed by various mechanisms during different stages of the rock history. Dennis (1972, p. 242) presents a photograph of the joint system in the Entrada Sandstone near Moab, Utah, showing a high ratio of joint length to joint spacing, which is a hallmark of syntectonic fracturing.

Dyer (1983) presents additional data, supporting this possibility. According to Dyer, systematic extensional joints in the Moab Member of the Late Jurassic Entrada Sandstone display a zoned character (see also joint zone by Hodgson 1961b). Subparallel joints tend to cluster in narrow zones, 0.5–2.0 m wide, where joint spacing ranges up to about 1.5 m (Dyer 1983, Fig. 2.3). Interzone spacing is regular (27 m) on the order of the Moab Member thickness. The longest joint measured reached about 27 m, but a joint zone may reach several kilometres in lateral extent.

Fractures are confined to the Moab Member. A photograph by Dennis (1972, Fig. 11.2) shows a straight horizontal fringe of echelons at the base of a vertical joint in a near-horizontal massive sandstone unit. Dyer also reports vertical cracks 1 to 4 m long in the echelon fringes at the lower and upper boundaries of the Moab Member. A photography by Muench and Pike (1974, p. 107) portraying a series of horizontal fringes along boundaries of the sandstone layers, shows such a fringe adjacent to a horizontal plume which bears similarity to a plume of a 345° set in N. Y. State (Fig. 4.9). All these observations imply horizontal fracture propagation in the Moab Member, similar to the large joints of the syntectonic 345° set cutting shales in New York State (Sect. 5.2.3). The horizontal fracture propagation of the syntectonic joints in the sandstones is quite distinct from the vertical propagation of the uplift joints (e.g. Fig. 5.20).

Dyer notes small lateral displacements along the joints in the Moab Member, but dip-slip was nowhere observed. This implies horizontal maximum and minimum principal stresses which, according to Dyer, have undergone controlled rotation (caused by movement of salt at depth), while every time a new joint zone developed the stress field was perturbed by older joint zones.

Dyer (1983, pp. 35–36) observes that the systematic joints are "generally sub-parallel to the axis of the Salt Valley anticline" and that there also is "a genetic link between the extensional fracturing and faulting and growth of the Windows anticline". Hence, syntectonic fracturing.

According to Dyer, the youngest joint set is related to the on-going collapse of the Salt Anticline and this "requires that the youngest systematic joint sets formed under near-surface conditions, probably with less than 200 meters of overburden". This agrees well with the occurrence of the highly developed fringes that he notes along the layer boundaries (see also Sect. 5.2.2.2).

5.3.6 Syntectonic Jointing in Association with Fault Termination

A syntectonic joint set associated with a vertical fault occurs in the Middle Eocene chalks at Wadi Na'im, near Beer Sheva. This set (briefly mentioned by Bahat 1987b) strikes 344° and occurs in a single layer which is 60 cm thick. The joints are confined to a zone about 20 m long and some 7 m wide, straddling both sides of a fault which strikes 318° along its straight portions. The fault and the joints occur in the same layer and the 344° joint set does not occur elsewhere. There is no connection between this set and the 344° set that cuts Lower Eocene chalks around Beer Sheva (Sect. 5.2.1.1). The fault's extremity (tip) is divided into three left-stepping curved en echelon segments with maximal vertical displacement of 20 cm (Fig. 5.13). Their size and overall relief gradually diminish towards the end where the fault disappears. The joints curve in sympathy with the curving of the fault. Field relationships indicate that set 344° preceded an adjacent single-layer joint set (striking 028°, Fig. 5.14), which is thought to be a product of the burial phase in the same rock layer. The joints of set 344° are generally longer than those of set 028° (longest measured fracture in set 344° is 10 m) and involve only few interactions within the set. Spacing varies from a few cm to a few tens of cm, depend-

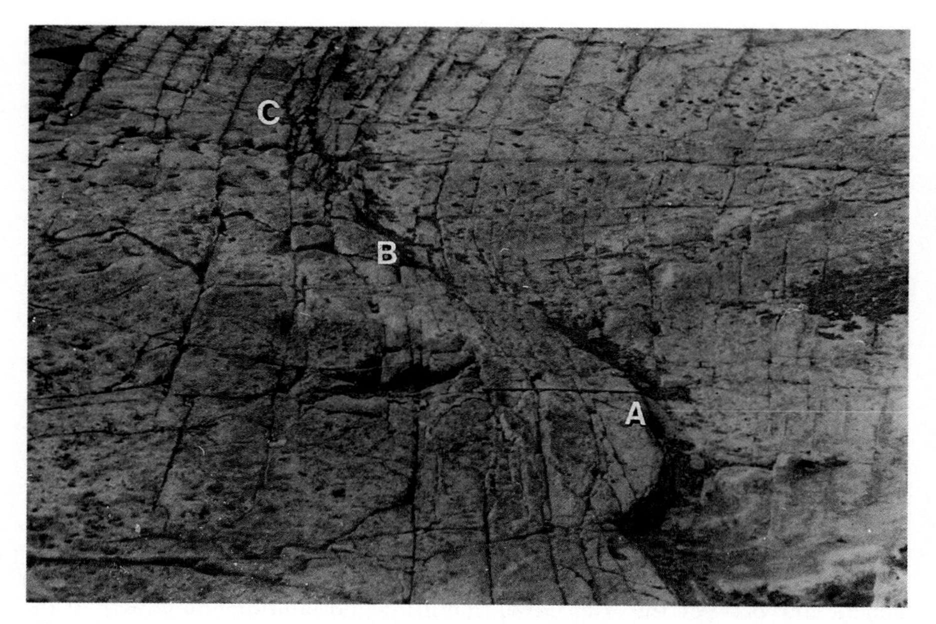

Fig. 5.13. A fault cutting platform 1 in Wadi Na'im near Beer Sheva (outcrop Y in Fig. 5.3 a) and dying out in a series of left-stepping, curved en echelon segments. A set of joints develop close to the fault (but nowhere else), and the individual joints bend in sympathy with the curving of the en echelon segments. Segments *B* and *C* are narrow stretches of splintered rock. 1-m scale crossing segment *A* is aligned 062°

ing on the curvature of the joints as a function of their location with respect to curvings of the fault. Opening is generally small (1 mm or less). No fracture markings could be examined on the surface of these joints, due to lack of exposures.

The above relationship between the 318° fault and the joint set 344° is probably not a good example of fault-joint interaction. The full length of the fault is not exposed, and it is difficult to establish the amount of horizontal displacement along the fault. Nevertheless, the situation illustrates a case of fault-joint interaction along the terminal portion of a fault. As such it is characterized by the following:

1. The jointed zone on either side of the fualt is notably restricted, the width being comparable to the length of the longest en echelon segment of the fault's extremity.
2. The mean strikes of the fault segments and the joints are at acute angles to each other (26°). However, at close proximity to the fault the joints may be parallel to the fault, crossing it at its curvings.
3. The joints near the fault show no displacement, and display neither opening nor filling.

Fig. 5.14. Chalk stratification and jointing in Middle Eocene chalks at Wadi Na'im, near Beer Sheva. Anastomose fractures of set 062° (at *centre* of picture) cut through platform (Fig. 5.3a) and the soft chalk bed above it, and continue into the cliff. Note the irregularity in the platform and straightness of these fractures in the cliff. Straight joints striking 028° at *left* produce an acute angle with the 062° fractures. *Arrows* mark the thickness of layer 5 (60 cm) at the cliff

4. Joint spacing varies as a function of distance from the fault and location with respect to fault curving. Closer spacing near a curving of the fault trace is associated with rock splintering along the fault.

5.3.7 Syntectonic Joints Related to Unfolding

The joint set oriented 062° which occurs in the Middle Eocene chalks (the Horsha Formation) near Beer Sheva fits the pattern of fold-linked syntectonic fracturing which is distinguishable from both earlier burial jointing and late uplift jointing, all identified along the same outcrop (Bahat 1987b).

Set 062° occurs along various parts of the creek Wadi Na'im in the Beer Sheva syncline, extending for hundreds of metres along two strath platforms flanking the creek. The joints undulate and zigzag along the platforms, forming straight vertical traces on the intervening cliff (Fig. 5.14). The 062° joints in the platforms are not evenly spaced. In some parts of the outcrop, normal distances between adjacent joints are several metres, while in other parts they form a bunch of subparallel joints several centimetres apart. Joint openings along the strike also vary from about 1 mm to several cm. The openings continue downward into irregular cracks (as revealed by opening with a chisel). The orientation 062° roughly corresponds to the strike of the Beer Sheva syncline, but no multi-layer joints similar to the 062° set occur normal to this direction. Set 062° cuts in the same manner through alternating indurated and soft chalks (Fig. 5.3 a), even though joints in the latter tend to be more closely spaced.

The average strikes of the single-layer set 028° which developed during the burial phase, and set 062°, as measured on the platforms, differ by 34°. Assuming that the rock was approximately isotropic in its mechanical properties, the extensional stresses that caused the fracturing most probably acted at different times. Set 062° seems to have developed over a long interval of time. Occasional joints of set 028° seem to be arrested at surfaces of 062° joints in the platforms, although generally set 028° is either cut or superposed by 062° fractures. On the other hand, set 062° cuts through the entire cliff above the strath platforms, where it postdates set 028° (Sect. 5.4.2.1). Typical, however, is the near-lack of lateral offsets along the joints of either set, demonstrating that the dilation of one set did not contribute to the opening of the other.

The distinctive characteristics of set 062° in the Middle Eocene of Wadi Na'im, as compared with the multi-layer joints typical of the Lower Eocene (Sect. 5.4.1), clearly suggest different mechanisms of fracture. The great irregularity and extensive length of set 062° sub-parallel to the synclinal axis, and the lack of multi-layer joints normal to this direction, suggest a process related to the synclinal structure. Jointing occurred as breakage in an extended buckling rebound at a time when the compression normal to the synclinal axis relaxed (Park 1983, Fig. 10.2 A).

Almost no fracture bifurcations have been observed in joints of the Beer Sheva area. This is particularly noteworthy for the long fractures of set 062°, where branching under conditions of constant stress would be stimulated by the increase of fracture length [Eq. (1.57)]. The implication is that jointing was generally slow, and occurred under conditions of low K_I intensity. No exposed surfaces of these joints are found in the platforms, and fracture surface markings contribute no clues. The continuation of set 062° further up on the adjacent cliff (Fig. 5.14), on the other hand, reveals distinct fracture surface morphologies (Sect. 5.4.2.1).

5.3.8 Syntectonic Discoidal Fracturing in Flint Clays

The discoidal fracturing encountered in the (Lower Jurassic) Ramon flint clays, known as the Mishhor Formation, is linked to the syntectonic phase. Arguments for this affinity are as follows:

1. The discoidal fractures (Sect. 3.1.7.2) were formed in flint clays that developed as alteration products, replacing previous soils (Goldbery 1982). Alteration took place under an erosional surface during the Early Jurassic. Hence, the discoids were not formed during sedimentation, and are not representatives of the early burial stage.
2. On the other hand, the general appearance of the discoids does not fit the description of uplift joints in terms of size, spacing and other properties (Table 5.9).
a) Cross-cutting relations and differences in the pigmentation of precipitates that coat the surfaces of various discoids possibly point to several fracturing episodes under different chemical regimes.
b) Adjacent discoids differ considerably in size, ranging from 10 to 500 mm, and their shapes vary from flat to conical.
c) Discoids occur commonly on vertical joints or with deviations up to 20° from the vertical.
d) Spacing of discoid-marked joints is generally small. In an exposure where discoids are clearly associated with a fault, spacing is 10 to 30 mm.
e) Fracture markings vary in size and shape on adjacent joints. In this respect the Ramon discoids resemble those of the Santonian chalks (Table 5.6).

Most joints in the Ramon flint clay strike in the range 306° (N54°W) to 332°. This range is similar to that of the major joint sets in the Santonian chalks in Israel (Sect. 5.3.1). This is the direction of the main stress exerted by the impingement of Europe and Africa during the Late Cretaceous and the Tertiary (de Sitter 1962; Garfunkel 1981). It should, however, be noted that this criterion is not exclusively syntectonic. Single-layer joints of the burial stage in Lower Eocene chalks also have a prominent NNW orientation (Sect. 5.2.1.2).

The implication is that in certain areas burial or syntectonic joints may be the products of the same causative forces at different times and under different effective stresses.

5.3.9 Characteristic Features of Syntectonic Joints

Following is a summation of properties that are present, to variable degrees, in most sets of syntectonic joints described above.

1. Fractures are vertical or somewhat inclined. The inclined joints appear to have been initially vertical and were only subsequently tilted, or they were formed during changing stress field conditions. The vertical joints imply post-tilting tectonics. The changing stress field may cause lateral displacements along the joints.

Table 5.6. Styles of fracture surface morphology on syntectonic joints

1 No.	2 Stype of fracture surface morphology	3 Rock description	4 Rock stratigraphy	5 Location	6 Joint inclination	7 Vertical size of marking (m)	8 Horizontal size of marking (m)	9 Special remarks	10 Source or figure number in text
1	Perfectly concentric undulations	Chalk	Santonian	Arad Israel	Vertical and slightly inclined	0.35	0.35	Common	Fig. 4.2a
2	Radial plumes, concentric undulations and concentric en echelon fringes	Chalk	Coniacian	Paris Basin	Vertical	1.5	1.5		Fig. 5.10b
3	Adjacent unrelated radial plumes and curved en echelons fringes	Chalk	Coniacian	Paris Basin	Vertical	1	1	Common	Fig. 5.10a
4	Plumes which reveal fracture front splitting	Chalk	Coniacian	Paris Basin	Vertical	1	1	Uncommon	Fig. 5.10d
5	Vicinal plumes	Chalk	Upper Cretaceous	Kent England	Vertical	0.4	0.3	Uncommon	
6	Cuspate-hackles	Chalk	Upper Cretaceous	Kent England	Vertical	0.6	0.6	Uncommon	Fig. 4.6
7	Straight plume	Siltstone	Devonian	Appalachian Plateau	Vertical	0.2	0.5 – 40	Common	Fig. 3.18

8	Curved undulations convexed sidewise	Shales	Devonian	Appalachian Plateau	Vertical	2	0.5	Uncommon	Fig. 4.9
9	Straight en echelon fringes along boundaries of thick layers	Sandstone	Jurassic	Utah	Vertical	0.5 – 3	tens		Dennis 1972, Fig. 11.2; Muench and Pike 1974, pp. 2 and 107
10	Radial plumes	Flint clays	Jurassic	Ramon Israel	Vertical and slightly inclined	up to 0.5	up to 0.6	Common discoids	Figs. 3.42; 3.43
11	Concentric angular undulations and concentric fringes	Chalk	Cenomanian	Carmel Israel	Vertical and slightly inclined	up to 0.4	up to 0.4	Uncommon discoids	Fig. 3.42

2. Fractures vary in size from several tens of mm (discoids in flint clay) to several m (in chalks), and up to several tens of m in thick sandstones. Syntectonic joints are generally unconfined to boundaries in thin layers, but in massive beds they are confined.
3. Spacing is generally narrow (several cm to several tens of cm) and it characteristically narrows more upon approaching a fault or a sharp bending of the layers (pointing to a possible underlying fault).
4. Joint opening and subsequent mineral precipitation are often considerable, and vein fillings of 5–20 mm thickness are common. In the Paris Basin mineralization in chalks occurs along hydraulic breccias that are associated with non-mineralized syntectonic joints (Coulon and Frizon de Lamotte 1988a).
5. Joint orientation often approximates the major regional stress directions but it may also correspond to local structures. It remains to be seen to what extent unpronounced joint orthogonality (Sect. 5.3.1.2) characterizes syntectonic joints.
6. Syntectonic joints are usually formed by more than a single fracturing event. Correspondingly, fracture markings vary considerably in size and shape, even on adjacent surfaces.
7. Fracture propagation is both radial and horizontal. Radial propagation produces circular rib markings, occasional circular fringes and radial plumes. Horizontal plumes and straight horizontal fringes found in thick sandstone formations, point to horizontal propagation.
8. Fracture markings are usually of small to medium size (10 mm to 2 m). However, in sandstone formations they may reach tens of m in length.
9. Horizontal fractures (sheets, Sect. 5.4.7) seem to cut syntectonic joints.
10. Vicinal plumes were observed on syntectonic joints (Sect. 5.3.3.1). However, vicinal plumes also appear on post-uplift joints (Sect. 5.5). A list of the more common FSM styles on syntectonic joints is presented in Table 5.6.

5.4 Uplift Jointing

Burial joints and uplift joints are easier to identify and distinguish in a given area, when one knows what to look for (Fig. 5.15a). Good criteria for distinction between these two fracturing phases are found in the Lower Eocene chalks around Beer Sheva. Middle Eocene chalks in this region also offer various aspects to uplift joints, often displaying FSM overprinting. A different approach to the distinction of uplift fracture types has been applied in the Appalachian Plateau (Engelder 1985), which is brought here for comparison.

5.4.1 Burial Joints and Uplift Joints in Lower Eocene Chalks Around Beer Sheva

In the Lower Eocene chalks of the Beer Sheva area, single-layer burial joints (Sect. 5.2.1) are occasionally traversed or displaced by normal faults. Vertical

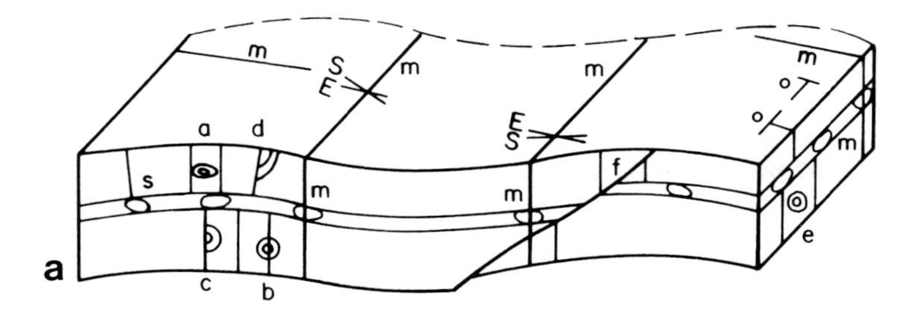

Fig. 5.15. **a** Fracture architecture in Lower Eocene chalks. Block diagram showing fracture in slightly folded beds (folding is exaggerated in diagram). Two chalk layers sandwich a bed of nodular chert at the *centre*. A single-layer joint *s* does not cross the chert bed, whereas a multi-layer joint *m* does. *E* and *S* are extensional and hybrid joints (Hancock 1985) respectively. Two reverse fracture sequences imply alternating episodes of jointing at *0*. A fault f crosses *s* and is cut by *m*. Fracture markings on surfaces of single-layer cross-fold joints describe: *a* Fracture initiation close to centre of bed. *b* A cross-fold fracture is crossed by a strike joint. *c* A cross-fold joint initiates as a strike joint. *d* A cross-fold joint initiates at a corner between layer boundary and a strike joint. A fracture marking *e* occurs on a single-layer strike joint. **b** Diagonal low-angle fault in Lower Eocene chalks (Fig. 5.29, station 58) displaced along a vertical fault (*parallel to the hammer handle*). The vertical fault belongs to a set of multi-layer joints

multi-layer joints and vertical faults that are associated with them, often traversing entire outcrops, may run alongside or across normal faults (Fig. 5.15b). These fractures are therefore considered to have been formed during or after uplift of the area. Moreover, the multi-layer joints (MLJ) differ from the early single-layer joints (SLJ) in other respects also as follows:

1. Spacing in SLJ is regular, ranging from 120 to 330 mm but commonly around 200 mm, while in MLJ it is irregular and wide, often ranging between 5 and 10 m.
2. SLJ are tightly closed, whereas MLJ have considerable openings (tens of mm to tens of cm). Correspondingly, joint mineralization along the former is slight (< 1 mm thickness), while in the latter, openings are mainly filled with chalk, calcite, gypsum or soil.
3. MLJ often have parallel "satellite" fractures flanking them at low spacings (Fig. 5.16a), 5–50 mm apart. SLJ do not display satellite joints.
4. MLJ overprint some of the earlier SLJ along both cross-fold and strike trends, cutting through chert nodules that were not affected by the single-layer fracturing (Fig. 5.16b).
5. Unlike SLJ, the MLJ surface bear no fracture markings. The above characteristics strongly suggest that the multi-layered joints developed in the chalks well after their diagenesis, and certainly not before consolidation of the chert nodules. The wide spacing is typical of fracturing that affects thick sections (e.g. Crosby 1982; Ladeira and Price 1981), as are the satellite joints. It appears that the Neogene tectonism and uplift of the region (Gvirtzman 1979) were ac-

Fig. 5.16a, b. Multi-layer joints in Lower Eocene chalks near Beer Sheva. **a** Alternating layers of chalk and thin chert beds (the second chalk layer from bottom is 90 cm thick), traversed by the multi-layer joint at *centre* of picture. Note the typical sub-parallel, closely spaced "satellite" fractures close to the main joint. **b** Many single-layer joints and a multi-layer fracture at the *centre* crossing the entire outcrop and cutting both chalk and chert beds. Due to its considerable opening, the multi-layer fracture is very evident. Note that the late MLJ overprints an early SLJ. Scale at bottom is 90 cm long (Fig. 5.29, station 4)

companied by fracturing (Price 1959) which overprinted the existing single-layer burial joints (Nickelsen and Hough 1967), as well as forming new, multi-layer, satellited and well-opening uplift joints.

5.4.2 Uplift Joints in Middle Eocene Chalks and Limestones from the Northern Negev (Southern Israel)

5.4.2.1 Large Joints

Large joints (6 m high) occur in massive unstratified chalks (an outcrop 8 km south of Beer Sheva in the highway to Nizzana) in the lower part of the Horsha Formation, striking NNW. The description of these joints corresponds with that of other uplift joints. The joints are straight and planar, spaced several metres apart and having openings of several cm filled with gypsum and carbonates. This set does not appear at sub-surface outcrops. One notes that small or medium-sized systematic joints resembling either burial or syntectonic joints from other outcrops in the area do not occur in these thick unstratified chalks. It appears that lack of conductive burial or tectonic conditions for fracture inhibited the growth of such joints.

The multi-layer joints from Wadi Na'im (see Sect. 5.3.7) near Beer Sheva that cut stratified chalks of the upper part of the Horsha Formation (Bahat 1987b) display variable strikes from 062 to 035°. Joints exposed in the platform differ from those in the cliff by their strike and morphological characters, suggesting different propagation phases. The MLJ in the cliff belong to the uplift stage, and they possess some uncommon but distinctive fracture markings: Firstly, there are en echelon cracks along vertical straight fringes (Fig. 5.17a) which are distinct from en echelon cracks that occur along horizontal straight fringes on single-layer joints (Fig. 3.11). This, by analogy suggests vertical propagation of the parent fracture. This interpretation, however, cannot be conclusive when compared with Fig. 5.20c which shows a horizontal fringe and en echelon cracks that propagated vertically under a parent fracture that propagated both sidewise and downward. Hence, this analogy seems to require some further studies. Secondly, radial plume-like markings diverge downward from an initiation point (Fig. 5.17b), suggesting downward fracture propagation.

At Ein Mor, near Sede Boqer (some 45 km south of Beer Sheva) the Upper Middle Eocene limestones of the Nizzana Formation are traversed by very long (tens of m or more) vertical joints, at wide spacings (tens of m). They form a set of average strike 055°, approximately normal to the direction of Tertiary regional compressive stress (Sect. 5.3.8). In this respect they may fit Engelder's description of release joints (Sect. 5.4.3). These joints are devoid of fracture markings.

5.4.2.2 Mark Overprinting

The term mark overprinting is used to describe the joint surface morphology that results when a late fracturing phase affects an existing joint surface, both

Fig. 5.17 a, b. Fracture markings on multi-layer joints cutting several layers in Middle Eocene chalks at Wadi Na'im near Beer Sheva (Fig. 5.29, station 40). **a** Note a series of en echelon cracks along a subvertical fringe at *left* near pen. The inferred direction of propagation of the echelons is *from right to left*, away from the joint shoulder, suggesting a vertical propagation of the parent fracture. **b** A multi-layer joint is shown with delicate striae which cut across several layers and radiate downward, towards the *left* indicating downward vertical propagation of the multi-layer joint

leaving distinct fracture markings. Although fracture overprinting is a common feature, mark overprints are rare. The importance of the latter lies, of course, in the high credibility that they add to the tectonic interpretation of the former. Two examples illustrate this:

1. A horizontal plume and en echelon fringes occur along a single-layer burial joint in a Lower Eocene chalk outcrop in the Elah Valley, some 50 km north of Beer Sheva. This joint is overprinted by concentric rib markings of downward convexity (Fig. 5.18 a) belonging to a downward propagating fracture which passed through the horizontal plume, most likely during the uplift stage. Since there were no lateral displacements during the overprinting the plume was not erased.
2. Mark overprinting is distinguished in an outcrop of Middle Eocene chalks (Fig. 5.18 b) near Shivta, some 32 km SW of Beer Sheva. The burial phase is represented by delicate horizontal plumes. Jointing of the uplift phase produced semi-elliptical concentric rib markings which overprinted the plumes on the first and third layers. The rib markings have downward convexity, denoting downward propagation.

The above-described cases of overprinting belong to two episodes of predominantly mode I fracturing which, according to Engelder (1987), should be distinguished from joints that have undergone frictional sliding under shear loading.

Mark overprinting indicates a multi-phase process, which may be considered as an alternative interpretation to the shielding effect. Occasionally, the presence of neighbouring single-layer and multi-layer joints in parallel orientations may be explained by a shielding single-phase process: "As a joint set evolves in a homogeneous rock, more and more joints are shielded, and the resulting population is characterized by many short joints and a few long joints" (Pollard and Aydin 1988) (see also Sect. 1.5.3).

5.4.3 Unloading Joints and Release Joints from the Appalachian Plateau, U. S. A.

Engelder (1985) distinguishes between two types of joints that form near the surface in response to thermal-elastic contraction accompanying erosion and uplift. These are unloading joints and release joints. The orientation of unloading joints is controlled by either residual or contemporary tectonic stress, the orientation of release joints is controlled by rock fabric. The development of these joints involves little or no abnormal pore pressure, either during burial or during the subsequent erosional events.

According to Engelder, the formation of unloading joints requires the removal of overburden equal to more than half their depth of burial at the time of initiation, depending on the change in Poisson ratio during lithification. "Estimates for the removal of overburden from the Appalachian Plateau of western New York vary from 500 m (Van Tyne 1983) to 2 km (conodont isograd index of Epstein et al. 1975). This denudation is compatible with the propaga-

Fig. 5.18 a, b. Fracture mark overprinting. **a** Fracture in Lower Eocene chalks from Elah Valley, central Israel. A horizontal plume in the upper part of the picture is cut by a series of rib markings which convex downward in the lower part of picture. Diameter of the *folded scale* at the bottom is 10 cm (see drawing in Fig. 4.14). **b** A three-stage jointing in Middle Eocene chalks near Shivta, south of Beer Sheva. Delicate horizontal plume of the first set *A* has been overprinted by a series of semi-elliptical rib markings *B* which convex downward, representing the second set. The third set *C*, orthogonal to the previous two, is visible near the hammer. A bulldozer scar marking *D* unrelated to the jointing cuts obliquely across the outcrop

tion of unloading joints at a few hundred metres to 1 km in depth". Engelder also surmises the existence of unloading joints in the crystalline basement at a still greater depth (1.6 km, Haimson and Doe 1983). He uses the basic equation proposed by Voight and St. Pierre (1974) to explain the horizontal stresses (σ_x and σ_y) during uplift for an isotropic elastic rock:

$$\sigma_x = \sigma_y = (\upsilon/1-\upsilon)\sigma_z + [\alpha E \Delta T(1-\upsilon)] \ , \tag{5.4}$$

where υ, σ_z, α, E and ΔT are Poisson ratio of the rock, vertical stress, thermal expansivity of the rock, Young's modulus and the temperature difference involved in the cooling of a homogeneous and a laterally confined half space. Based on this equation, Engelder concludes that rocks of high E may develop tensile stresses with the removal of much less overburden, and hence jointing is possible in a crystalline medium at greater depths than in more compliant sedimentary rocks.

Though the conditions for unloading are well-argued, two difficulties remain: first, the application of Eq. (5.4) does not entirely satisfy geological conditions. Glass experimentalists (Shand 1958) distinguish in restrained glass between a thermal stress σ_t developed under steady-state cooling conditions where

$$\sigma_t = E\alpha\Delta T/2(1-\upsilon) \ , \tag{5.5}$$

and one under transient (sudden cooling) conditions where

$$\sigma_t = E\alpha\Delta T/(1-\upsilon) \ . \tag{5.6}$$

Consequently, in a (very slow) steady-state geological process the thermal stress component should be the one used in Eq. (5.5) or even smaller, and not that of Eq. (5.6).

Secondly, uplift joints are probably restricted to the upper few hundred metres (Sect. 5.8), where thermal gradients are of small significance. At these depths the temperature differences are small and the near surface temperatures are affected by the flowing groundwater and undergo both diurnal and seasonal changes.

According to Engelder (1985), unloading joints in the Appalachian Plateau are orthogonal to the contemporary tectonic stress field. Release joints, which postdate active folding, are parallel to the fold axes, normal to the direction of compression. "On erosion these joints open in much the same manner as release joints in a triaxial compression experiment. Here the orientation of the joint may be controlled by some rock fabric such as solution cleavage planes rather then the contemporary tectonic stress at the time of propagation". Engelder observes that release joints (strike joints) in the Appalachian Plateau postdated the Alleghanian Orogeny, as indicated by abutting relationships within the deeper parts of the Devonian clastic section. See comparison of joint type characteristics by Engelder (1985) to present terminology in Table 5.7.

Table 5.7. Genetic joint terminology

Engelder (1985)		Present	
Joint type	Joint characteristics	Joint type	Joint characteristics
Tectonic	Form due to high pore pressure during tectonic compaction at depths of less than 3 km	Burial	Propagate in multi-stages mostly during the diagenetic processes and may initiate at very shallow depths (tens 4 m)
Hydraulic	Caused by abnormal pore pressure under restricted pore water circulation at depths in excess of 5 km	Syntectonic	Develop in response to intense deformation generally, post-date diagenesis and pre-date uplifting
Unloading	Propagate after more than half of their overburden has been removed and controlled by either a contemporary tectonic stress or a residual stress, but little or no abnormal pore pressure	Uplift	Form during the unloading processes
Release	Propagate like unloading joints in response to the removal of overburden during erosion but are fabric controlled and form parallel to the axes of folds after folding		
		Post-uplift	Result from geomorphological processes

5.4.4 Sequential Formation of Uplift Joints

Uplift joints may be formed through several fracturing stages, sometimes quite complex. A sequence of several fracturing stages is observed in the western scarp of Mt. Abu Azayat in eastern Sinai, Egypt (Fig. 5.19a, b). Early tectonics produced two large E-W joints, marked n and s, which are about 50 m apart and some 60 m high. A large N-S joint (parallel to photo) is contained between joints n and s and is therefore evidently of later age. Its lower part is vertical, with exceptionally large fracture markings, designated t and u. The upper part of this joint t is inclined backward to the wall. The large joint may have started as an inclined fracture before propagating vertically downward (Johnson 1970, p. 381). The horizontal fractures above and below the person at m were the next development. If they were earlier than the N-S central joint, the latter probably would not cross them downward. The upper horizontal joint h is thought to be either older than or contemporaneous with the lower one near m (see Sect. 5.4.5.2). The many closely spaced joints at the top of the outcrop (above the large joint) could have developed at any time. Note, however, the small spacing in these joints, which may signify that the closely spaced joints developed first, during early uplift. A short series of arrest lines on these joints

Fig. 5.19 a, b. A "giant" marking on a joint surface in a Palaeozoic sandstone formation at Mt. Abu Azayat, Sinai. **a** *u* undulations; *t* striae; *n* northern vertical joint; *s* southern vertical joint; *h* horizontal joint; *m* Ron Bogoch is standing at the foot of the mountain. **b** A fracture marking with hackles pointing downward on the E-W *n* fracture

(above t) shows these markings to be confined between pre-existing E-W joints, suggesting that at least some of the E-W joints at the top of the outcrop preceded the N-S ones. A large fracture marking on joint n has downward-pointing hackles on its lower side (Fig. 5.19b). Hence, two large fracture markings on mutually orthogonal joints both imply downward fracture propagation (see also undulations u in Fig. 5.19a). The debris at the foot of the outcrop suggest that the fracturing process is continuing, supplying fragments at a faster rate than the desert alluvial process can remove them. All this goes to indicate that the tensional uplift jointing in the massive Palaezoic sandstones is a prolonged process, which probably started during the great Neogene uplifts (Garfunkel 1978) and continues to present times (for further interpretation, see Bahat 1980a).

5.4.5 Vertical Propagation of Uplift Joints

Fracture propagation during the burial phase is usually parallel to the bedding, as is suggested by the common occurrence of horizontal plumes (Sect. 5.2.4). What about fracture propagation during the uplift phase?

5.4.5.1 Observations and Theories of Crustal Stress

Studies by Hoek and Brown (1978), Jamison and Cook (1979) and Jaeger and Cook (1979, p. 376) show that "the average values of the horizontal components of the crustal stress range from about a half to more than three times that of the vertical stress at the same depth", and the average of principal horizontal stresses at the crustal surface is compressional. Jaeger and Cook suggest that if denudation is rapid, the vertical stress would diminish elastically, in proportion to the decreasing weight of the overburden. Elastic relaxation of the horizontal stress components, however, would be only $\upsilon/(1-\upsilon)$ times that of the vertical stress because initially $\sigma_x = \sigma_y = (\upsilon/1-\upsilon)\,\sigma_z$ [Eq. (5.3)]. Consequently, average horizontal stresses at the surface would be only slightly smaller than those which prevailed at depth before denudation. More mechanisms for large horizontal stresses at shallow depths were proposed by other authors (e.g. Seagar 1964; Engelder and Sbar 1984).

These observations, however, are in conflict with the great ubiquity of shallow joints which by all criteria should be regarded as tensional rather than compressional. Perhaps a mechanism which prevails over lithostatic or compressional states of stress in the crust (McGarr 1988) is required to explain this conflict. Several authors suggest mechanisms that may lead to tension during uplift. According to Price (1959), the erosional removal of overburden that accompanies uplift is associated with a rapid decrease of the total lateral stresses at shallow depths to the extent that they become tensile. Price estimates the average magnitude of the tensile stresses to be equal to half the decrease in vertical load. This involves lateral expansion by elastic stretching at the surface, which results in vertical joints. Voight and St. Pierre (1974) elaborate on the tensile stresses that develop due to gravitational unloading, in conjunction with the effect of thermal stresses that result from the cooling of the elevated rock.

Haxby and Turcotte (1976) suggest that the stress induced by erosional unloading may be broken down into three components: removal of overburden pressure results in compressional stress, whereas the stresses due to uplift on a sphere and the thermal stresses are tensional. Since the latter dominate, erosion leads to predominantly tensional stresses. Jaeger and Cook (1979, p. 382) visualize a concentration of tensile stress at the top of a hemi-spherical hill (presumably due to upbulging/bending). Bending mechanisms were contemplated as early as 1896 by van Hise and much later by Bahat (1979a).

Is there a way to test the various proposed mechanisms? An uplift has been described as "a broad and gentle epeirogenic increase in the elevation of a region without a eustatic change in sea level" (Press and Siever 1982). It would intuitively be expected that such an elevation be accompanied by a vertical stress gradient, which in turn should affect jointing. Do joints present evidence to support this expectation?

5.4.5.2 Vertical Fracture Propagation

Engelder (1985) describes cross-fold joints that may be traced vertically over much of the 50-m exposure in shales of the Genesee Group at Taughannock Falls, Ithaca State Park, New York. Joint growth in the shales is not impeded by bedding interfaces, and the uncontained joints propagate so that their vertical dimension is as large or larger than their horizontal (strike-parallel) dimension (Engelder et al. 1987), suggesting that the vertical gradient in stress was about equal to the horizontal gradients or greater. This is in contrast to cross-fold joints cutting bedded siltstone-shale sequences (Fig. 3.18), where fracture growth is impeded by bedding interfaces, and the horizontal dimension is much larger than the vertical dimension, implying a horizontal stress gradient greater than the vertical gradient. Engelder draws an analogy to observations from petroleum geology (Nolte and Smith 1981) and concludes that fractures will propagate along the gradient of decreasing least stress or, failing that, propagate in the direction of no gradient, rather than propagate along the gradient of increasing least stress. Because the least principal stress decreases in a vertical direction by virtue of there being less overburden in that direction, vertical fracture propagation would be upward. Hodgson (1961b) has also envisioned upward joint propagation: "Joints form early in the history of a sediment and are produced successively in each new layer of rock as soon as it is capable of fracture".

More consistent with observations of fracture surface morphology is a vertical downward propagation of uplift joints. Downward fracture propagation of large uplift joints in Palaeozoic sandstones from Sinai is described in the previous section (Fig. 5.19). Vertical or sub-vertical downward propagation of uplift joints occurs also in Maastrichtian marls (Fig. 5.20a), Lower Eocene chalks (Fig. 5.18a), Middle Eocene chalks (Figs. 5.17a, b, 5.18b and 5.20b) and sandstones from Central Australia (Fig. 5.20c) (in the latter sidewise propagation is prominent). Similar evidence from fracture markings on uplift joints has been observed on other exposures (not presented here). Characteristic to

a
b
c

Fig. 5.20 a–e. Uplift joints that propagated vertically downward. **a** A large uplift joint with concentric rib markings and a curved fringe with en echelon cracks in its lower part in Maastrichtian marls, southern Israel, is examined by Ludwig Avraham. This fracture is orthogonally cut by a later set of joints having an average spacing of about 1 m. **b** Itamar Pelly is standing near a series of convexing rib marks on a joint that cuts Middle Eocene chalks at Elah Valley. **c** A series of convexing rib marks at the *top* and a horizontal fringe of en echelon cracks at the *bottom* on a joint surface approximately 30 m high cutting the Devonian Mereenie sandstone, King's Canyon, Northern Territory Australia. (Courtesy of Ramon Loosveld). **d** Remnants of two radial plumes with origins at the upper parts of two adjacent joints, and two vertical plumes having the shape of inverted palm tree that propagated downward, on two adjacent joints cutting Jurassic sediments in the Ramon, Central Negev. **e** Drawing of **d**; *a, b, c, d* are four distinct plumes, *e* is the fringe of *b*. *Bar* = approximately 1 m

all these fracture markings are downward convex arrest lines, with or without curved or vertical fringes of en echelon cracks. Fracture markings on joints in Palaeozoic shales from the eastern U. S. A. also point to downward vertical propagation, (see pictures in Kulander et al. 1979, Figs. 21, 29, 32, 87 and 89 and by Pollard and Aydin (1988, Fig. 13 B). Lateral divergence of rib markings is also known to occur (Fig. 5.21). In places, vertical downward propagation is indicated by a vertical plume (Fig. 5.20 d, e). Sub-horizontal plumes on uplift joints (Fig. 3.54) are rare.

What can this strong evidence for downward joint propagation teach us about the mode of uplift fracture growth? Perhaps the most important implication is that these joints grow in response to a stress gradient in which the least principal stress decreases downward and not upward. Such a gradient is best explained by lateral expansion or a bending mechanism which causes greatest strain at the surface of the crust, where the greatest tangential tension exists. Thus, fracturing initiates at the surface, and as elevation progresses, compensation by erosion gradually exposes the rock from greater depths to this maximal tension.

An analogous situation prevails in a lava lake which solidifies during cooling, when columnar joints initiate at the surface and propagate from the region of maximal cooling rate [and hence maximal tensile stress, Eq. (5.5)] downward to the interior (Ryan and Sammis 1978), where the cooling rate is smallest and the resulting stresses are minimal. An interpretation of downward

Fig. 5.21. A divergent rib marking on a remnant of a large multi-layer joint. Ribs radiate sidewise from the left and downward. Upper Senonian chalk near Tsefat, Upper Galilee. Scale at *left* = 1 m

propagation of joints from the surface was offered by Lachenbruch (1961, 1962) on grounds of fracture mechanics.

If the mechanism of bending by upward bulging and downward fracture propagation is accepted, without excluding other vertical propagating fracture modes in other cases (see for instance Fig. 3.3), certain implications are unavoidable. Fracturing that initiates at the surface would be considerably influenced by topography and other surface conditions. The compensation between regional and local stresses would affect the orientation of joints. Consequently, on a regional scale, sets of uplift joints would display greater azimuth spreading than sets of burial joints. This seems to agree with Engelder and Oertel's (1985) observation from the Devonian Catskill Delta that "when viewed in the form of rose diagrams, one set of tectonic joints has smaller circular variance ($<5°$) than a set of unloading joints ($>10°$)".

5.4.6 Orthogonal Joint Sets

Orthogonal or sub-orthogonal ($90 \pm 10°$) relations are quite common in sets of vertical jointing, which occur among single-layer burial joints as well as among uplift joints. Price (1959) suggests that mutually orthogonal sets are formed sequentially during uplift, because as soon as fracturing takes place the tensile stresses are released and the direction of least principal stress may change. However, when the lateral stresses are initially the same, orthogonal sets of joints may possibly develop practically simultaneously. The same holds true if one considers effective stresses [Eq. (5.1)].

Fracture surface morphology may shed light on the conditions that favour joint orthogonality, as shown in Fig. 4.11. Here, two orthogonal joint surfaces are marked by a similar pattern of rib markings. The two patterns, at right angles to each other, are so symmetric that they appear as if divided by a mirror plane. The symmetry of the fracture marking indicates orthogonal and probably identical tensile stresses and simultaneous fracturing.

5.4.7 Fracture Markings on Horizontal Surfaces

Johnson (1970, p. 381) explains sheeting (horizontal jointing sub-parallel to the surface) as a process that follows overburden removal under conditions of plane strain (Sect. 1.2.4) when there is no horizontal deformation and vertical tensile strain is maximal. He describes fractures in the granite of Fletcher Quarry near Westford, Massachusetts, that are essentially horizontal at depth, but those fractures that occur near the ground surface may bend upward or downward.

Horizontal fracture markings, which are mostly the product of vertical tension, occur on several types of surface: on flat horizontal joints, curved horizontal joints, and on bedding interfaces.

Horizontal and curved joints occur close together in the ceiling of a bell-shaped cavern excavated in the massive chalk of the Middle Eocene Tsor'a Formation, some 40 km north of Beer Sheva, Israel. One fracture marking on a horizontal joint shows hackles that radiate at the periphery of a concentric series of arrest lines (Fig. 4.4), suggesting intense fracturing. Divergent plumes appear on the curved fractures (Fig. 3.24).

Markings on bedding planes appear to reflect processes associated with vertical strains that separate layers that have been "welded" together due to overburden pressure. Two typical marking styles are observed: a combination of concentric undulations with radial plumes (Fig. 5.22a) which resemble many markings on vertical joints, and faint radial plumes that can easily be missed (Fig. 5.22b). Presumably, horizontal joints that often occur in massive rock formations do not develop in layered rocks where vertical strain is accommodated by layer separation.

5.4.8 Common Features of Uplift Joints

Uplift joints occur, for unknown reasons, either without markings, or their surfaces are decorated with various styles and sizes of non-annular rib markings. They rarely show those horizontal plumes and annular undulations that are characteristic of burial joints, or annular undulations which occur on syntectonic joints. Some joint features and some joint surface markings are typical, and occasionally diagnostic. The following spatial features are typical of uplift joints:

1. Uplift joints may be vertical, horizontal, and occasionally bended. They often are large (several m to tens or perhaps hundreds of m), and cross layer boundaries. Multi-staged uplift joints may occur at a single exposure, representing progressive stages of uplift (Fig. 5.19). Vertical uplift joints occasionally cut tilted syntectonic joints (in the Judean Desert).
2. Joint spacing is usually wide (several m).
3. Joint opening is generally large (several cm). Mineral infilling varies from slight (as in burial joints) to thick (as in syntectonic joints).
4. Orthogonal jointing is common.
5. Sets of uplift joints have generally wider scatter around an azimuth than have sets of burial joints.
6. Uplift joints are often associated with vertical faults (Fig. 5.15b). Displacements along vertical uplift joints, which thus become vertical faults, are not rare.

The following fracture markings are typical of uplift joints:

1. Rib markings are generally large (Figs. 5.19 and 5.20c) compared to rib markings on burial joints (Figs. 5.1b and 5.5).
2. Rib markings are quasi-circular, quasi-elliptical or semi-elliptical (compare quasi-circular markings on an uplift joint in Palaeozoic sandstone, Fig. 3.23 to circular undulations on a syntectonic joint in Senonian chalk, Figs. 3.22 and 4.2a).
3. The long axes of elliptical markings often deviate from the horizontal, and may be vertical (Fig. 5.23) or inclined (Fig. 5.19).

Fig. 5.22 a, b. Inter-layer markings on horizontal surfaces in Lower Eocene chalks, vertical overhead views. **a** Concentric rib markings (*near pen*). Part of the marking is hidden by the remaining upper layer, in the *lower part* of the picture. **b** Delicate radial striae

Fig. 5.23. Asymmetric concentric undulations forming an elliptic pattern, with large sub-vertical diameter, on a joint surface formed in Palaeozoic sandstones. Wadi Tweibeh, northern Sinai. *Hammer* in lower part of picture gives scale

4. Typically, fracture markings initiate at the centre or upper part of a joint, rarely in the lower part.
5. Uplift joints generally propagate downwards and sidewise rather than upward. This is indicated by radial striae that fan downward (Fig. 5.17b), by downward-convexity of rib markings (Fig. 5.18a), and by the downward convexing fringes of en echelon cracks (Fig. 5.20a). Uplift joints have inclined, vertical or horizontal straight fringes (Figs. 5.17a and 5.20c) compared with almost exclusively horizontal fringes in burial joints (Fig. 3.11).
6. Overprinting of fracture markings may occur when uplift joints propagate along pre-existing joints (Fig. 5.18).

5.5 Post Uplift Jointing

Post-uplift joints are regular sets of joints that can best be characterized by geomorphological criteria. They usually reflect local conditions rather than regional tectonics and may be caused by various fracture mechanisms.

Inclined fractures that occur in sets appear to be formed by post-uplift late processes. Engelder and Sbar (1984) correlate an alignment of inclined joints in Precambrian crystalline rocks of the Adirondack Mountains, New York, with the local topography. Johnson (1970, p. 385), on the other hand, considers high-angle fractures that intersect horizontal joints in Fletcher Quarry, Massachusetts, to be the result of quarrying.

Peculiar joints occur north of Beer Sheva, near Tel Shoqet. They cross the local Lower Eocene chalks at 55 to 75° angles and correspond to the local topography. The jointed outcrop is close to a vertical road cut, and the joints produce an acute angle between the joint dips and the road cut. The surface morphology of these joints (Fig. 5.24) is quite different from the fracture markings of uplift joints and from blasting scars (Fig. 3.60). They resemble vicinal plumes that occur on vertical syntectonic joints in having a common centre and propagating separately to different directions (Sect. 3.1.2.1). The vicinal markings on the inclined surfaces cut two or three layers. The difference between vertical uplift joints and inclined joints with vicinal plumes suggests different fracture mechanisms. The correspondence of the Tel Shoqet joints to the topography seems to reflect a sub-recent event.

Fig. 5.24. A joint surface, dipping approximately 65° and cutting through several alternating chalk layers and chert beds in the Lower Eocene near Beer Sheva (Fig. 5.29, station 2). Two vicinal plumes are observed on the joint surface (see drawing, Fig. 3.4)

Vertical uplift joints that cut Maastrichtian marls in the Tsin Valley, southern Israel, seem to propagate upward and cut younger sediments, possibly in response to latter-day stresses. This feature that may be classified as post-uplift jointing needs to be further investigated. FSM styles on uplift and post-uplift joints are exemplified in Table 5.8.

5.6 Comparison of Jointing During the Three Major Tectonic Phases: a Summary

The evolution of joints can be correlated with the three phases of rock history: burial, syntectonic deformation and uplift. The three corresponding joint categories can be identified and distinguished on the basis of (1) the spatial characters of the joint plane, and (2) the fracture markings on the joint surface.

5.6.1 Spatial Characters of the Joint Plane

5.6.1.1 Strike

Joints of burial stage occur in oriented sets (Fig. 5.25). Grouping of these joints often presents difficulties, but many investigators (e.g. Engelder and Geiser 1980) have recognized a regional consistency (of tens or hundreds of km) in the alignment of joint sets along particular azimuths (Figs. 5.25; 5.26). Note the pronounced orthogonality relationship in Fig. 5.26.

Syntectonic joints may be produced by both regional and local stresses. Joints in Santonian chalks in southern Israel, are oriented NW-NNW, corresponding to regional compression along this direction. In some outcrops the alignment of the NW trending syntectonic joint sets is quite similar to that of the burial phase (compare Fig. 5.8 to Fig. 5.26). Syntectonic joints often occur in close proximity to folds (Fig. 5.6), which exert local influences. Likewise, they occur in association with and are occassionally confined to the close vicinity of faults (Fig. 5.13). The strike of such sets is determined primarily by local stresses, and may show no affinity to the directions of regional stress.

Uplift joints often occur in distinct sets (Fig. 5.27a). Orthogonal relationships are common. Occaionally, uplift fractures overprint existing burial joints, but more often they develop along new directions. Some sets of uplift joints around Beer Sheva are aligned with previous single-layer burial joints, though other sets deviate considerably from these directions (compare Fig. 5.28 to Fig. 5.26 and Fig. 5.25). Figure 5.28 is related to Fig. 5.29.

5.6.1.2 Dip

Burial joints in a given outcrop mostly dip uniformly, they are vertical in undeformed strata. Syntectonic joints, both vertical and those that deviate from verticality often occur in close proximities. Uplift joints, which reflect

Table 5.8. Styles of fracture surface morphology on uplift and post-uplift joints

1 No	2 Style of fracture surface morphology	3 Rock description	4 Rock stratigraphy	5 Location	6 Joint inclination	7 Vertical size of marking (m)	8 Horizontal size of marking (m)	9 Special remarks	10 Interpreted process	11 Source or figure number in text
1	Radial striae and hackles	Granite	Precambrian	Sinai	Non-in situ block, assumed vertical	0.5	0.5		Neogene uplifts	Fig. 4.20a
2	"Giant" elliptical undulation	Sandstone	Palaeozoic	Sinai east	Elliptical long axis subhorizontal	10	25		Neogene uplifts	Fig. 5.19
3	Asymmetric concentric undulations	Sandstone	Palaeozoic	Sinai north	Vertical	2	0.7		Neogene uplifts	Fig. 5.23
4	Compound-circular undulations	Sandstone	Palaeozoic	Sinai north	Vertical	4	3.5		Neogene uplifts	Fig. 4.2b
5	Asymmetric non-concentric divergent undulations	Chalk	Senonian	Upper Galilee	Vertical	2	9		Uplifting stage	Fig. 5.21
6	Asymmetric concentric undulations and curved en echelon fringe	Marl	Maasstrichtian	Central Negev	Vertical	3	4		Uplifting stage	Fig. 5.20a
7	Partial concentric undulations	Indurated chalk	Middle Eocene	Beit Govrin	Vertical	4	3		Uplifting stage	Fig. 5.20b
8	Concentric semi-elliptical undulations	Indurated chalk	Middle Eocene	Shivta area	Vertical	0.4	2	Mark over-printing	Uplifting stage	Fig. 5.18b
9	Concentric undulations, delicate plume and radial hackles	Indurated chalk	Middle Eocene	Beit Govrin	Horizontal fracture		2×1.4	Sheet	Uplifting stage	Fig. 4.4
10	Vicinal plumes	Indurated chalk	Lower Eocene	Beer Sheva	Inclined	0.8	0.4	Inclination similar to topography	Sub-Recent post-uplift	Fig. 5.24

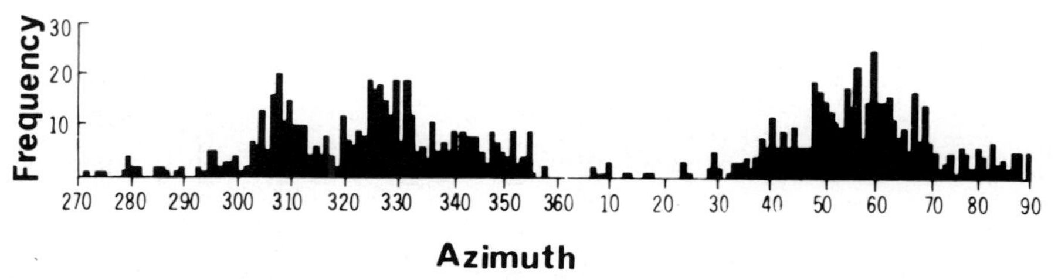

Fig. 5.25. Strike orientation histograms of some 1330 burial joints in Lower Eocene chalks around Beer Sheva. Median directions of the cross-fold and strike fractures as in Fig. 5.26

post-folding and post-faulting fracturing episodes, are generally vertical, and may appear in close association with vertical faults.

5.6.1.3 Height

Burial joints are mostly confined to single layers. Occasionally, they cross more than one bed, possibly because at the time of fracturing layer separation did not yet exist. Burial joints may therefore be no higher than several cm in thin clay beds, and seldom exceed 1 m in thicker layers.

Syntectonic joints are likely to develop a long time after burial, possibly because the ambient pore pressures are not sufficient for initiation of early fracturing. This development seems mostly to occur in unstratified thick layers (10 m or more) and less frequently in thin (0.5 m or less) stratified beds. Such joints cut across entire unstratified or badly stratified exposures.

Uplift joints are characteristically high, cutting through entire outcrops, stratified as well as unstratified. As a rule they propagate downward, more rarely these joints propagate upward or sidewise.

5.6.1.4 Length

Burial joints are mostly between 20 and 200 cm long (Fig. 3.17 a–c). Syntectonic fractures range from several cm to several tens of m (Figs. 3.42 and 3.18),

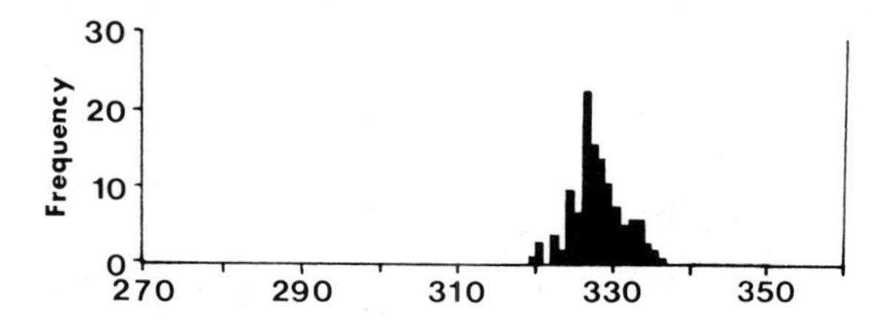

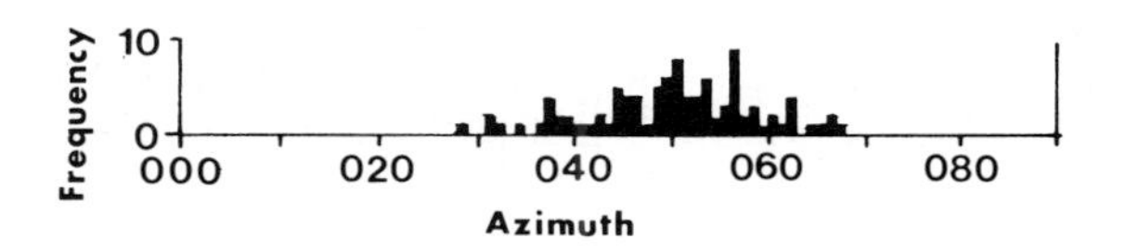

Fig. 5.26. Strike orientation histogram of single-layer, cross-fold burial joints around 328°, and strike fractures around 058° in a Lower Eocene chalk outcrop near Beer Sheva (Fig. 5.29, station 20)

Fig. 5.27 a, b. Uplift joints in southern England. **a** Vertical fractures oriented 290° cut the entire chalk outcrop at Captain Digby beach, Margate, south England. **b** In Campton Bay, Isle of Wight, south England, the layers are tilted close to vertical (see dip of the chert beds). The rock is traversed by joints which parallel the bedding, as well as by late horizontal joints and by inclined joints sub-parallel to the cliff face. *Picture width* 2 m

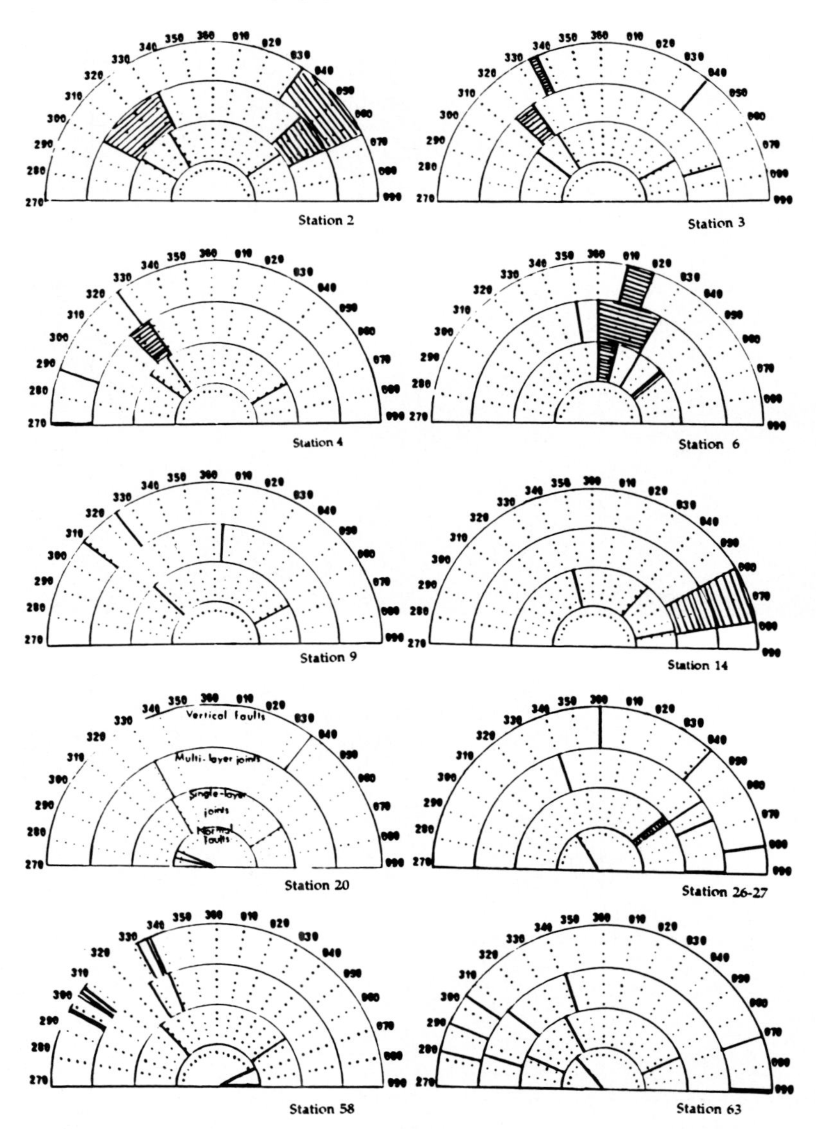

Fig. 5.28. Rose diagrams comparing strikes of four failure structures in ten stations from the Lower Eocene chalks near Beer Sheva. Average strikes for normal faults, single-layer joints, multi-layer joints and vertical faults are shown as heavy lines in successive arcs for each diagram (see details in station 20). In some stations ranges of strikes are shown by *strips* rather than lined. *Station numbers* correspond to Fig. 5.29

and may even attain greater length, as in the Entrada sandstone, Utah. Uplift joints generally range from several m up to tens (hundreds) of m in length.

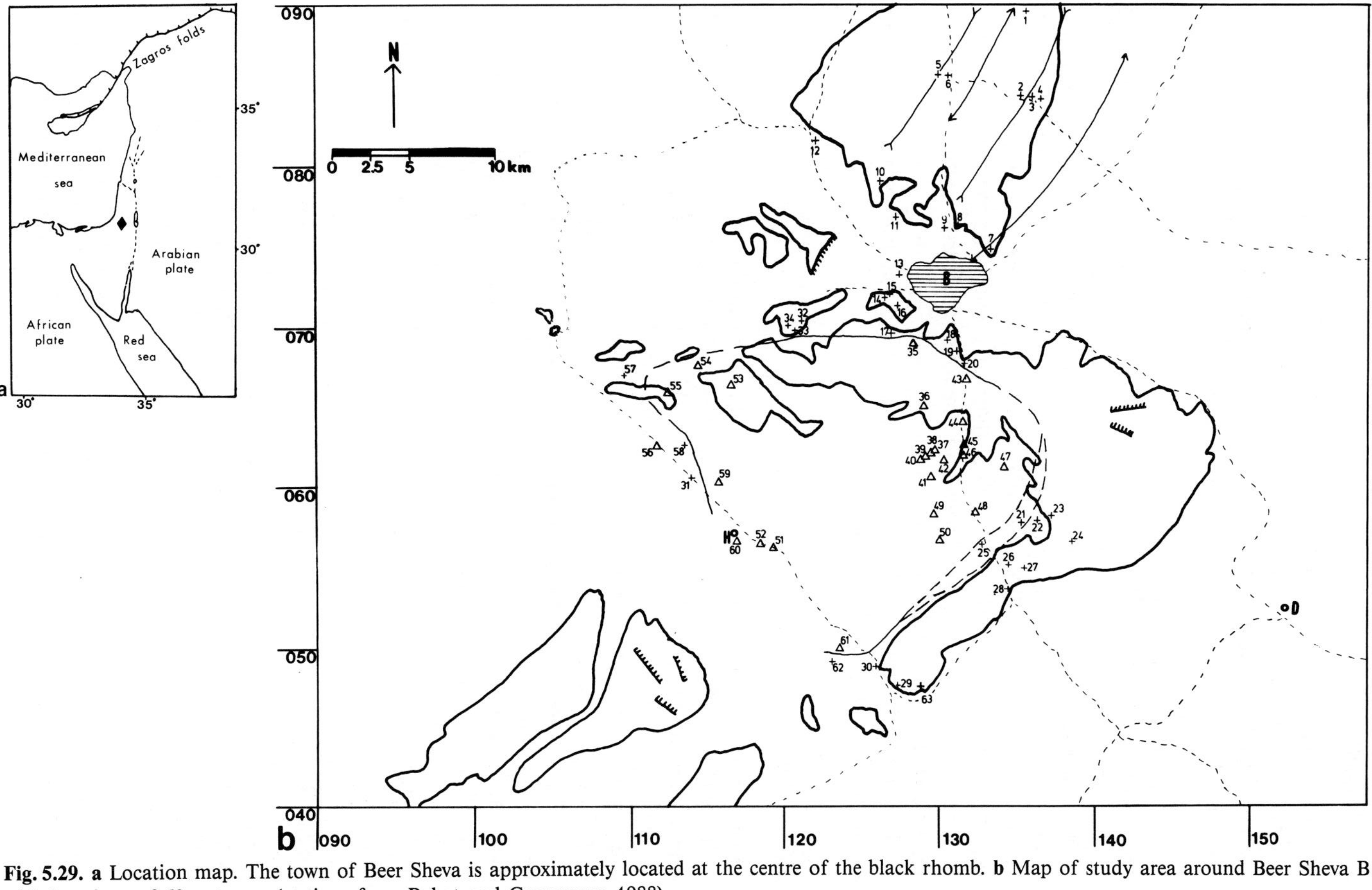

Fig. 5.29. a Location map. The town of Beer Sheva is approximately located at the centre of the black rhomb. **b** Map of study area around Beer Sheva B with locations of 63 outcrops (stations from Bahat and Grossmann 1988)

5.6.1.5 *Joint Shape*

Burial joints generally develop under uniform regional or areal (more localized) stress conditions. Accordingly, they are usually smooth, straight and planar (Fig. 3.10), although occasionally they are bent along the strike. The fractures are mostly square or rectangular, not circular. Their shape is dictated by layer boundaries and previous burial joints.

Syntectonic joints also develop under either regional or areal stresses. The regional stress produces flat joints. The results of areal stronger stresses may be bunches of joints in which single joints are either straight and flat or irregular in shape (Fig. 5.14). Discoids occur in a variety of shapes (Figs. 3.42; 3.43), they generally are categorized as syntectonic joints. Large joint surfaces, either syntectonic or uplift-type, occasionally display bends either along the dip (Fig. 5.19) or along the strike. The shape of syntectonic joints is influenced to some extent by layer boundaries and previous structures. Syntectonic joints may be square, rectangular or circular.

The shape of uplift joints may be influenced by topography when they start at the surface (see next section).

5.6.1.6 *Aspect Ratio (Length/Height)*

Burial joints, when confined between layer boundaries, propagate either radially (Fig. 5.1 b) or horizontally (Fig. 3.10). The aspect ratio differs accordingly, from about 1 in the former up to about 5 in the latter.

Syntectonic joints often are larger than burial joints. The aspect ratios of these joints vary from about 1 to about 10, but occasionally they are higher (Fig. 3.18).

No aspect ratios can be determined for large multi-layer joints that extend beyond the outcrop. For the many uplift joints that seem to start at the surface, they are assumed to be downward-propagating hemicircular fractures (Fig. 2.31), presenting an aspect ratio of about 2 with maximal length at the surface. However, larger aspect ratios are inferred for joints in Middle Eocene limestones near Sede Boqer (Sect. 5.4.2) and possibly also for some sandstone formations (Hodgson 1961 b).

5.6.1.7 *Spacing*

In burial joints spacing varies systematically as a function of layer thickness. A common spacing ranges from 10 to 60 cm. There are also systematic differences in spacing between cross-fold and strike joints when they occur in the same layer at the same location (Table 5.4). Spacing is a little wider in the strike joints than in the cross-fold joints.

Syntectonic joint spacing in the Santonian chalks (Menuha formation) of the northern Negev and Judean Desert shows a connection between spacing and the proximity of post-genetic deformations. Spacing diminishes from tens of cm to several mm near zones of strong bending (Fig. 5.6), or it may change in concordance with its proximity to a fault or the curvature of the fault (Fig. 5.13). Dyer (1983) describes narrow joint zones, 0.5 – 2.0 m wide, in the Moab Member of the Entrada Sandstone, Utah where spacing is from 0 to about 1.5 m.

Unlike burial joints, spacing in uplift fractures does not vary systematically with thickness, but ranges apparently unsystematically from 1 to 10 m or more. In uplift fractures spacing is not influenced by the proximity of faults.

5.6.1.8 The Ratio of Joint Length to Joint Spacing

The ratio of joint length to joint spacing, L/S, varies characteristically according to joint type. In burial joints it is low, varying between 2 and 10. Syntectonic joints are often quite long and spacing in these fractures is usually small to very small (Figs. 5.7; 5.6b, respectively). Therefore, the L/S ratio is high (about 20) to extremely high (>100). Discoids deviate, of course, from this rule. In uplift joints, L/S values generally range between those of the burial and syntectonic stages.

5.6.1.9 Fracture Width and Mineralization

Burial joints are formed under conditions of low lateral stresses and increasing overburden pressure, and are generally closed (Fig. 5.5). Water circulation through the joints is restricted and there is little or no mineral precipitation along the joints during the burial stage. Joint filling occurs along cross-fold burial joints of the 335° set in the Devonian siltstone from the Appalachian Plateau, with plumes that appear inscribed into the mineral film, which is less than 1 mm thick. Almost no mineralization occurs along burial joints of the Lower Eocene chalks around Beer Sheva.

Syntectonic joints develop in response to considerable lateral stresses, often in association with folding and/or faulting which may be accompanied by high pore pressures. The result may be fairly pervasive fluid circulation and crack opening, accompanied by mineral fillings along the fractures. Fillings of 1 to 20 mm are common. Some syntectonic joints (set Ia at Watkins Glen) lack mineralization.

Uplift joints develop under tensional conditions. Therefore they tend to be open (several cm or even several tens of cm), and to be mineralized or filled with soil. In some areas, lack of mineralization implies opening above the water table.

5.6.1.10 Joint Behaviour with Respect to Local Tectonic Structures

Since burial joints develop fairly early in the rock history, they generally predate other structures like folds or faults and bear no genetic relation to them.

Syntectonic joints, on the other hand, have genetic relations to local tectonic structures, and features such as orientation, spacing and shape often vary with distance from these structures.

Uplift joints generally post-date the earlier tectonic structures, and disregard them. They cut in similar style through series of thin layers and single thick layers, characteristically with wide spacings, unaffected by intervening structures. There are, however, instances where uplift joints overprint previous joints of the same orientation. Topography and surficial stresses may also influence the behaviour of shallow uplift joints.

5.6.1.11 Influence of Regional Structures and the State of Crustal Stress on Joint Behaviour

As mentioned, burial joints may often be divided into cross-fold and strike sets (e.g. Engelder and Geiser 1980). The former sets trend sub-orthogonal, the latter sets sub-parallel to the regional fold axis. However, deviations up to about 20° from orthogonality and parallelism appear to be more the rule than the exception, representing variations in the directions and magnitudes of stresses in the region (Engelder and Geiser 1980; Bahat 1986a; Hancock et al. 1987). The ubiquity of burial joints imply extensional conditions of fracture [Eq. (5.3)].

Syntectonic joints appear to be influenced by local structures and by lithologic factors, at least as much as by the main directions of regional stress. The unpronounced orthogonality in syntectonic joints (Fig. 5.8) possibly implies that these fractures develop by combined effects of horizontal stress and lithostatic stress state (McGarr 1988) following an inelastic relaxation of the deviatoric stress from the burial state of stress expressed by Eq. (5.3).

A strict correlation of uplift joint orientation with regional stresses may be misguided. On a regional scale some uplift joints conform to the regional patterns and others deviate considerably from them (Fig. 5.28). Local influences are often the more decisive. Uplift joints quite possibly develop tensionally in lithostatic or compressional states of stress (Sect. 5.4.5.1). Some distinctive features of burial, syntectonic and uplift joints are summarized in Table 5.9.

5.6.2 Fracture Surface Markings

5.6.2.1 Plumes

Straight horizontal to sub-horizontal, wavy and rhythmic plumes (Fig. 3.2a–c) are common on burial joints, less frequent on syntectonic joints, and rare on uplift joints. Radial plumes (Fig. 3.2d) are common on burial and syntectonic joints, and rare on uplift joints. Vertical plumes occasionally appear on uplift joints (Fig. 5.20d).

5.6.2.2 Undulations

Annular undulations are common on burial (Fig. 5.1b) and syntectonic joints (Figs. 3.22 and 4.2a), rare on uplift joints. Non-annular undulations (e.g. Fig. 3.23) are uncommon on burial and syntectonic joints, common on uplift

Table 5.9. Distinguishing features of burial, syntectonic and uplift joints

	Division into sets	Conjugate sets	Orthogonal sets	Dips in an outcrop	Height	Maximum length	Shape (flatness)
Burial joints	Generally fairly well defined	Positively established	Positively established	Uniform	Several cm to 1 m, being confined to single layers	Several m	Straight and flat
Syntectonic joints	Varies from good to poor definitions	Not established	Question-able	Occasionally non uniform	Generally several m, but may also be confined to single massive layers (25 m thickness)	Tens m	Vary from flat to conic or ir-regular and anastomotic
Uplift joints	Varies from very good to poor definitions	Not established	Positively established	Uniform	Up to tens of m, unconfined to layers	Many tens m	Generally straight and flat, but may be somewhat folded

	Aspect ratio (length/height)	Spacing	Ratio of joint length to joint spacing L/S	Joint opening and mineralization	Response to local structures	Response to regional structures
Burial joints	From 1 to 5	Regular and systematic, generally tens cm	Very low, be-tween 2 to 10	Joints are close and mineralization is generally poor	Unaffected, occa-sionally being cut by later faults	Directly affected
Syntectonic joints	Generally high (>10) but may also be low or very high	Characteristically irregular, varies from several mm to several tens cm	Ranges from high 20 to very high $\gg 100$	Joints are open and mineralization is considerable	Affected, joint orientation, is in-fluenced by existing structures	Directly and in-directly affected
Uplift joints	May vary from 1 to very high ($\gg 10$)	Varies from regular to ir-regular, ranges from several m to several tens m	Intermediate between burial and syntectonic ranges	Joints are open and mineralization and soil fillings are considerable	Generally unaffected but may overprint previous joints, may be influenced by topography	Generally af-fected, but occa-sionally the rela-tionship is com-plicated

joints. Divergent undulations (Fig. 3.24) are uncommon on burial joints, rare on syntectonic joints, and fairly common on uplift joints.

5.6.2.3 Transitional Markings

Transitional markings (Fig. 3.6b) which probably result when plumes and undulations are synchronously formed on the same surface (Bahat 1987a) occur occasionally on burial joints and are rare on uplift joints.

Intriguing transitional markings have been observed on certain joints. In contrast to analogous markings on burial joints, which have corresponding symmetries where the plume axis approximately coincides with the radii of the rib curves, on some syntectonic joints and fractures in limestone concretions the two axes seem to be independently oriented, as if formed separately.

5.6.2.4 Maximal Size

Fracture markings are generally not more than 1 m high on burial joints. Size is limited by layer boundaries, and the development of burial joints seems to be inhibited in thick layers. Generally, markings on burial joints reaches several m in length (Fig. 3.17). Both height and length of fracture markings may reach tens of metres on syntectonic joints (in sandstone formations) and on uplift joints (Fig. 5.19).

5.6.2.5 Fracture Initiation and Direction of Propagation

Circular rib markings may initiate anywhere on burial joints: at layer boundaries, edges of previous joints (Fig. 3.21) or within the layer. They propagate radially. Plumes initiate in similar fashion, and generally propagate subparallel to the layer boundaries.

On syntectonic joints, circular rib markings initiate either within the layer or from edges of previous joints. Initiation at layer boundaries is rare. Fracture propagation is mostly radial. Plumes also initiate within the layer, but they mostly propagate radially. Vertical propagation of plumes is rare.

Undulations are common on uplift joints. Characteristically, they start closer to the upper than to the lower layer boundary, but not at the edges of previous joints. They propagate radially downward (Figs. 5.19, 5.20). Rare vertical plumes indicating vertical downward (Fig. 5.20d) or upward propagation (Fig. 3.3), and rare horizontal plumes implying horizontal propagation (Fig. 3.54) are known as well.

5.6.2.6 En Echelon Cracks

Horizontal plumes on burial joints commonly occur in association with en echelon planar cracks, forming straight fringes along the lower and/or upper layer boundaries.

En echelon cracks in curved fringes are fairly common in syntectonic joints in Upper Cretaceous chalks at Chepy, northeast France, and occasionally in chalks of Israel. They may mark the peripheries of circular undulations on syntectonic discoids. Straight horizontal fringes of en echelons occur along boundaries of massive beds in the Entrada sandstones, Utah.

On uplift joints, en echelons appear in semi-circular fringes along the lower parts of ribs (Figs. 5.20), as well as in vertical sets (Fig. 5.17). The more distinctive fracture surface features on burial, syntectonic and uplift joints are summarized in Table 5.10.

5.7 Propagation, Opening, Mineralization and Joint Intensity of the Various Joint Types

5.7.1 Propagation, Opening and Mineralization

Joint propagation and joint opening do not necessarily coincide. Fracture markings develop during propagation, and at this stage the two matching sides of the joint usually remain intact. Opening of the joint may or may not occur at a later stage, and the time interval between propagation and opening can vary from zero to indefinite. Growth of crystals orthogonal to the walls of a joint indicates mineralization that accompanies opening (Ramsay 1980). On the other hand, random orientation of crystals may imply post-opening mineralization.

In burial joints, opening and mineral precipitations are generally slight (< 1 mm) or absent (Fig. 5.5). Any opening belongs to a later phase. The time interval between opening and mineralization can be very long. No mineralization appears on burial joints around Beer Sheva; except for some random, non-uniform iron oxide stainings on joints in Middle Eocene chalks, evidently of post-fracturing age. In the Appalachian Plateau, a thin film of precipitate occurs on C-type plumes (see Fig. 3.7), faithfully preserving the fine relief of the fracture markings. It is not known whether the film was contemporaneous with or subsequent to initial fracturing. Secondary fracturing may cut through the mineral precipitate (Ramsay 1980).

Narr and Burruss (1984) have investigated the genesis of burial fractures in reservoir carbonates from Little Knife Field, North Dakota. These planar joints range from 15 to 46 cm in height, with less than 1 mm separation. Narr and Burruss consider that "predating and/or coincident with dolomitization is an early episode of fracture formation, possibly induced by the volumetric changes that might have accompanied dolomitization". These authors also contemplate the possibility that fracture post-dated dolomitization. Hence, the time lapse between burial jointing and mineralization is peculiar to each case. Derre et al. (1986) report on multiple observations of granites in which joints originated long before the opening and infilling.

In syntectonic joints, fracture propagation is distinguishable from joint opening. In the Coniacian chalks of the Chepy quarry in the Paris Basin, joint opening virtually did not occur, and it is generally smaller than 1 mm. In the

Santonian chalks from Israel, on the other hand, joint opening is slight in some outcrops and quite large (1 cm or more) in others, and so are lateral displacements along the joints. Opening and shearing in the Santonian chalks operated during the long syntectonic phase, which included repeated tectonic events, that produced the Syrian Arc. Mineralogical and geochemical data on fillings from these Santonian syntectonic joints (Sect. 5.3.1.1) and from uplift joints in Eocene outcrops in the Negev (Issar et al. 1988, see below, this section) suggest similarities of the infilling processes, implying that opening and mineralization of the syntectonic joints was much later than the fracturing itself (see also Sect. 5.3.1).

The iron oxides that stain surfaces of discoids in the Ramon flint clays display various colours (Sect. 5.3.8), which may testify to several episodes. A single staining episode would expectedly produce more uniform colors. Tillman (1983) describes syntectonic fractures in the Appalachian region, filled by veins which display a diversified mineralization sequence, formed in several episodes. Altogether, syntectonic joints display varied time sequences of fracture propagation, fracture opening and mineral precipitation, with evidence for repetition of events.

Table 5.10. Distinguishing features of fracture markings on burial, syntectonic and uplift joints

	Horizontal plumes	Radial plumes	Vertical plumes	Wavy or rhythmic plumes	Circular annular undulations	Circular nonannular undulations	Divergent undulations
Burial joints	Characteristic on conjugate and coaxial sets	Fairly common	Absent	Common	Characteristic on coaxial sets	Occasional	Uncommon
Syntectonic joints	Uncommon	Common	Uncommon	Uncommon	Characteristic	Occasional	Uncommon
Uplift joints	Uncommon	Uncommon	Occasional	Rare	Very rare	Characteristic	Characteristic

Belfield et al. (1983) investigated the Miocene Monterey Formation along the Santa Barbara Channel coastline of California. They found that "fractures developed during burial and prior to regional folding are insignificant both in terms of volumetric importance and as hydrocarbon transporting conduits, compared to the fractures interpreted as having formed during regional folding". The distinction between burial and syntectonic fracturing and their respective opening and volumetric importance in the Monteray Formation correspond quite well with the analogous differences in the Israeli Eocene and Santonian chalks.

Uplift joints in the Beer Sheva area are thought to be formed due to the Tertiary uplift. They are typically large multi-layer joints. Multi-layer joints that cut through Eocene chalks are often mineralized, with a 1-mm film of fine-grained carbonate (aragonite and calcite) lining the joint planes, and coarse sparitic gypsum filling the rest of the vein. The carbonate lining can perhaps be roughly co-eval with the jointing process. The rest of the fill appears to be derived from the leaching of eolian soils during the Late Pleistocene (Issar et al. 1988). Thus the latest opening, if not most of it, can be dated in this case to a sub-recent event. The same conclusion applies to some syntectonic joints cutting the Santonian chalks (Sect. 5.3.1.1).

Sub-vertical arrest lines	Transitional markings (plumes and undulations)	Maximum size (m)		Marking initiation	Propagation styles	En echelon fringes	Hackles
		height	length				
Common	Fairly common	1	3	Layer boundaries, inside the layer, surface of previous joint	Axially horizontal, radial, wavy, rhythmic	Occasionally horizontal, parallel to layer boundaries	Very rare
Common	Fairly common	10	40	Inside the layer, layer boundaries, surface of previous joint	Radial, axially horizontal, vicinal, propagation from multinucleation sites	Occasionally concentric, or curved convexed to all directions	Rare
Uncommon	Uncommon	50	50	Inside the formation or previous joints	Concentric with strong emphasis downward	Occasional in curved rims at lower parts of marking, convexed downward	Occasional

Burial joints and syntectonic joints are interpreted as rock failure due to pore pressure of the internal fluid, which is also the driving force that causes joint propagation. Uplift joints are the product of tensional stresses generated by the relaxation of confining pressures, upbulging, erosional unloading and other surface-related processes.

5.7.2 Joint Intensity

Joint intensity is an expression of joint area per unit rock volume, given as cm^2 joint surface per cm^3 of rock. There are additional definitions of this term (Kulatilake 1988). According to Wheeler and Dixon (1980) "Joint intensity affects bearing strength and slope stability of rock masses and is a consideration in mine design. Intensity affects the ability of a rock mass to transmit and hold fluids such as ground water, pollutants, oil, gas and perhaps some mineralizing solutions".

Joint intensity is a useful formal parameter in all above applied fields, but it could be a more meaningful index if assessed separately for each of the three major joint categories. Morphologic differences between the various joint types (Sect. 5.6.1) suggest, for example, that uplift joints have potentially superior drainage properties compared to burial sets.

Following this trend of thought, a model of the hydraulic conductivity in the (almost) unfolded Eocene rocks from the Beer Sheva syncline (Sects. 5.4.1 and 5.4.2) should possibly consist of three horizontal compartments for the successive rock units. The following considerations apply to these compartments: (1) lower and upper hydraulic constraints, with perhaps local stochastic changes in the hydraulic conductivity, should serve as the mathematical boundaries on the heterogeneous fracture system of the Mor Formation (Fig. 5.15 a). (2) The unstratified part of the Horsha Formation consists of chalks with a rather uniform distribution of joints in the upper part of the formation and relatively uniform distribution of lags in its lower part. This may justify the postulation of two average values for the required fluid transport parameter. (3) Quite a heterogeneous fracture system occurs in the stratified upper part of the Horsha Formation. Liquid flow in these rocks is facilitated by burial, syntectonic and uplift joints (Fig. 5.14). A mathematical approach similar to the one suggested for the Mor Formation is applicable to the stratified part of the Horsha Formation. Horizontal layer boundaries (Fig. 5.22) and horizontal fractures (Figs. 3.6 a) are considered to be essential postulates in the analysis of each compartment and of the combined compartments.

5.8 Approximate Maximum Depths at Which Various Joint Types Develop

Hubbert (1951) suggests that the vertical stress at a given depth is equal to overburden pressure and that there is a critical depth for every material below which the horizontal stress is compressive. Consequently, absolute tension is impossible below very shallow depths in the crust. Griggs and Handin (1960) consider tensile stresses to be unlikely to occur at depths greater than a few hundreds of metres. Secor (1965), on the other hand, suggests that tension fractures can develop at increasingly greater depth in the Earth as the ratio of fluid pressure to overburden weight λ approaches unity, and the hydrostatic fluid pressure distribution (where λ is between 0.4 and 0.5) will permit the development of tension fractures to a depth of several thousand metres. Building on Secor's theory, Engelder (1985) infers the formation of tectonic joints at depths between 0.5 km and 2.5 km, and hydraulic joints at 5 km depth (Sect. 5.2.3). Engelder and Oertel (1985) suggest that unloading joints (Sect. 5.4.3) in the Appalachian Plateau occur only within the upper 0.5 km of the crust. Price (1959) and Nur (1982), on the other hand, claim that large joints that result from uplift may penetrate as brittle fractures down to several kilometres into the crust.

Burial joints develop in extension during subsidence and they may form by pore pressure at considerable depth, provided that the liquid pressure can build up and transmit sufficient effective stresses to induce fatigue cracking. Narr and Burruss (1984) estimate the depth of fracturing at 3 km or deeper. Future research may determine to what extent the presence or absence of horizontal en echelon fringes can help to determine the thickness of overburden during fracture (Sect. 5.2.2.2).

Syntectonic joints generally develop by extension during episodes of tectonic deformation. Like burial joints, they may develop at considerable depths (several km). Syntectonic joints spaced at several cm are abundant at the bottom of the Glacier Gorge along the creek near Grindelwald in the Alps. These fractures cut the Oehrli Upper Jurassic limestone around the base of the Eiger Mountain at an elevation of about 1000 m. This mountain rises to an elevation of about 4000 m. Some syntectonic joints however (e.g. set 062°, Sect. 5.3.7) are thought to have developed during stress relaxation, possibly associated with uplifting. Such fracturing may have occurred at very shallow depths (tens of m).

Uplift joints generally develop due to tension close to the surface. When the extent of uplift is limited, uplift joints may be confined to very shallow depths. In Middle Eocene chalks south of Beer Sheva, uplift joints seem to be significantly fewer at depths of 50 m compared with their frequency at surfacial outcrops. It appears that mesoscale uplift joints do not penetrate deeper than a few hundred metres. Post-uplift joints develop at the surface and are influenced by geomorphologic processes.

Thus, large mesoscale joints that are encountered at considerable depths are unlikely to be due to uplift fracturing; unless the extent of uplift was especially great, they are more likely syntectonic joints.

5.9 Fracture Interaction

Neighboring joints display a wide variety of geometric interrelation styles, from straight parallel fractures that do not interact at all through adjacent fractures whose tips curve toward each other, to anastomosing crack complexes of high connectivity. A special case of fault-joint interaction is presented in Section 5.3.6.

5.9.1 Extents of Joint Interaction

Sets of uplift joints consist mostly of widely spaced, straight joints that do not interact. Joints of set 344° that are closely associated with a vertical fault do not interact, in spite of their close spacing and their considerable curvature (Fig. 5.13). Joints that cut Santonian chalks in Israel (Fig. 5.6) and Upper Cretaceous chalks in the Paris Basin, even though closely spaced, are straight and almost do not interact. Common to these three examples are their syntectonic characteristics (Sect. 5.3.3.2).

Cross-fold joints from the Appalachian Plateau, New York State (Sect. 5.2.3) occasionally interact with each other (Bahat 1983, Fig. 7). The fracture interactions within the Ib and Ia sets are generally suborthogonal in style (Fig. 5.30b). Asymptotic (Fig. 5.30a) and tip-to-plane interactions were also observed between Ia joints.

No interaction occurs among the burial joints of the Lower Eocene chalks around Beer Sheva (Sect. 5.2.1). In set 328° this is not surprising, considering the short lengths of these joints and their blocky structure (Fig. 5.5). However, lack of interaction is also evident in the longer joints of sets 309 and 344°.

Joints of the burial set 028° from the Middle Eocene chalks near Beer Sheva (Sect. 5.2.2) interact quite intensively. A study of some 160 terminations of non-coplanar joints from this set (Bahat 1987b) shows that the linkage of

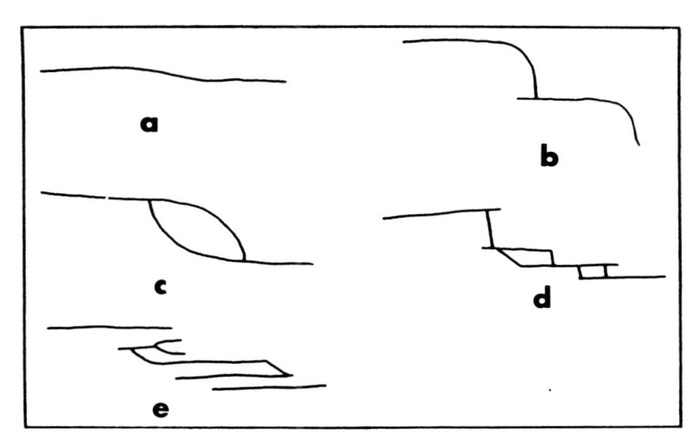

Fig. 5.30a–e. Four fracture interaction types of joints in Middle Eocene chalks near Beer Sheva. **a** Asymptotic abutting curve. **b** Suborthogonal abutting curve. **c** Interaction between two oppositely propagating cracks along curved trajectories (often termed tip-to-plane). **d** Straight to suborthogonal, occasionally oblique crack connections between parallel joints. Connections may be single or multiple. **e** Five overlapping en echelon cracks. Some cracks do not form contacts, others are connected by straight fractures. The two latter interactions represent variations of the straight interaction style

fractures by interaction may basically be grouped into four types (Fig. 5.30). The relative frequencies of these interaction styles in chalks (Table 5.11) shows that the straight end member is the most common among the various linkage styles.

The common occurrence of straight linkages (Fig. 5.30d) seems to imply that the en echelon cracking, which is very common in this set, is not sufficient to release all the energy stored in the rock, and some retained stress expresses itself by a delayed fracture.

It is quite intriguing to find two non-coplanar joints that link tip-to-plane, but the plumes on their surfaces show propagation in opposite directions. That is, on the joint closer to the observer in Fig. 3.34 the plume initiated at or close to the contact between the two fractures and propagated away from the contact. This perhaps indicates that on rare occasions the tip-to-plane linkage is a consequence of crack initiation rather than crack termination.

Set 062° from the Middle Eocene near Beer Sheva (Sect. 5.3.7) displays anastomosing geometry. This set appears as interacting cracks that form clusters of braided zones rather than distinct individual fractures.

Table 5.11. Frequencies of different interaction styles of fracture terminations between non-coplanar joints of set 028° in chalks

Fracture termination style		Platform 1		Platform 2		Combined data	
		Frequency	(%)	Frequency	(%)	Frequency	(%)
A1	Asymptotic curve abutting (Fig. 5.30a)	7	8	7	9	14	9
A2	Sub-orthogonal curve abutting (Fig. 5.30b)	12	13	10	14	22	14
A3	Interaction along curve trajectories (Fig. 5.30c)	8	9	4	5	12	7
B1	Sub-orthogonal connection by single traverse crack (Fig. 5.30d)	30	34	28	37	58	35
B2	Sub-orthogonal connection by two traverse cracks (Fig. 5.30d)	5	6	3	4	8	5
B3	Sub-orthogonal connection by an oblique crack (Fig. 5.30d)	7	8	5	7	12	7
C	Overlapping en echelon cracks (Fig. 5.30e)	20	22	18	24	38	23
Total		89	100	75	100	164	100

Platforms 1 and 2 are diagramatically shown in Fig. 5.3a.

5.9.2 Conditions of Joint Interaction

The wide spectrum of joint interaction styles described above exemplifies mostly fracture in chalk. Thus, it is obvious that lithology has no influence on the style or intensity of interaction, which appears to depend more on other geological conditions and peculiarities of the stress field. It seems that common to the non-interacting joints that cut the Lower Eocene chalks around Beer Sheva (Sect. 5.2.1.3) and the Santonian chalks in the Judean Desert (Sect. 5.3) was a persistent regional NNW compression. This compression was coaxial or at low angle to the strike of the joints and resulted in considerable stress differences (compared to the rock strength). Similar conditions but on a much smaller scale were determined in connection with set 344° that developed in association with a vertical fault (Sect. 5.3.6). The horizontal compression was coaxial with the direction of the 344° oriented joints (unpublished study). In the Paris Basin, large horizontal stress differences resulted in the regional Oligo-Miocene extensional events and the jointing described in Section 5.3.2.

Pore pressure has no influence on the advanced propagation of uplift joints subject to extension (Bahat and Rabinovitz 1988). Although interaction between uplift joints (viewed from overhead) has not been observed in Israeli chalks, this does not discount interaction altogether, since uplift joints generally propagate vertically. Therefore interactions should be looked for in vertical sections. A possible example of such an interaction is the bending of the large N-S joint in Fig. 5.19.

No stress conditions associated with the fractures of set 028° that exhibit intensive interaction have been determined. Possibly, pore pressure that was exerted via joint rotation (Sect. 5.2.4) under general conditions of small stresses and small stress differences at shallow depths (Bahat 1989) has promoted crack interaction. The joint rotation process in the Appalachian Plateau (Engelder and Geiser 1980) contributed to interaction of cross-fold joints. Compared to the intensity of fracture interaction in set 028°, the interactions in sets Ib and Ia were moderate, possibly due to the great depths at which these joints were formed.

Fig. 5.31. a Schematic illustration of propagation behaviors of two non-coplanar parallel cracks. Inner tips of both cracks are *X* and *X'*, and outer tips are *Y* and *Y'*. Stage a describes cracks before overlapping. **b** Shows propagation from *X* and *X'*. At *c* overlap is attained, propagation stops at *X* and *X'* and proceeds from *Y* and *Y'*. Stage *d* represents a finite body. After the arrival of *Y* and *Y'* at the edges of specimen, interaction from *Z* and *Z'* proceeds. (Yokobori et al. 1971, Fig. 26). **b** Two parallel cracks caused by indentation (at *two dashed rings*) are simulated by stress wave. This simulation induces fracture propagation (at = 300 m/s). At the *centre*, prior to overlap, the cracks deviate slightly away from each other, then they grow towards each other and join. Distance between the cracks is 25 μm. Distance between indentations is 140 μm. (Swain and Hagan 1978). **c** The interaction behaviour of two faults *A* and *C* in the Gregory Rift, Kenya. Each fault consists of two oppositely propagating segments along curved trajectories. The angular deviation γ of the two segments and the K_{II}/K_I ratio for each fault follow the technique by Swain and Hagan (1978). The *X* and *Y* segments are represented by *crosses* and *dots*, respectively. (Bahat 1984)

a

b

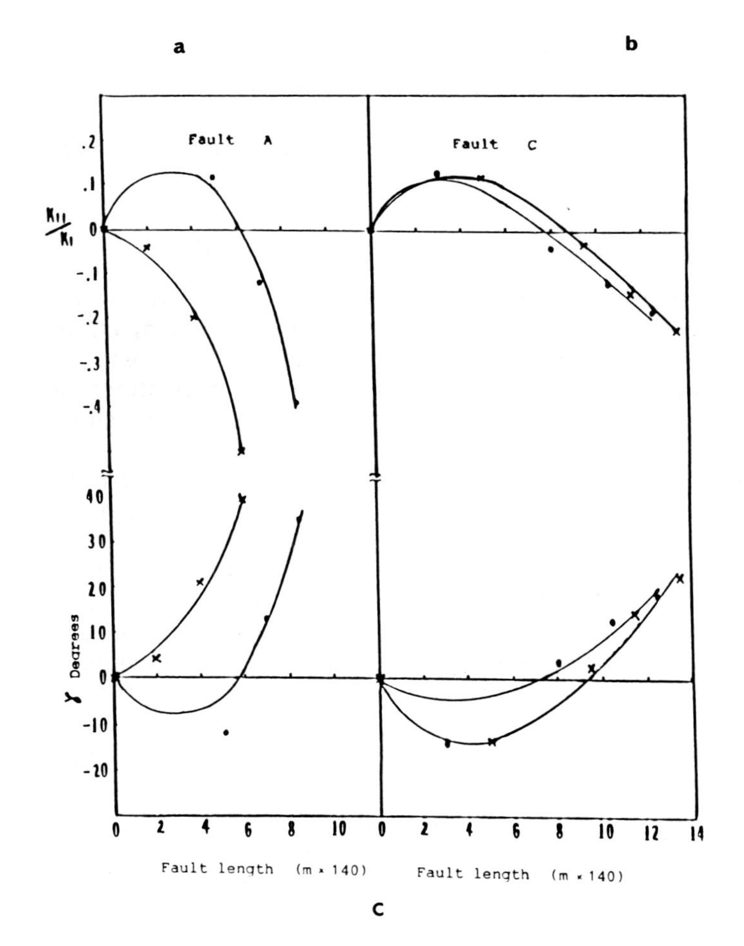
Fault A
Fault C
K_{II}/K_I
.2
.1
0
-.1
-.2
-.3
-.4
40
30
20
10
0
-10
-20
γ Degrees
0 2 4 6 8 10
0 2 4 6 8 10 12 14
Fault length (m × 140)
Fault length (m × 140)

c

The asymptotic abutting curve (Fig. 5.30a) results when a straight crack propagates toward a non-coplanar opposing second crack. When overlapping is about to occur, the propagating crack curves at a very small angle towards the stationary crack. The suborthogonal abutting curve (Fig. 5.30b) results under similar conditions, but the propagating crack curves around at a large angle, almost perpendicularly, to the stationary crack. These two interaction styles imply slow propagation. One non-coplanar fracture can be thought of as a free boundary at which the shear stress is zero (and the principal stresses are parallel and perpendicular to the boundary). Therefore, the abutting fracture (the younger one) tends to meet the existing surface either asymptotically or perpendicularly. Rapid fracture propagation, on the other hand, would have been diverted from small or large angles due to the various influences of shear and due to inertia.

The anastomose fracture pattern of set 062° reflects strong uniaxial tension at shallow depths that resulted from the release of the compressional stresses orthogonal to the synclinal axis (Sect. 5.3.7).

Thus, fracture interaction in chalk does not occur in burial joints that developed in extension in an orthorhombic symmetry (Wilson 1982), in syntectonic joints associated with intense deformation and in uplift joints. On the other hand, fracture interaction is intense among burial joints that developed at shallow depths resulting from joint rotation and among syntectonic joints formed due to tension.

5.9.3 Interaction Mechanisms of Experimental Cracks and Faults

The interaction between two propagating opposing cracks along curved trajectories (Figs. 5.30c and 5.31a) under tensile loading was investigated by Lange (1968) and Yokobori et al. (1971). The numerical method proposed by Yokobori et al. is in good agreement with their experimental observations on cellophane paper. They show that crack curving may occur under tensile loading due to local shear stresses close to the tip of the interacting cracks even without the influence of remote shear (Fig. 5.31a). Yokobori et al. determine the variation of the K_I and K_{II} stress intensities (Sect. 1.1.4.1) in the region of the overlapping cracks at various stages as the tips overlap.

Kalthoff's (1973) model (Fig. 2.40) may also be applicable for describing the curving paths of interacting non-coplanar cracks. As overlapping is attained, the two curves seem to repel each other, bending slightly away from each other, reflecting $K_{II}/K_I > 0$ condition (left side of Fig. 2.40). On overlapping, the cracks grow towards each other (right side of Fig. 2.40). These crack deviations occur as the shear senses are reversed in order to maximize the tensile stress at the two crack tips. Swain and Hagan (1978) observed similar interactions between non-coplanar cracks that link tip-to-plane in soda lime glass. Just prior to overlap, the cracks deviate slightly away from each other and then they converge strongly (Fig. 5.31b). Swain and Hagan show that the angular deviation from straight line growth γ for these cracks may be determined from Eq. (2.13, Fig. 2.40).

Bahat (1984) examined small faults from the Gregory Rift in Kenya assuming that their interaction followed the principles outlined by Swain and Hagan (1978). Figure 5.31c shows variations in γ and K_{II}/K_I [calculated by Eq. (2.13)] for two non-coplanar faults that have the morphology shown in Fig. 5.30c. The two deviating "cracks" X and Y of fault C reflect close to idealized conditions: γ shows initial divergent trends of the two cracks that then converge strongly. Commensurately, the K_{II}/K_I ratios indicate opposite shear senses as they cross the zero line. Fault A, on the other hand, shows a deviation from the idealized interaction: crack X does not diverge away from crack Y, and therefore, there is no negative γ. Also, there is no sign change in the K_{II}/K_I ratio. Hence, this analysis may help to elucidate the interaction mechanism and minor changes in shear versus tensile conditions close to the tips of faults. Such an analysis may also be applicable to joint interaction.

5.9.4 Extensional Branching of Faults

Fault tectonofractography is beyond the scope of this book, which primarily concerns jointing. Two exceptions are the brief reference to fault interaction in the previous section and the following discussion on branching (Sects. 1.4.4 and 2.1.4) of faults.

Kalthoff's (1973) observations of the angular relationships in branching (Sect. 2.1.4.3) were applied to macroscale fracture analyses. Faults produced during the 1968 earthquake at the Coyote Creek fault in California were intensively branched. The angular behaviour of the branching fractures in eight forks followed Kalthoff's theory unusually well (Bahat 1982). Kalthoff's angular relationships were generally followed by the branching faults in the Gregory Rift of Kenya (Bahat 1984). It was, however, found that most initial branching angles were $>28°$ and after some propagation the branches generally attracted each other and continue on sub-parallel planes. Crack initiations with branching angles lower than 28° were rare. Preston's (1935) measurements of forking angles (Sect. 2.1.4.3) have been compared by Bahat to branching angles of 44 faults in the Gregory Rift. A maximum, close to 45°, of the fault branching angles was interpreted as an indication of fracture by tension, a conclusion which corresponded with other observations.

A comparison of the branchings of faults in Kenya and in South California (Fig. 5.32a, b) with braided streams near the edges of melting glaciers (e.g. Press and Siever 1982, p. 168) and with the pattern of a theoretical modular tree (Fig. 5.32c) reveals that, possibly due to areal dictates, initial forks of large angles, which later produce further forks of smaller angles, are more common than the reverse sequence of forks that enlarge their initial small angles as they develop.

Rice's (1984) rebranching rate [Eq. (2.12), Sect. 2.1.4.2] may or may not be applied to dynamic fracturing which has to obey the rules of fracture mechanics. But in other systems, such as in plants, where branching serves the purposes of liquid transportation and mechanical support, it was observed, for

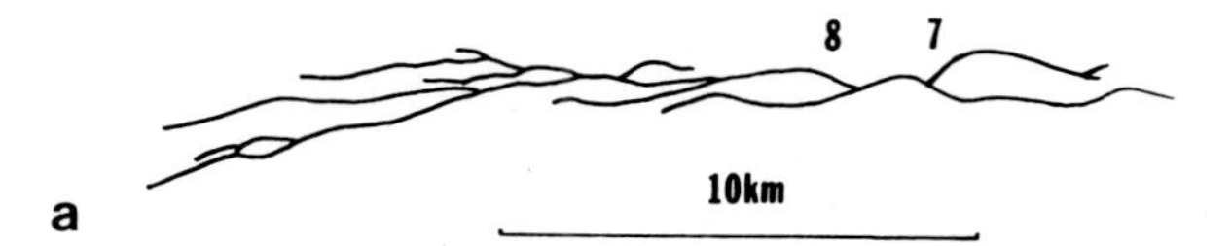

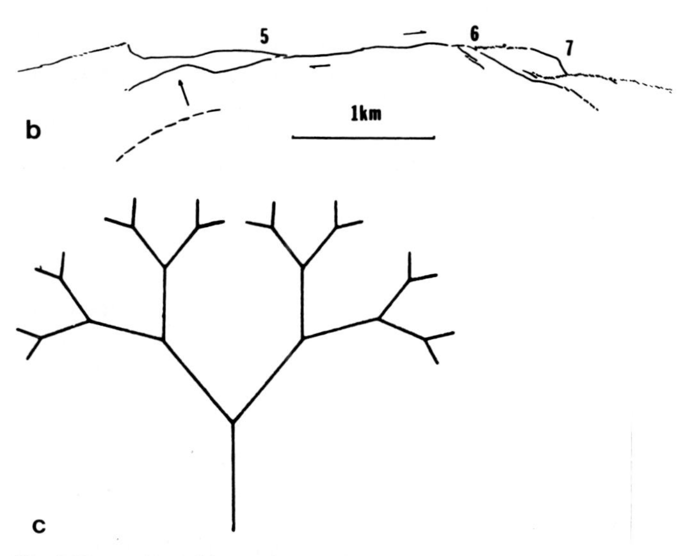

Fig. 5.32 a – c. Branching styles and fracture interactions. **a** Complex branching of faults from the Gregory Rift in Kenya (Nos. *7, 8); **b*** Complex branching (No *5*) and two additional forks (Nos. *6* and *7*) from the Koyote Creek fault, South California. The *dashed line* designates approximately E-W folding, which implies approximately local N-S compression that affected branching of No. 5. The direction of compression and the location of branching "deformation" is shown by an *arrow*. Note the differences in scale. (After Bahat 1982, 1984). **c** A modular tree. Similar but smaller units are added at the tip of each branch. This is repeated three times, starting from the large *Y* at the bottom of the sketch. (Stevens 1974)

example by Stevens (1974), that later branches turned out to be shorter and more numerous than early ones (Fig. 5.32c).

Suggestions pertaining to (1) the importance of bifurcation as an indicator of extensional faulting along the San Andreas fault system, to (2) the dominant tensile mode of faults at the surface, which is distinct from the shear mode at depth, and to (3) the local uplifts which cannot be explained by compression accompanying the strike-slip faulting were made by Bahat (1982) and Zoback et al. (1987). Some of the fractographic features outlined in this section merit additional studies.

References

Achenbach JD (1974) Elastodynamic stress intensity factors for a bifurcating crack. In: Sih GC, Van Elst HC, Broek D (eds) Prospects of fracture mechanics. Noordhoff, pp 319–336

Aharoni E (1976) The surface and subsurface structure of southern Judean Desert, Israel. Thesis, Hebrew Univ, Jerusalem, 100 p (in Hebrew)

Alexander KM, Taplin JH (1962) Concrete strength, paste strength, cement hydration and the maturity rule. Aust J Appl Sci 13:277–284

Alexander KM, Taplin JH (1964) Analysis of the strength and fracture of concrete based on unusual intensitivity of cement-aggregate bond to curing temperatures. Aust J Appl Sci 15:160–170

Alexander KM, Wardlow J, Gilbert DJ (1965) Aggregate-cement bond, cement paste strength and the strength of concrete. Proc Conf Struct of concrete and its behaviour under load, Lond, pp 59–91

Anderson EM (1936) The dynamics of the formation of cone-sheets, ring-dikes and caldron-subsidences. Proc R Soc Edinb 56:128–157

Anderson H (1969) Stress intensity factors at the tips of a star-shaped contour in an infinite tensile sheet. J Mech Phys Solids 17:405–417

Anthony SR, Congleton J (1968) Crack branching in strong metals. Metal Sci J 2:158–160

Atkinson BK (1984) Subcritical crack growth in geological materials. J Geophys Res 89:4077–4144

Atkinson BK (1987) Introduction to fracture mechanics and its geophysical applications. In: Atkinson BK (ed) Fracture mechanics of rock. Academic Press, Lond, pp 1–26

Atkinson BK, Avdis V (1980) Fracture mechanics parameters of some rockforming minerals determined using an indentation technique. Int J Rock Mech Min Sci Geomech Abstr 17:383–386

Atkinson BK, Meredith PG (1987) The theory of subcritical crack growth with applications to minerals and rocks. In: Atkinson BK (ed) Fracture mechanics of rock. Academic Press, New York, pp 111–166

Aydin A, DeGraff JM (1988) Evolution of polygonal fracture patterns in lava flows. Science 239:471–476

Bache HH, Nepper-Christensen P (1965) Observations on strength and fracture in lighweight and ordinary concrete. Proc Conf Struct of concrete and its behaviour under load, London, pp 93–108

Bahat D (1977a) Thermally-induced wavy Hertzian fracture. J Am Ceram Soc 60:118–120

Bahat D (1977b) Prehistoric Hertzian fracture of chert. J Mater Sci Lett 12:616–620

Bahat D (1978) Hertzian fracture, a sound physical basis for the explanation of carbonatite structure. Proc Int Colloq, Orleans, France. Earth Sci Measurements, Mem B. R. G. M. 91:275–283

Bahat D (1979a) Theoretical considerations on mechanical parameters of joint surfaces based on studies on ceramics. Geol Mag 116:81–92

Bahat D (1979b) On the African rift system, theoretical and experimental study. Earth Plan Sci Lett 45:445–452

Bahat D (1979c) Interpretation on the basis of Hertzian theory of a spiral carbonatite structure at Homa mountain, Kenya. Tectonophysics 60:235–246

Bahat D (1980a) A "giant" plumose marking in paleozoic sandstone in Sinai. Tectonophysics 68:T1–7

Bahat D (1980b) Secondary faulting, a consequence of a single continuous bifurcation process. Geol Mag 117:L373–380

Bahat D (1980c) Plumose markings on joints surfaces from Cenomanian, Senonian and Eocene chalks in Israel. Isr J Earth Sci 29:171–174
Bahat D (1982) Extensional aspects of earthquake induced ruptures determined by an analysis of fracture bifurcation. Tectonophysics 83:163–183
Bahat D (1983) New aspects of rhomb structures. J Struct Geol 5:591–601
Bahat D (1984) Fracture interaction in the Gregory Rift, East Africa. Tectonophysics 104:47–65
Bahat D (1985) Low angle normal faults in Lower Eocene chalks near Beer-Sheva, Israel. J Struct Geol 7:613–620
Bahat D (1986a) Joints and en echelon cracks in middle Eocene chalks near Beer-Sheva, Israel. J Struct Geol 8:181–190
Bahat D (1986b) Criteria for the differentiation of en echelons and hackles in fractured rocks. Tectonophysics 121:197–206
Bahat D (1987a) Correlation between styles of fracture markings and orientation of cross-fold joints. Tectonophysics 136:323–333
Bahat D (1987b) Jointing and fracture interactions in middle Eocene chalks near Beer-Sheva, Israel. Tectonophysics 136:299–321
Bahat D (1988a) Early single-layer and late multi-layer joints in the lower Eocene chalks near Beer-Sheva, Israel. Annal-Tecton 2:3–11
Bahat D (1988b) Fractographic determination of joint length distribution in chalk. Rock Mech Rock Engin 21:79–94
Bahat D (1989) Fracture stresses at shallow depths during burial. Tectonophysics 169:59–65
Bahat D, Engelder T (1984) Surface morphology on cross-fold joints of the Appalachian Plateau, New York and Pennsylvania. Tectonophysics 104:299–313
Bahat D, Grossman HNF (1988) Regional jointing and paleostresses in Eocene chalks around Beer-Sheva. Isr J Earth Sci 37:1–11
Bahat D, Rabinovitch A (1988) Paleostress determination in a rock by fractographic method. J Struct Geol 10:193–199
Bahat D, Sharpe MR (1982) Dependence of Hertzian fracture angle on Poisson's ratio and indentation techniques. J Mater Sci 17:1167–1170
Bahat D, Leonard G, Rabinovitch A (1982) Analysis of s-symmetric fracture mirrors in glass bottles. Int J Fract 18:29–38
Ball A, Payne BW (1976) The tensile fracture of quartz crystals. J Mater Sci 11:731–740
Ball MJ, Landini DJ, Bradt RC (1984) Fracture mist region in a soda-lime-silica float glass. In: Mecholsky JJ Jr, Powell SR Jr (eds) Fractography of ceramic and metal failures. ASTM STP 827, Philadelphia, pp 110–120
Bankwitz P (1965) Über Klüfte 1. Beobachtungen im Thüringischen Schiefergebirge. Geologie 14:241–253
Bankwitz P (1966) Über Klüfte 2. Die Bildung der Kluftfläche und eine Systematik ihrer Strukturen. Geologie 15:896–941
Bankwitz P (1978) Über Klüfte 4. Aspekte einer bruchphysikalischen Interpretation geologischer Rupturen. Z Geol Wiss 3:301–311
Bankwitz P, Bankwitz E (1984) Die Symmetrie von Kluftoberflächen und ihre Nutzung für eine Paläospannungsanalyse. Z Geol Wiss 12:305–334
Bansal GK (1976) Effect of flaw shape on strength of ceramics. J Am Ceram Soc 59:87–88
Bansal GK (1977) On fracture mirror formation in glass and polycrystalline ceramics. Phil Mag 35:935–944
Bansal GK, Duckworth WH (1977) Fracture stress as related to flaw and fracture mirror sizes. J Am Ceram Soc 60:304–310
Barenblatt G (1962) Mathematical theory of equilibrium cracks in brittle fracture. Adv Appl Mech 7:55
Barquins M, Ghalayini K, Petit JP (1989) Branchement des fissures sous compression uniaxiale. CR Acad Sci Paris 308:899–905
Barton DC (1933) Surface fracture system of South Texas. Am Assoc Pet Geol Bull 17:1194–1212
Barton CC (1983) Systematic jointing in the Cardium sandstone along the Bow River, Alberta, Canada. Ph. D. Thesis Univ Microfilms Int 302 pp
Barton NR (1973) Review of a new shear strength criterion for rock joints. Eng Geol 7:287–332

Beachem CD (1975) The effects of crack tip plastic flow directions upon microscopic dimple shapes. Metall Trans A 6:377–383

Beall GH, Chyung K, Stewart RL, Donaldson KY, Lee HL, Baskaran S, Hasselman DPH (1986) Effect of test method and crack size on the fracture toughness of a chain silicate glass-ceramic. J Mater Sci 21:2365–2372

Becker GF (1893) Finite homogeneous strain, flow and rupture of rocks. Bull Geol Soc Am 4:13–90

Belfield WC, Helwig J, La Pointe P, Dahleen WK (1983) South Ellwood oil field, Santa Barbara Channel, Ca. A Monterey formation fractured reservoir. In: Isaacs CM, Garrison RE (eds) Petroleum generation and occurrence in the Miocene Monterey Formation, California. Pacific Section SEPM

Bentor YK, Vroman A (1954) A structural contour map of Israel, 1:250000 with remarks on the dynamical interpretation. Bull Res Counc Isr 4:125–135

Berry MV (1976) Waves and Them's theorem. Adv Phys 25:1–26

Berry JP (1960a) Some kinetic considerations on the Griffith criterion for fracture. 1: Equation of motion at constant force. J Mech Phys Solids 8:194–206

Berry JP (1960b) Some kinetic considerations on the Griffith criterion for fracture. 2: Equations of motions at constant deformation. J Mech Phys 8:207–216

Bhowmick AK (1986) Fractographic investigation of tensile and fatigue fracture surfaces of rubber containing inhomogeneous inclusions. J Mater Sci Lett 5:1042–1044

Bluhm JI (1969) Fracture arrest. In: Liebowitz H (ed) Fracture. An advanced treatise. Academic Press, New York 5:1–63

Brace WF, Bombolakis EG (1963) A note on brittle crack growth in compression. J Geophys Res 68:3709–3713

Brace WF, Silver E, Hadley K, Goetze C (1972) Cracks and pores: a closer look. Science 178:162–163

Broek D (1982) Elementary engineering fracture mechanics. 3rd edn, Nijhoff, Boston, 469 pp

Brown SR (1989) Transport of fluid and electric current through a single fracture. J Geophys Res 94:9429–9438

Bullen FP, Henderson F, Wain HL (1970) Crack branching in heavily drawn chromium. Phil Mag 21:689–699

Bullock RE, Kaae JL (1979) Size effect on the strength of glassy carbon. J Mater Sci 14:920–930

Buist DS (1980) Columnar sandstone, Island of Bute, Scotland. Geol Mag 117:381–384

Carlsson J, Dahlberg L, Nilsson F (1973) Experimental studies of the unstable phase of crack propagation in metals and polymers. In: Sih GC (ed) Proc Int Conf Dynamic Crack Propagation. Noordhoff, Leyden, pp 165–181

Carman CM, Katlin JM (1966) Low cycle crack propagation characteristics of high strength steels. J Basic Eng 66:792–800

Carmichael ISE, Turner FJ, Verhoogen J (1974) Igneous petrology. Mc Graw Hill, New York, 739 pp

Carter CS (1971) Stress corrosion crack branching in high-strength steels. Eng Fract Mech 3:1–13

Cegla J, Dzulyuski S (1967) Experiments on feather fracture in sediments. Annal Soc Geol Pologne 38:489–497

Charles RJ (1958a) Static fatigue of glass 1. J Appl Phys 29:1549–1553

Charles RJ (1958b) Static fatigue of glass 2. J Appl Phys 29:1554–1560

Charles RJ (1958c) Dynamic fatigue of glass. J Appl Phys 29:1657–1662

Charles RJ (1961) A review of glass strength. In: Burke JE (ed) Prog Ceram Sci 1:1–38

Charles RJ, Hillig WB (1962) The kinetics of glass failure by stress corrosion. In: Symposium sur la résistance mechanique du verre et les moyens de l'améliorer compte rendu (Symp mechanical strength of glass and ways of improving it, 1961). Union Scientifique du Verre, Charleroi, Belgium, pp 511–527

Cina B, Kaatz T (1976) Learning from experience of the stress-corrosion failure of high-strength aluminium alloy forgings. STP 610, ASTM, Philadelphia, USA, pp 61–71

Clark ABJ, Irwin GR (1966) Crack propagation behaviors. Exp Mech 6:321–329

Cleveland JJ (1977) The critical grain size for microcracking in the Pseudobrookite structure. M. Sc. Thesis, Pennsylvania State Univ

Congleton J (1973) Practical applications of crack-branching measurements. In: Sih GC (ed) Proc Int Conf Dynamic crack propagation. Noordhoff, Leyden, pp 427–438
Congleton J, Petch NJ (1967) Crack branching. Phil Mag 16:749–760
Conway JC Jr, Mecholsky JJ Jr (1989) Use of crack branching data for measuring near-surface residual stresses in tempered glass. J Am Ceram 72:1584–1587
Corten HT, Gallagher JP (1971) Fracture toughness Proc 1971 National Symposium on Fracture Mechanics Part II. ASTM STP 514, Philadelphia, 191 pp
Cotterell B (1965) On brittle fracture paths. Int J Fract Mech 1:96
Cotterell B (1972) Brittle fracture in compression. Int J Fract Mech 8:195–208
Coulon M, Frizon de Lamotte D (1988a) Les craies éclatées du secteur d'Orney (Marne, France): le résultat d'une brechification par fracturation hydraulique en contexte extensif. Bull Soc Geol Fr 8:177–185
Coulon D, Frizon de Lamotte D (1988b) Les extensions cénozoiques dans l'est du Bassin de Paris: mise en évidence et interpretation. CR Acad Sci Paris 307:1113–1119
Crosby WO (1882) On the classification and origin of joint structures. Proc Boston Soc Nat Hist 22:72–85
Das S, Scholz CH (1981) Theory of time-dependent rupture in earth. J Geophys Res 137:6039–6051
Davis MW, Wasylyk JS, Southwick RD (1983) On the static fatigue limit for soda-lime-silica glass. Glasstech Ber 56:117–124
Deere DU (1963) Technical description of rock cores for engineering purposes. Felsmech Ing Geol 1:16–22
De Freminville MCh (1907) Caractères des vibrations accompagnant le choc déduits de l'examen des cassures. Rev Met 4:837–884
De Freminville MCh (1914) Recherches sur la fragilité – l'éclatement. Rev Met 11:971–1056
DeGraff JM, Aydin A (1987) Surface morphology of columnar joints and its significance to mechanics and direction of joint growth. Geol Soc Am 99:605–617
Dennis JG (1972) Structural geology. Ronald, New York, 532 pp
Derre C, Lecolle M, Roger G, Tavares De Freitas Carvalho J (1986) Tectonics, magmatism, hydrothermalism and sets of flat joints locally filled by Sn-W aplite-pegmatite and Quartz veins; southeastern border of the Serra de Estrela granitic massif (Beira Baixa, Portugal). Ore Geol Rev 1:43–56
De Sitter LU (1962) Structural development of the arabian shield in Palestine. Geol Mijnbouw 41:116–124
Doll W (1975) A molecular weight dependent fracture transition in polymethylmethacrylate. J Mater Sci 10:935–942
Doremus RH (1973) Glass science. Wiley, New York, 349 pp
Doremus RH (1976) Cracks and energy-criteria for brittle fracture. J Appl Phys 47:1833–1836
Doremus RH (1980) Importance of crack tip radii in fracture and fatigue of glass. J Non-Cryst 39:493–496
Doremus RH, Johnson WA (1978) Depths of fracture-initiating flows and initial stages of crack propagation in glass. J Mater Sci 13:855–858
Dugdale DS (1960) Yielding of steel sheets containing slits. J Mech Phys Solids 8:100
Dyer JR (1983) Jointing in sandstones, Arches National Park, Utah Ph. D. Thesis Stanford, California, Stanford Univ, 202 pp
Engelder T (1982) Is there a genetic relationship between selected regional joints and contemporary stress within lithosphrere of North America? Tectonics 1:161–177
Engelder T (1985) Loading paths to joint propagation during cycle: an example of the Appalachian Plateau, USA. J Struct Geol 7:459–476
Engelder T (1987) Joints and shear fractures in rock. In: Atkinson BK (ed) Fracture mechanics of rock. Academic Press, Lond, pp 27–69
Engelder T, Geiser P (1980) On the use of regional joint sets as trajectories of paleostress fields during the development of the Appalachian Plateau, New York, J Geophys Res 85:6319–6341
Engelder T, Geiser P (1984) Near-surface in situ stress. 4. Residual stress in the Tully limestone Appalachian plateau, New York. J Geophys Res 89:9365–9370

Engelder T, Oertel G (1985) Correlation between abnormal pore pressure and tectonic jointing in the Devonian Catskille Delta. Geology 13:863–866
Engelder T, Sbar ML (1984) Near-surface in situ stress: introduction. J Geophys Res 89:9321–9322
Engelder T, Geiser P, Bahat D (1987) Alleghanian deformation within shales and siltstones of the Upper Devonian Appalachian Basin Finger Lakes district, New York. Geol Soc Am Continental Filed Guide, Northeastern Sect, pp 113–118
Epstein AG, Epstein JB, Harris LB (1975) Conodont color alteration – an index to organic metamorphism. US Geol Surv Open File Rep 54:75–379
Erdogan F (1968) Crack propagation theories. In: Liebowitz H (ed) Fracture: an advanced treatise. Academic Press, New York, pp 49–590
Erdogan F, Sih GC (1963) On the crack extension plates under plane loading and transverse shear. J Basic Eng 85:519–527
Ernsberg FM (1977) Mechanical properties of glass. In: Götz J (ed) Glass 1977. Proc 11th Int Congr glass. North Holland, Amsterdam, pp 295–321
Ernstson K, Schinker M (1986) Die Entstehung von Plumosekluftochenmarkierungen und ihre tektonische Interpretation. Geol Rundsch 75:301–322
Evans AG (1972) A method for evaluating the time-dependent failure characteristics of brittle materials and its application to polycrystalline alumina. J Mater Sci 7:1137–1146
Evans AG (1973) Strength degradation by projectile impacts. J Am Ceram Soc 56:405–409
Evans AG, Faber KT (1984) Crack growth resistance of microcracking brittle materials. J Am Ceram Soc 67:255–260
Evans AG, Tappin G (1972) Effects of microstructure on the stresses to propagate inherent flaws. Proc Brit Ceram Soc 20:275–297
Evans AG, Wiederhorn SM (1974) Proof testing of ceramic materials – an analytical basis for failure prediction. Int J Fract 10:379–392
Field JE (1971) Brittle fracture: its study and application. Contemp Phys 12:1–31
Finnie I, Saith A (1973) A note on the angled crack problem and the directional stability of cracks. Int J Fracture 9:484–486
Flexer A (1971) Late Cretaceous paleogeography of northern Israel and its significance of the levant geology. Palaeogeogr Palaeoclimatol Palaeoecol 10:293–316
Fraissinet C, Frizon de Lamotte D, Coulon M (1988) Estimation de paleoprofondeur. 12 ème Réunion Sci Terre, Lille Abstr, 56 pp
Frank FC, Lawn BR (1967) On the theory of Hertzian fracture. Proc R Soc A 299:291–306
Frechette VD (1972) The fractology of glass. In: Pye L (ed) Introduction to glass science. Plenum Press, New York, pp 433–450
Frechette VD (1984) Markings on crack surfaces of brittle materials: a suggested unified nomenclature. In: Mecholsky JJ Jr, Powell SR Jr (eds) Fractography of ceramic and metal failures, ASTM STP 827, Philadelphia, pp 104–109
Frechette VD, Michalske TA (1978) Fragmentation in bursting glass containers. Bull Am Ceram Soc 57:427–429
Freiman SW (1980) Fracture mechanics of glass. Glass Sci Tech 5:21–77
Freiman SW (1984) Effects of chemical environments on slow crack growth in glasses and ceramics. J Geophys Res 89:4077–4114
Freimann SW, McKinney KR, Smith HL (1974) Slow crack growth in polycrystalline ceramics. Fract Mech Ceram 2:659–676
Freiman SW, Gonzalez AC, Mecholsky JJ Jr (1979) Mixed-made fracture in Soda-Lime glass. J Am Ceram Soc 62:206–208
Friedman M, Logan JM (1970a) The influence of residual elastic strain on the orientation of experimental fractures in three quartzose sandstones. J Geophys Res 75:387–405
Friedman M, Logan JM (1970b) Microscopic feather fractures. Geol Soc Am Bull 81:3417–3420
Friedman M, Logan JM (1973) Lüder's bands in experimentally deformed standstone and limestone. Geol Soc Am Bull 84:1465–1476
Fuller KNG, Fox PG, Field JE (1975) The temperature rise at the tip of fast moving cracks in glassy polymers. Proc R Soc Lond A 341:537–557

Gallois RW (1965) British regional geology – the Wealden district 4th edn. Nat Environ Res Counc Inst Geol Sci, Lond, 101 pp
Garfunkel Z (1978) The Negev regional synthesis of sedimentary basins. IAS 10th Int Congr, Guideb 1:33–110
Garfunkel Z (1981) Internal structure of the Dead Sea leaky transform (rift) in relation to plate kinetics. Tectonophysics 80:81–108
Garfunkel Z, Bartov Y (1977) The tectons of the Suez Rift. Bull Geol Surv Isr 71:44
Gilman JJ (1958) Creation cleavage steps by dislocation. Trans Metall Soc AIME 212:310–315
Glucklich J (1963) Fracture on plain concrete. J Eng Mech Div, Am Soc Civil Eng Proc 6:127–138
Goldbery R (1982) Structural analysis of soil microrelief in palaeosoils of the lower Jurassic "Laterite derivative facies" (Mishhor and Ardon formation) Makhtesh Ramon, Israel. Sediment Geol 31:119–140
Gramberg J (1965) Axial cleavage fracturing, a significant process in mining and geology. Eng Geol 1:31–72
Gran RJ, Orazio FD Jr, Parvis PC, Irwin GR, Hertzberg RW (1971) AFFDL-TR pp 70–149
Green PJ, Nicholson PS, Embury JD (1977) Crack shape studies in brittle porous materials. J Mater Sci 12:987–989
Grenet L (1899) Mechanical strength of glass. Bull Soc Enc Indust Nat Ser 5:838–848
Griggs DT, Handin J (1960) Observations on fracture and hypothesis of earthquakes. In: Griggs DT, Handin J (eds) Rock deformation. Geol Soc Am Mem 79:347–364
Griffith AA (1920) The phenomena of rupture and flow in solids. Phil Trans R Soc Lond Ser A 221:163–198
Griffith AA (1924) The theory of rupture. Proc First Int Conf Appl Mechanics, Delft, pp 55–63
Gulati ST (1980) Cracks kinetics during static and dynamic loading. J Non-Cryst 38 and 39:475–480
Gulati ST, Helfinstine JD, Beall GH (1986a) Strength and fracture properties of multiple chain silicates: part 1. Collected pap 14th Int Cong Glass, pp 39–46
Gulati ST, Helfinstine JD, Beall GH (1986b) Strength and fracture properties of multiple chain silicates: part 2. Collected pap 14th Int Cong Glass, pp 47–54
Gvirtzman G (1979) Geology and geomorphology of the Beer-Sheva valley. In: Grados Y, Stern E (eds) The Beer-Sheva book. Keter, Jerusalem, pp 333–355 (in Hebrew)
Hagan JT, Swain MV, Field JE (1979) Fracture-strength studies on annealed and tempered glasses under dynamic conditions. Phil Mag 39:743–756
Hahn GT, Rosenfield AR (1965) Local yielding and extension of a crack under plane stress. Acta Metall 13:293–306
Haimson BC, Doe TW (1983) State of stress, permeability and fractures in the Precambrian granite of northern Illinois. J Geophys Res 88:7355–7372
Hall J (1843) Geology of New York, Part 4. Survey of the Fourth Geological District. Albany, NY, 683 pp
Hallbauer DK, Wagner H, Cook NGW (1973) Some observations concerning the microscopic and mechanical behaviour of quartzite specimens in stiff, triaxial compression tests. Int J Rock Mech Min Sci, Geomech Abstr 10:713–726
Hamil BM, Sriruang S (1976) A study of rock fracture induced by dynamic tensile stress and its application to fracture mechanics. In: Strens RGJ (ed) The physics and chemistry of minerals and rocks. Wiley, New York, pp 151–196
Hancock PL (1985) Brittle microtectonics: principles and practice. J Struct Geol 7:437–457
Hancock PL, Al-Kadhi A (1978) Analysis of mesoscopic fractures in the Dhurma-Nisah segment of the central Arabian Graben System. J Geol Soc 135:339–347
Hancock PL, Al-Kadhi A (1982) Significance of arcuate joint sets connecting oblique grabens in Central Arabia. Mitt Geol Inst ETH, Univ Zürich, N.F. 239:128–131
Hancock PL, Al-Kadhi A, Barka AA, Bevan TG (1987) Aspects of analyzing brittle structures. Annal Tectonicae 1:5–19
Haneman D, Pugh EN (1963) Tear marks on cleaved Germanium surfaces. J Appl Phys 34:2269–2272
Haxby WF, Turcotte DL (1976) Stresses induced by the addition of removal of overburden and associated thermal effects. Geol 4:181–185

Hayati G (1975) The engineering geology of some chalks in Israel. Ph. D. Thesis, Israel Inst Technol, 1837 pp (in Hebrew)

Healey JT, Mecholsky JJ Jr (1984) Scanning electron microscopy techniques and their application to failure analysis of brittle materials. In: Mecholsky JJ Jr, Powell SR Jr (eds) Fractography of ceramic and metal failures. ASTM STP 827, Philadelphia, pp 157–181

Heard HC (1960) Transition from brittle to ductile flow in Solenhofen Limestone as a function of temperature, confining pressure and interstitial fluid pressure. In: Griggs D, Handin J (eds) Rock deformation. Geol Soc Am Mem 79:193–226

Hertz H (1881) Contact of elastic solids. J Reine Angew Math (Crelle) 93:156–171

Hertzberg RW (1976) Deformation and fracture mechanics of engineering materials. Wiley, New York, 605 pp

Hertzberg RW, Mills WJ (1976) Character of fatigue fracture surface micromorphology in the ultra-low growth rate regime. Fractography – Microscopic Cracking Processes. ASTM STP 600, Philadelphia, pp 220–234

Hoagland RG, Hahn GT, Rosenfield AR (1973) Influence of microstructure on fracture propagation in rock. Rock Mech 5:77–106

Hobbs BE, Means WD, Williams PF (1976) An outline of structured geology. Wiley, New York, 571 pp

Hodgson RA (1961 a) Classification of structures on joint surfaces. Am J Sci 259:493–502

Hodgson RA (1961 b) Regional study of jointing in comb ridge-Navajo Mountain area, Arizona and Utah. Bull Am Assoc Petrol Geol 45:1–38

Hodkinson PH, Nadeau JS (1975) Slow crack growth in graphite. J Mater Sci 10:846–856

Hoek E, Brown ET (1978) Trends in relationship between measured in situ stresses and depths. Int J Rock Mech Min Sci, Geomech Abstr 15:211–215

Holland AJ (1936) The effect of sustained loading on the breaking strength of glass. Glass Rev 12:133–135

Holloway DG (1973) The physical properties of glass. Wykeham, Lond, 220 pp

Holloway DG (1986) The fracture behaviour of glass. Glass Tech 27:120–133

Holzhausen GR, Johnson AM (1979) Analyses of longitudinal splitting of uniaxially compressed rock cylinders. Int J Rock Mech Min Sci Geomech Abstr 16:163–177

Honeycombe RWK (1984) The plastic deformation of metals. Arnold, Lond, 33 pp

Horton RE (1945) Erosional development of streams and their drainage basins: hydrological approach to quantitative morphology. Geol Soc Am Bull 56:275–370

Hsieh C, Thomson R (1973) Lattice theory of fracture and crack creep. J Appl Phys 44:2051–2063

Hsu TC, Slate FO, Sturman GM, Winter G (1963) Microcracking of plain concrete and the shape of the stress-strain curve. J Am Conc Inst 60:209–223

Hubbert MK (1951) Mechanical basis for certain familiar geologic structures. Bull Geol Soc Am 62:355–372

Inglis CE (1913) Stresses in a plate due to the presence of cracks and sharp corners. Trans Inst Naval Architects 55:219–242

Ingraffea AR (1977) Discrete fracture propagation in rock: laboratory test and finite element analysis. Ph. D. Dissertation, Univ Colorado, Boulder, 347 pp

Irwin GR, Jonassen F, Reep WP, Bayless RT (1948) Fracture dynamics. In: Fracturing of metals. Am Soc Metal, pp 147–166

Irwin GR (1957) Analysis of stresses and strains near the end of a crack traversing a plate. J Appl Mech 24:361–364

Irwin GR (1958) Fracture. Encyclopedia Phys 6:551

Irwin GR (1960) Fracture mode transition for a crack traversing a plate. J Basic Eng 82:417–425

Irwin GR (1962) Crack-extension force for a part-through crack in a plate. J Appl Mech 29:651–654

Irwin GR (1964) Structural aspects of brittle fracture. Appl Mater Res 3:65–81

Irwin GR (1977) Comments on dynamic fracturing. In: Hahn GT, Kanninen MF (eds) Fast fracture and crack arrest. ASTM STP 627, Philadelphia, pp 7–18

Irwin GR, Wells AA (1965) A continuum-mechanics view of crack propagation. Metall Rev 10:223–270

Irwin GR, Kies JA, Smith HL (1958) Fracture strengths relative to onset and arrest of crack propagation. Proc ASTM 58, Philadelphia, pp 640–657
Issar A, Bahat D, Wakshal E (1988) Occurrence of secondary gypsum veins in joints in chalks in the Negev, Israel. Catena 15:241–247
Jaeger JC, Cook NGW (1979) Fundamentals of rock mechanics. Chapman and Hall. Lond, 593 pp
Jamison DB, Cook NGW (1980) A note on measured values for the state of stress in the earth's crust. J Geophys Res 85:1833–1838
Johnson JW, Holloway DG (1966) On the shape and size of the fracture zones on glass fracture surfaces. Phil Mag 14:731–743
Johnson JW, Holloway DG (1968) Microstructure of the mist zone on glass fracture surfaces. Phil Mag 17:899–910
Johnson KL, O'Connor JJ, Woodward AC (1973) The effect of the indenter elasticity on the Hertzian fracture of brittle materials. R Soc Lond Proc A 334:95–117
Johnson RM (1970) Physical processes in Geology. Freeman, San Francisco, 577 pp
Kalthoff JF (1973) On the propagation direction of bifurcated cracks. In: Sih GC (ed) Dynamic crack propagation. Noordhoff, Leyden, pp 449–458
Kamdar MH (1977) Mechanism of embrittlement and brittle fracture in liquid metal environments. Fracture 1:387–405
Kanninen MF, Popelar CH (1985) Advanced fracture mechanics. Oxford Univ Press, New York, 563 pp
Kaplan MF (1959) Flexural and compressive strength of concrete as affected by the properties of coarse aggregates. J Am Conc Inst 55:1193–1208
Kaplan MF (1961) Crack propagation and the fracture of concrete. J Am Conc Inst 58:591–610
Karper MJ, Scuderi TG (1964) Modulus of rupture of glass in relation to fracture pattern. Ceram Bull 43:622–625
Kendall PF, Briggs H (1933) The formation of rock joints and the cleat of coal. R Soc Proc 53:164–187
Kennedy WJ, Garrison RE (1975) Morphology and genesis of nodular chalks and hardgrounds in the upper Cretaceous of southern England. Sedimentology 22:311–386
Kerkhof F (1973) Wave fractographic investigations of brittle fracture dynamics. In: Sih GC (ed) Dynamic crack propagation. Noordhof, Leyden, pp 3–35
Kerkhof F (1975) Bruchmechanische Analyse von Schadensfällen an Gläsern. Glastechn Ber 48:112–124
Kerkhof VF, Müller-Beck H (1969) Zur Bruchmechanischen Deutung der Schlagmarken an Steingeräten. Glastechn Ber 42:439–448
Kies JA, Sullivan AM, Irwin GR (1950) Interpretation of fracture markings. J Appl Phys 21:716–720
Kirchner HP (1978) The strain intensity criterion for crack branching in ceramics. Eng Fract Mech 10:283–288
Kirchner HP, Conway JC Jr (1987) Comparison of the stress-intensity and Johnson and Holloway criteria for crack branching in rectangular bars. J Am Ceram Soc 70:565–569
Kirchner HP, Gruver RM (1973) Fracture mirrors in alumina ceramics. Phil Mag 30:1433–1446
Kirchner HP, Gruver RM (1974) Fracture mirrors in polycrystalline ceramics and glass. Plenum Press, New York, pp 309–320
Kirchner HP, Kirchner JW (1979) Fracture mechanics of fracture mirrors. J Am Ceram Soc 62:198–202
Kirchner HP, Gruver RM, Sotter WA (1975) Use of fracture mirrors to interpret impact fractures in brittle materials. J Am Ceram Soc 58:188–191
Kirchner HP, Gruver RM, Sotter WA (1976) Fracture stress-mirror size relations for polycrystalline ceramics. Phil Mag 33:775–780
Kirchner HP, Gruver RM, Swain MV, Garvie RC (1981) Crack branching in transformation toughened Zirconia. J Am Ceram Soc 64:529–533
Kitagawa H, Yuuki R, Ohira T (1975) Crack morphological aspects in fracture mechanics. Eng Fract Mech 7:515–529
Knott JF (1973) Fundamentals of fracture mechanics. Halsted, New York, 273 pp

Kobayashi AS, Ramulu M (1985) Dynamic fracture mechaniss. J Aeronaut Soc India 37:1 – 19
Kobayashi AS, Wade BG (1973) Crack propagation and arrest in impacted plates. In: Sih GC (ed) Proc Int Conf Dyn Crack Propagation, Noordhoff, Leyden, pp 663 – 677
Kohn B, Eyal M (1981) History of uplift of the crystalline basement of Sinai and its relation to opening of the Red Sea as revealed by fission track dating of apatites. Earth Planet Sci Lett 52:129 – 141
Koifman L, Flexer A (1975) Possibilities of underground storage for crude oil in Israel. Isr Inst Petrol Energy, 217 pp
Krohn DA, Hasselman D (1971) Relation of flow size to mirror in the fracture of glass. J Am Ceram Soc 54:411
Kulatilake PH (1988) State of the art in stochastic joint geometry modeling. Rock Mechanics, Proc 28th US Symp, Tucson
Kuszyk JA, Bradt RC (1973) Thermal expansion anisotropy-grain size effects in Magnesium Dititanate. J Am Ceram Soc 56:420 – 423
Kulander BR, Dean SL (1985) Hackle plume geometry and propagation dynamics. In: Stephanson O (ed) Fundamentals of rock joints. Proc Lulea Univ Technology, Sweden, pp 85 – 94
Kulander BR, Barton CC, Dean SC (1979) The application of fractography to core and outcrop fracture investigations. Rep to U. S. D. O. E. Morgantown Energy Technol Cent METC/SP-79/3, 174 pp
Lachenbruch AH (1961) Depths and spacing of tension cracks. J Geophys Res 66:4273 – 4291
Lachenbruch AH (1962) Mechanics of thermal contraction cracks and icewedge polygons in permafrost Geol Soc Am Spec Pap 70
Ladeira FL, Price NJ (1981) Relationship between fracture spacing and bed thickness. J Struct Geol 3:179 – 183
Laird C, Grosskreutz JC (1967) The influence of metallurgical structure of the mechanisms of the fatigue crack propagation. In: Fatigue crack propagation. ASTM, Philadelphia, pp 131 – 180
Lange EF (1968) Interaction between overlapping parallel cracks; a photoelectric study. Int J Fract Mech 4:287
Lawn BR, Wilshaw TR (1975 a) Fracture of brittle solids. Cambridge Univ Press, Lond, 204 pp
Lawn BR, Wilshaw TR (1975 b) Review indentation fracture: principles and applications. J Mater Sci 10:1045 – 1081
Lawn BR, Wiederhorn SM, Johnson HH (1975) Strength degradation of brittle surfaces: blunt indenters. J Am Ceram Soc 58:428 – 432
Lawn BR, Roach DH, Thomson RM (1987) Thresholds and reversibility in brittle cracks: an atomistic surfaceforce model. J Mater Sci 22:4036 – 4050
Leonard G (1979) Physical aspects (qualitative and quantitative) of failure in glass containers. M. Sc. Thesis, Ben Gurion Univ Negev, Beer-Sheva, Dep Geol Mineral, 90 pp
Leopold LB, Langbein WB (1962) The concept of entropy in landscape evolution. US Geol Surv, Prof Pap 500 A:1 – 20
Levengood WC (1958) Effect of origin flaw characteristics on glass strength. J Appl Phys 29:820 – 826
Low JR Jr (1959) A review of the microstructural aspects of cleavage fracture. In: Averbach BL, Felbeck DK, Hahn GT, Thomas DA (eds) Fracture. M. I. T. Press, Cambridge, Massachusetts, 646 pp
Lutton RJ (1971) Tensile fracture mechanics from fracture surface morphology. In: Clark CB (ed) Dynamic rock mechanics. 12th US Symp on rock mechanics, pp 561 – 571
Luyckx SB (1980) Fractography: a tool for failure analysis. In: Engineering applications of fracture analysis, Pergamon Press, Oxford, 428 pp
Lynch SP (1979) Mechanisms of fatigue and environmentally assisted fatigue. In: Fatigue mechanisms, ASTM STP 675, Philadelphia, pp 174 – 213
Macculoch J (1829) On a prismatic structure in standstone induced by artificial heat; and on certain prismatic rocks found in nature, including the columnar sandstone of Dunbar. Q Sci 2:247 – 265
MacCurdy E (1955) The note books of Leonardo da Vinci. Reynal and Hitchcock, New York, 655 pp

Marshall DB, Lawn BR, Mecholsky JJ Jr (1980) Effect of residual contact stresses on mirror/flaw-size relations. J Am Ceram Soc 63:358–360
Mayer FM (1972) The effect of different aggregates on the compressive strength and modulus of elasticity of normal concrete. Beton 22:61–62
McClintok FA, Walsh JB (1962) Friction on Griffith cracks in rocks under pressure. Proc 4th US Nat Cong Appl Mech, pp 1015–1021
McGarr A (1988) On the state of lithospheric stress in the absence of applied tectonic forces. J Geophys Res 93:609–617
McMahon CJ Jr, Vitek V, Kameda J (1981) Mechanics and mechanisms of intergranular fracture. In: Chell GG (ed) Developments in fracture mechanics 2. Appl Sci Publ, Lond, pp 193–246
Means WD (1979) Stress and strain. Springer, Berlin Heidelberg New York, 399 pp
Mecholsky JJ, Freiman SW (1979) Determination of fracture mechanics parameters through fractographic analysis of ceramic. In: Freiman SW (ed) Fracture mechanics applied to brittle materials. ASTM, Philadelphia, 220 pp
Mecholsky JJ, Rice RW, Freiman SW (1974) Prediction of fracture energy and flaw size in glasses from measurements of mirror size. J Am Ceram Soc 57:440–443
Mecholsky JJ, Freiman SW, Rice RW (1976) Fracture surface analysis of ceramics. J Mater Sci 11:1310–1319
Mecholsky JJ Jr, Rice RW (1984) Fractographic analysis of biaxial failure in ceramics. In: Mecholsky JJ, Powell SR Jr (eds) Fractography of ceramic and metal failures. ASTM STP 827, Philadelphia, pp 185–193
Michalske TA (1979) Dynamic effects of liquids on crack growth loading to catastrophe failure in glass. Ph. D. Thesis Alfred Univ, pp 1–75
Michalske TA (1983) The stress corrosion limit: its measurement and implications. In: Bradt RC, Hasselman DPH, Lange FF (eds) Fracture mechanics of ceramics. Plenum, New York, 5:277–289
Michalske TA (1984) Fractography of slow fracture in glass. In: Mecholsky JJ Jr, Powell SR Jr (eds) Fractography of ceramic and metal failures. ASTM STP 827, Philadelphia, pp 121–136
Michalske TA, Bunker BC (1984) Slow fracture model based on strained silicate structures. J Appl Phys 56:2686–2693
Michalske TA, Bunker BC (1987) The fracturing of glass. Sci Am 257:78–112
Michalske TA, Frechette VD (1980) Dynamic effects of liquids on crack growth leading to catastrophic failure in glass. J Am Ceram Soc 63:603–609
Michalske TA, Freiman VD (1982) A molecular interpretation of stress corrosion in silica. Nature (Lond) 295:511–512
Michalske TA, Varner JR, Frechette VD (1978) Growth of cracks partly filled with water. In: Bradt RC, Hasselman DPH, Lange FF (eds) Fracture mechanics of ceramics. Plenum, New York, 4:639–649
Mimran Y (1977) Chalk deformation and large-scale migration of calcium carbonate. Sedimentology 24:333–360
Mimran Y (1985) Tectonically controlled freshwater carbonate cementation in Chalk. Soc Econ Paleont Mineral 36:371–379
Mimran Y, Michaeli L (1986) The transaction of a major fault into secondary faults in chalk and its effects on the mechanics of the host rock. Isr J Earth Sci 35:114–123
Mould RE (1952) The behavior of glass bottles under impact. J Am Ceram Soc 35:230–236
Mould RE (1967) The strength of inorganic glasses. Fundamental Phenomena Mater Sci 4:119–147
Mould RE, Southwick RD (1959) Strength and statistic fatigue and abraded glass controlled ambient conditions: 2 Effect of various abrasions and the universal fatigue curve. J Am Ceram Soc 42:582–592
Mukai Y, Watanabe M, Murata M (1978) Fractographic observation of stress corrosion cracking of AISI 304 stainless steel in boiling 42 percent magnesium chloride solution. In: Strauss BM, Cullen WH Jr (eds) Fractography in failure analysis. ASTM, Philadelphia, 390 pp
Murgatroyd JB (1942) The significance of surface marks on fractured glass. J Soc Glass Tech 26:155–171

Murray CD (1926) The physiological principle of minimum work applied to the angle of branching of arteries. J Gen Physiol 9:835–841
Murray CD (1927) A relationship between circumference and weight in trees and its bearing on branching angles. J Gen Physiol 10:725–739
Muench D, Pike D (1974) Anasazi – ancient people of the rock. American West Publ Co, 191 pp
Nadai A (1950) Theory of flow and fracture of solids. Mc-Graw Hill, New York, 572 pp
Nadeau JS (1974) Subcritical crack growth in vitreous carbon at room temperature. J Am Ceram Soc 57:303–306
Narr W, Burrus RC (1984) Origin of reservoir fractures in little knife field, North Dakota. AAPG Bull 68:1087–1100
Nepper-Christensen P, Nielsen TPH (1969) Modal determination of the effect of bond between coarse aggregate and mortar on the compressive strength of concrete. Proc Am Conc Inst 66:69–72
Nicholson R, Ejiofor IB (1987) The three-dimensional morphology of echelon and sigmoidal mineral-filled fractures: data from North Cornwall. J Geol Soc Lond 144:79–83
Nicholson R, Pollard DD (1985) Dilation and linkage of echelon cracks. J Struct Geol 7:583–590
Nickelsen RP, Hough ND (1967) Jointing in the Appalachian plateau of Pennsylvania. Geol Soc Am Bull 78:609–630
Nolte KG, Smith MB (1981) Interpretation of fracturing pressures. J Petrol Technol 33:1767–1772
Nur A (1982) The origin of tensile fracture lineaments. J Struct Geol 4:31–40
Nye JF (1952) The mechanics of glacier flow. J Glaciol 2:82–93
Obreimoff JW (1930) The splitting strength of mica. Proc R Soc A 27:290–297
Orowan E (1948) Fracture and strength of solids. Rep Prog Phys 12:185
Orowan E (1949) Fracture and the strength of solids. Rep Progr Phys 12:185–232
Orowan E (1974) Origin of the surface features of the moon. Proc R Soc Lond A 336:141–163
Orr L (1972) Practical analysis of fractures in glass windows. ASTM, Philadelphia, pp 21–23, 47
Padovani ER, Shirey SB, Simmons G (1982) Characteristics of microcracks in amphibolite and granulite facies grade rocks from southerneastern Pennsylvania. J Geophys Res 87:8605–8630
Paris PC, Sih GC (1965) Stress analysis of cracks. In: Brown WF Jr (Symposium chairman) Fracture toughness testing and its applications. ASTM STP 381, Philadelphia, pp 30–81
Park RG (1983) Foundations of structural geology. Chapman and Hall, New York, 135 pp
Parker JM (1935) Regional systematic jointing in gently dipping sedimentary rocks. Ph. D. Thesis, Cornell Univ Library, New York
Parker JM (1942) Regional systematic jointing in slightly deformed sedimentary rocks. Geol Soc Am Bull 53:381–408
Parker JM (1969) Jointing in south-central New York: discussion. Geol Soc Am Bull 80:923–926
Patterson LW, Solberger JB (1979) Water treatment of flint. Lithic Technol 8:50–51
Payne BW, Ball A (1976) The determination of crack velocities in anisotropic materials by the analysis of lines. Phil Mag 34:917–922
Petit J, Barquins M (1987) Formation de fissures en milier confine dans le PMMA (polymetachrylate de methyle): modèle analogique de formation de structures tectoniques cassantes. CR Acad Sci 305:55–60
Picard L (1943) Structure and evolution of Palestine with comparative notes on neighbouring countries. Geol Dep, Hebrew Univ, Jerusalem, 187 pp
Pohn HA (1981) Joint spacing as method of locating faults. GLGYB 9:258–261
Pollard DD, Aydin A (1988) Progress in understanding jointing over the past century. Geol Soc Am Bull 100:1181–120
Pollard DD, Segall P, Delaney PT (1982) Formation and interpretation of dilatant echelon cracks. Bull Geol Soc Am 93:1291–1303
Poncelet EF (1958) The markings on fracture surfaces. J Soc Glass Tech 42:279–288
Press F, Siever R (1982) Earth. 3rd edn, Freeman, San Francisco, 613 pp
Preston FW (1926) A study of the rupture of glass. J Soc Glass Tech 10:234–269
Preston FW (1931) The propagation of fissures in glass and other bodies with special reference to the split-wave front. J Am Ceram Soc 14:419–427
Preston FW (1932) The form of cracks in bottles. J Am Ceram Soc 15:171–175

Preston FW (1935) Angle of forking of glass cracks as an indicator of the stress system. J Am Ceram Soc 18:175–176
Preston FW (1938) The term "striation" and certain other terms used in glass technology. Am Ceram Soc Bull 18:12–20
Preston FW (1952) Internal surface damage to bottles. GLI 33:639–644
Price NJ (1959) Mechanics of jointing in rocks. Geol Mag 96:142–167
Price NJ (1966) Fault and joint development in brittle and semi-brittle rock. Pergamon Press, New York, pp 110–161
Price NJ (1974) The development of stress systems and fracture patterns in undeformed sediments. Int Soc Rock Mech Cng Proc, pp 487–496
Proctor BA, Whitney I, Johnston JW (1967) Strength of fused silica. Proc R Soc Ser A Lond 297:534–557
Pugh EM, Heine-Geldern R, Foner S, Mustchler EC (1952) Glass cracking caused by high explosives. J Appl Phys 23:48–53
Quakenbush CL, Frechette VD (1978) Crack front curvature and glass slow fracture. J Am Ceram Soc 61:402–406
Rabinovitch A (1979) A note on the fracture-branching criterion. Phil Mag 40:873–874
Rabinovitch A, Bahat D (1979) Catastrophe theory: a technique for crack propagation analysis. J Appl Phys 50:231–234
Raggat HG (1954) Markings on joint surfaces in the Anglesea member of Demon's Bluff formation, Anglesea, Victoria. AAPG Bull 38:1808–1810
Ramsay JG (1967) Folding and fracturing of rocks. Mc-Graw Hill, New York, 568 pp
Ramsay JG (1980) The crack-seal mechanism of rock deformation. Nature (Lond) 284:135–139
Ramulu M, Kobayashi AS (1983) Dynamic crack curving – a photoelastic evaluation. Exp Mech 23:1–9
Ramulu M, Kobayashi AS (1985) Dynamic fracture toughnesses of reaction-bonded silicon nitride. J Am Ceram Soc 66:151–155
Ramulu M, Kobayashi AS, Kang BSJ (1982) Dynamic crack curving and branching in line-pipe. J Pressure Vessel Technol 104:317–322
Randall PN (1966) Plain strain crack toughness testing of high strength metallic materials. ASTM STP 410, Philadelphia, pp 88–129
Ravi-Chandar K, Knauss WG (1984a) An experimental investigation into dynamic fracture: 2, microstructural aspects. Int J Fract 26:65–80
Ravi-Chandar K, Knauss WG (1984b) An experimental investigation into dynamic fracture: 3. On steady-state crack propagation and crack branching. Int J Fract 26:141–154
Ravi-Chandar K, Knauss WG (1984c) An experimental investigation into dynamic fracture: 4. On the interaction of stress waves with propagation cracks. Int J Fract 26:189–200
Reed DA, Bradt RC (1984) Fracture mirror-failure stress relations in weathered and unweathered window glass panels. Am Ceram Soc Commun 67:227–229
Rice JR (1968) A path-independent integral and the approximate analysis of strain concentrations by notches and cracks. J Appl Mech 35:379–386
Rice RW (1974) Fracture topography of ceramics. In: Frechette VD, La Course WC and Burdick VL (eds) Surfaces and interfaces of glass and ceramics. Plenum Press, New York, pp 439–472
Rice RW (1979) The difference in mirror-to-flaw size ratios between dense glasses and polycrystals. J Am Ceram Soc 62:533–535
Rice RW (1984) Ceramic fracture features, observations, mechanisms and uses. In: Mecholsky JJ, Powell SR Jr (eds) Fractography of ceramic and metal failures. ASTM, Philadelphia, pp 1–51
Rice RW, Freiman SW, Becher PF (1981) Grain-size dependence of fracture energy in ceramics: 1, experiment. J Am Ceram Soc 64:345–349
Richter H (1974) Rissfrontkrümmung und Bruchflächenmarkierung im Übergangsbereich der Bruchgeschwindigkeit. Glastechn Ber 47:146–147
Ritchie RO (1985) Crack tip shielding in fracture and fatigue: intrinsic vs. extrinsic toughening. Second international conference on fundamentals of fracture. Oak Ridge Nat Lab, pp 90–91
Ritchie RO, Thompson AW (1985) On macroscopic and microscopic analyses for crack initiation and crack growth toughness in ductile alloys. Metall Trans A 16A:233–248
Roberts DK, Wells AA (1954) The velocity of brittle fracture. Engineering 178:820–821

Roberts JC (1961) Feather-fracture and the mechanics of rock jointing. Am J Sci 259:481–492
Roesler FC (1956) Brittle fractures near equilibrium. Proc Phys Soc 69B:981–982
Rummel F (1987) Fracture mechanics approaching. In: Atkinson BK (ed) Fracture mechanics of rock. Academic Press, Lond, pp 217–239
Ryan MP, Sammis CG (1978) Cyclic fracture mechanisms in cooling basalt. Geol Soc Am Bull 89:1295–1308
Sailors RH, Corten HT (1972) Relationship between Material Fracture Toughness using Fracture Mechanics and Transition Temperature Tests. ASTM STP 514, Part II, pp 164–191
Schardin H (1959) Velocity effects in fracture. In: Averbach BL, Felbeck DK, Hahm GT, Thomas DA (eds) Fracture. Wiley, New York, pp 297–330
Schardin H, Struth W (1937) Neuere Ergebnisse der Funkenkinematographie. Z Tech Physik 18:474
Scholz CH (1972) Static fatique of quartz. J Geophys Res 77:2104–2114
Schultz G (1974) Ice age lost. Anchor/Doubleday, Garden City, New York 342 pp
Seagar JS (1964) Pre-mining lateral pressures. Int J Rock Mech Min Sci 1:413–419
Secor D Jr (1965) Role of fluid pressure in jointing. Am J Sci 263:633–646
Secor D Jr (1969) Mechanics of natural extension fracturing at depth in the earth's crust. In: Baer AJ, Norris DK (eds) Research in tectonics. Geol Surv Canada Paper, pp 3–47
Segall P (1984) Formation and growth of extensional fracture sets. Geol Soc Am Bull 95:454–462
Segall P, Pollard DD (1983) Joint formation in granitic rock of the Sierra Nevada. Geol Soc Am Bull 94:563–575
Shainin VE (1950) Conjugate sets of en echelon tension fractures in the Athens limestone at Riverton, Virginia. Bull Geol Soc Am 61:509–517
Shand EB (1954) Experimental study of fracture of glass: 2 Experimental data. J Am Ceram Soc 37:559–572
Shand EB (1958) Glass engineering handbook. Mc-Graw Hill, New York, 484 pp
Shand EB (1959) Stress behaviour of brittle materials. Am Ceram Soc Bull 38:653–660
Shand EB (1965) Strength of glass-the Griffith method revised. J Am Ceram Soc 48:43–49
Shand EB (1967) Breaking stresses of glass determined from fracture surfaces. Glass Ind 47:190–194
Sheldon P (1912) Some observations and experiments on joint planes. J Geol 20:53–79
Shepherd J, Creasey JW, Rixon LK (1982) Comment on joint spacing as a method of locating faults. Geol 10:282
Shetty DK, Bansal GK, Rosenfield AR, Duckworth WH (1980) Criterion for fracture-mirror boundary formation in ceramics. J Am Ceram Soc 63:106–108
Shinkai N, Sakata H (1978) Fracture mirrors in Columbia Resin CR-39. J Mater Sci 13:415–420
Sih GC (1973) A special theory of crack propagation. In: Sih GC (ed) Mechanics of fracture. Noordhoff, Leyden 1:21–45
Sih GC, Paris PC, Erdogan F (1962) Crack tip, stress-intensity factors for plane extension and plate bending problems. J Appl Mech 29:306–312
Simmons DA, Richardson EV (1961) Forms of bed roughness in alluvial channels. Am Soc Civil Eng Proc 87:87–105
Simmons G, Richter D (1976) Microcracks in rocks. In: Strens RGJ (ed) The physics and chemistry of minerals and rocks. Wiley, New York, pp 105–137
Smekal A (1936) Die Festigkeitseigenschaften spröder Körper. Natur Wiss 15:106–188
Smekal A (1940) Ultraschalldispersion und Bruchgeschwindigkeit. Phys Zeit 41:475–480
Smekal A (1950) Verfahren zur Messung von Bruchfortpflanzungsgeschwindigkeiten an Bruchflächen. Glastech Ber 23:57–67
Smekal A (1953) Zumm Bruchvorgang bei sprödem Stoffverhalten unter ein- und mehrachsigen Beanspruchungen. Oesterr Ing-Arch 7:49–70
Smith FW, Emery AF, Kobayashi AS (1967) Stress intensity factors for semicircular cracks. J Appl Mech 34E:953–959
Sneddon IN (1951) Fourier transforms. Mc-Graw Hill New York, 542 pp
Solomon M, Hill PA (1962) Rib and hackle marks on joint faces at Renison Bell Tasmania: a preliminary note. J Geol 70:493–496

Sommer E (1967) Das Bruchverhalten von Rundstäben aus Glas im Manteldruckversuch mit überlagerter Zugspannung. Glastech Ber 40:304–307

Sommer E (1969) Formation of fracture 'lances' in glass. Eng Fract Mech 1:539–546

Soroka I (1977) Properties and behavior of aggregate and concrete made of portland cement. Technion, Haifa, 460 pp (in Hebrew)

Speidel MO (1971) Branching of stress corrosion cracks in aluminium alloys. In: Scully JC (ed) The theory of stress corrosion cracking in alloys. NATO, pp 449–471

Stearns DW (1968) Certain aspects of fracture in naturally deformed rocks. In: Riecker RE (ed) Rock, mechanics seminar, vol 1, Spec Rep Terrestrial Sci Lab Air Force, Cambridge Res Labs, Bedford, Mass, pp 27–116

Steinitz G (1974) The deformational structures in the Senonian bedded cherts of Israel. Ph. D. Thesis, Hebrew Univ, 126 pp (in Hebrew)

Stevens PS (1974) Patterns in nature. Little and Brown USA, 240 pp

Stewart RL, Chyung K, Taylor MP, Cooper RF (1986) Fracture of SiC fiber/glass-ceramic composites as a function of temperature. In: Bradt RC, Hasselman DPH, Lange FF (eds) Fracture mechanics of ceramics. Plenum, New York 7:33–51

Stoyanov S, Davobski C (1986) Morphology of fracturing in zones of oblique extension. Experimentals results and geological implications. Bulg Acad Sci 19:3–22

Streit R, Finnie I (1980) An experimental investigation of crack path directional stability. Exp Mech 20:17–23

Sugden DE, John BS (1976) Glaciers and landscapes. Arnold, London, 376 pp

Swain MV, Hagan JT (1978) Some observations of overlapping interacting cracks. Eng Fract Mech 10:299–304

Swain MV, Lawn BR, Burns SJ (1974) Cleavage step deformation in brittle solids. J Mater Sci 9:175–183

Syme Gash PJ (1971) Surface features relating to brittle fracture. Tectonophysics 12:349–391

Teague JM Jr, Blau HH (1956) Investigations of stresses in glass bottles under internal hydrostatic pressure. J Am Ceram Soc 39:229–259

Terao N (1953) Sur une relation entre la resistance à la rupture et le foyer d'éclatement du verre. J Phys Soc Jap 8:545–549

Theocaris PS (1972) Complex stress intensity factors at bifurcated cracks. J Mech Phys Solids 20:265–279

Thomson RM (1986) Physics of fracture. In: Ehrenrich H, Turnbull D (eds) Solid state physics. Academic Press, New York 39:1–129

Thomson R, Hsieh C, Rana R (1971) Lattice trapping of fracture cracks. J Appl Phys 42:3154–3160

Tillman JE (1983) Exploration for reservoirs with fracture-enhanced permeability. Oil Gas J 81:165–179

Tipper CF (1957) The study of fracture surface markings. J Iron St Inst 185:4–9

Tipper CF, Hall EO (1953) The fracture of alpha iron. J Iron St Inst 175:9–15

Tomer A (1988) Metals through the microscope. Ministry Defence, Israel, 205 pp

Tschegg EK (1983) Mode 3 and mode 1 fatigue crack propagation behaviour under tortional loading. J Mater Sci 18:1604–1614

Tsirk A (1981) On a geometrical effect on crack front configuration. Int J Fract 17:185–188

Tuttle OF (1949) Structural petrology of planes of liquid inclusions. J Geol 57:331–356

Van Hise CR (1896) Principles of North-American Pre-Cambrian geology: US geology. US Geol Surv 16th Annual Rep, pp 581–874

Van Mier JGM (1986) Fracture of concrete under complex stress. Heron 31:1–90

Van Tyne A (1983) Natural gas potential of the Devonian black shales of New York. E Geol 5:209–216

Varner JR, Frechette VD (1971) Fracture marks associated with transition – region behavior of slow cracks in glass. J Appl Phys 42:1983–1984

Voight B, St Pierre BHP (1974) Stress history and rock stress. Proc 3rd Rock Mech Congr ISRM. National Academy of Sciences Washington, DC 2:580–582

Wachtman JB Jr (1974) Highlights of progress in the science of fracture of ceramics and glass. J Am Ceram Soc 57:509–518

Wallner H (1939) Linienstrukturen an Bruchflächen. Z Physik 114:368–37
Warren R (1978) Measurement of the fracture properties of brittle solids by Hertzian indentation. Acta Metall 26:1759–1769
Wedel AA (1932) Geological structures of the Devonian strata of south-central New York, N.Y. State. Mus Bull 294:74
Wells AA (1963) Application of fracture mechanics at and beyond general yielding. BRWJA 10:563–570
Westergaard HM (1939) Bearing pressures and cracks. Trans Am Soc Mech Eng 61A:49–53
Wheeler RL, Dixon JM (1980) Intensity of systematic joints: methods and application. Geol 8:230–233
Wiederhorn SM (1967) Influence of water vapor on crack propagation in soda-lime glass. J Am Ceram Soc 50:407–414
Wiederhorn SM (1969) Fracture surface energy of glass. J Am Ceram Soc 52:99–105
Wiederhorn SM (1972) A chemical interpretation of static fatigue. J Am Ceram Soc 55:81–85
Wiederhorn SM (1974) Subcritical crack growth in ceramics. In: Bradt RC, Hasselman DPH, Lange FF (eds) Fracture mechanics of ceramics. Plenum Press, New York 2:613–646
Wiederhorn SM (1978) Mechanisms of subcritical crack growth in glass. In: Bradt RC, Hasselman DPH, Lange FF (eds) Fracture mechanics of ceramics. Plenum Press, New York 4:549–580
Wiederhorn SM, Bolz LH (1970) Stress corrosion and static fatigue of glass. J Am Ceram Soc 53:543–548
Wiederhorn SM, Johnson H (1971) Effect of pressure on the fracture of glass. J Appl Phys 42:681–684
Wiederhorn SM, Johnson H (1973) Effect of electrolyte PH on crack propagation in glass. J Am Ceram Soc 56:192–197
Wiederhorn SM, Lawn BR (1979) Strength degradation of glass impacted with sharp particles: 2. Tempered surfaces. J Am Ceram Soc 62:66–70
Wiederhorn SM, Johnson H, Diness AM, Heuer A (1974a) Fracture of glass in vacuum. J Am Conc Inst 58:591–61
Wiederhorn SM, Evans AG, Fuller ER, Johnson H (1974b) Application of fracture mechanics to space-shuttle windows. J Am Ceram Soc 57:319–323
Wilkins BJS (1980) Slow crack growth and delayed failure of granites. Int J Rock Mech Mining Sci 17:365–369
Williams JG, Ewing PD (1972) Fracture under complex stress – the angled crack problem. Int J Fract Mech 8:441–446
Williams ML (1957) On the stress distribution at the base of a stationary crack. J Appl Mech 24:109–114
Wilshaw TR (1971) The Hertzian fracture test. J Phys D 4:1567–1581
Wilson G (1982) Introduction to small-scale geological structures. Dep Geol, Imperial Coll Sci Technol, Lond
Wise DU (1964) Microjointing in basement, Middle Rocky mountains of Montana and Wyoming. Geol Soc Am Bull 75:287–306
Woldenberg MJ (1969) Spatial order in fluvial systems: Horton's laws derived from mixed hexagonal hierarchies of drainage basin areas. Geol Soc Am Bull 80:97–111
Woldenberg MJ (1970) A structural taxonomy of spacial hierarchies. Colston Pap 22:147–175
Wolock J, Kies JA, Newman SB (1959) Fracture phenomena in polymers. In: Aberbach BL, Felback DK, Hahn GT, Thomas CA (eds) Fracture. Wiley, New York, pp 250–264
Woodworth JB (1895) Some features of joints. SCI 2:903–904
Woodworth JB (1896) On the fracture system of joints, with remarks on certain great fractures. Boston Soc Nat Hit Proc 27:63–184
Yoffe EH (1951) The moving Griffith crack. Phil Mag 42:739–750
Yokobori T, Uozomi M, Ichikawa M (1971) Interaction between non-coplanar parallel straggered elastic cracks. Rep Res Inst Strength Fract Mater 7:25–47
Yukawa S, Timo DP, Rubio A (1969) Fracture design practices for rotating equipment. In: Liebowitz H (ed) Fracture Academic Press, New York, pp 65–157
Zandman F (1954) Étude de la deformation et de la rupture des matières plastiques. Publ Sci Tech Ministère l'Air, Paris, France

Zapffe CA, Landgraf FK, Worden CO (1948) Transgranular cleavage facets in cast molybdenum. Met Prog 54:328–331

Zoback MD, Zoback ML, Mount VS, Suppe J, Eaton JP, Healy JH, Oppenheimer D, Reasenberg P, Jones L, Raleigh CB, Wong IG, Scotti O, Wentworth C (1987) New evidence on the state of stress of the San Andreas fault system. Science 238:1105–1111

Subject Index

abrasion 8, 37–38, 123
acceleration 46, 61
acetylcellulose 203
acicular 52–53
acoustic 30
ad infinitum 73, 177
ageing 42, 99
aggregates 54
aircraft 9, 99–100
allotropic 100
alloys 27, 99
  aluminium 99
alternate cycles 95
alumina 42, 104
aluminosilicate 41
ammonia 35
anastomosing networks 232
angle, s
  between conjugate macroscopic shear fractures 203
  between conjugate sets of bands 203
  between the Lüders' bands 203
  bond 35
  branching 17, 114–116, 323
  cone 20, 116, 118, 137
  crack 20, 185–187
  deviation from straight line growth 322
  dihedral 245
  i 241, 245
  interface 153–154, 156, 226–227
  kink 229–230
  m 153–154, 226, 245
  median 153–154, 226–227
  of dilation 177
  of internal friction 29, 189, 247
  of rotation 81
  of total internal reflection 156
  relationships in plumes 223, 226
  that the crevasses make 191
  twist 172–174, 177, 232, 246
anisotropy 51, 53
annealing 63, 104, 120, 130
annular undulations 167, 201, 262, 298, 310
anticline 183, 197, 275
applied
  force 18, 39
  fractography 119–120
aragonite 260, 315
arcade 175
archaeologists 67
arenite 197
argillaceous quartzite 55
asperities 54
ASTM 131
asymptotic 23, 33, 48, 176, 217, 318, 320
atomic
  spacing 7, 33, 43
  stress-strain curve 6
austenite 100
autobrecciated chert 199

baffle 126–128, 130
ball-and-socket 142
barb, s
  branching 144, 148, 150, 183
  maximum intensities 224
  radial 190
  rectilinear 190
barium titanate 42
basalts 157, 195, 197
basement 289
bending 49, 55, 61, 63, 65, 79, 84, 89, 105–106, 114–115, 123–124, 126, 128, 130, 188, 202, 209, 217, 241, 261, 282, 293, 296–297, 308, 320, 322
  biaxial 74
biharmonic equation 10
bilateral
  asymmetric mirror 106, 110
  plume 156–157, 159, 179, 201, 212, 224–226
biomicrites 259
biotite 232
bitumene 67–69
block
  diagram 146, 159, 170–172, 177–178, 237, 283
  tectonism 246

blunt
  flaw 111
  indentor 21
body-centered 95, 100
borehole 134–137
borosilicate 39, 41, 235
bottom centre 126
boundary 16, 33, 41, 46, 52, 56, 82, 89, 91, 93, 101, 103, 110, 137, 152–154, 159, 161, 165, 170, 174, 177, 195, 206–207, 216–217, 226, 231–232, 248, 264, 283, 312, 320
branching
  angle 17, 114–116, 323
  barb 144, 148, 150, 183
  conic 116
  criteria for 49–50
  geometry of 112
  microbranching 109, 112, 116
  of faults 323–324
Brazilian test 55, 202
breakdown 172–174, 175–177, 179, 212, 215, 233–234, 265
bridges 70, 119, 172, 176–177
brittle 9, 20–22, 24, 31, 33, 36, 51–52, 56, 65, 67, 73, 87, 95–96, 99–101, 103–104, 114, 117–178, 135, 164, 196, 202–203, 229, 317
bruises 67
bubble 91–92
buckling rebound 278
budding 88, 112–113
Burgers vector 31, 33, 73
butterfly shape 129

cadmium 99–100
calcite 259–260, 263, 284, 315
calcium oxide 55
Campanian 199, 260–261
Campanian-Maastrichtian-Paleocene 262
canasite 52–53
carbonate 131, 195, 215, 248, 285, 313, 315
carbonatite extrusions 21
cast iron 63
catastrophic growth 4, 42
cavitation scarp 72, 93, 95
celestite 260
cellulose acetate 67, 91–92
cement matrix 54
Cenomanian 185, 186, 209
chain-silicate 52
chalk, s
  exfoliation in 199, 201
  Lower Grey 267
  Middle White 267
  soft 278
Champagne wine 123
checklist 120
chert 163, 166, 217, 241, 243, 260–262, 283–284, 301, 305
chevron 53, 99–100, 132, 144, 152–154, 156
chilling 63, 77
chromium 67, 100, 116–117
circular
  cone 137
  crack 129, 134
  front 157
  hole 2, 16, 191
classification 118–119, 139, 142, 144, 160, 167, 188, 211, 237–239
clays 163, 185, 203, 259, 279, 314
cleaved flake 7
cliff 248, 277–278, 285, 305
CO_2 121, 131
coalesce 33, 100, 223
coating 146
Coconino Sandstone 203
COD 1, 24, 32
collar-split 123
columnar
  fans 196–197
  joint tiers 197
  joints 157, 159, 195–199, 211, 217–218, 220, 296
  quartzite 197
  sandstone 197
combined
  markings 166
  modes 19
compression
  diametral 202
  multiaxial 55
  triaxial 203, 289
  uniaxial 2, 14, 16, 202, 248
concave 67, 123, 129, 155, 164, 167, 183, 195, 206
conchoidal 65, 118, 139–142, 159, 237
concrete 1, 6, 13, 17, 22, 29, 51, 54–56
concretions 199
cone
  angle 20, 116, 118, 137
  circular 137
  elliptical 137
  Hertzian 20–21, 137
  shaped crack 20
confinement to layer boundaries 211
Coniacian 265, 267, 313
conic
  branching 116
  discoids 218
conjugate 170, 180, 204, 241, 245, 258

constant
  displacement 12, 18
  geometric 45, 108
  stress rate loading 45
  velocities 82
consumer 120, 123–124, 128
contact damage 20, 123
contortion 262
controlled laboratory conditions 202
convergent fan 260
convex 67, 129, 143, 150, 155, 164, 177, 183, 195, 207, 218, 222, 249, 255, 287–288, 295–296, 300
copper 192
cords 67
coring-induced fractures 205–206, 209
corrosion pit 96
Coulomb criterion 29
Coyote Creek fault 323
crack, s
  angle 20, 185–187
  Arrest 51, 81, 119, 144
  circular 129, 134
  closure 17, 78, 99, 248
  cone-shaped 20
  critical edge 12
  critical internal 12
  deepening 42
  deflection 52
  dimensions of en echelon 223, 232–233
  edge 9, 11, 18
  elliptical 2, 16
  extension force 12, 17–18, 48, 60
  front 9, 12, 19, 69, 71, 73, 79–82, 84, 86, 91–93, 95–96, 103
  geometry 12
  growth limit 41
  increment of crack growth per cycle 43
  interaction 51, 233, 320
  internal 9, 11, 58, 76
  kinking of the 49
  leading 129
  lengths 173
  mud 157, 159, 211
  opening 28, 36, 309
  opening displacement 24, 26, 31–32
  parent 14, 112, 171–172, 178
  penny shaped 18, 60
  peripheral 127
  profiles 34
  radial 66, 109, 129
  radius 45–46, 134
  ring 21
  rounded (blunt) crack tip 43
  secondary 14, 16, 29–30, 61, 91–92, 115
  shielding 53–54
  size 1, 7, 9, 11, 24, 42
  tip singularity 34
  tip zone 22
  width 2, 173, 232
Cretaceous 143, 182, 197, 258–259, 265, 267, 273, 279, 313, 318
critical
  edge crack 12
  flaw length 103
  internal crack 12
  length 11, 96
  strain energy release rate 12–13
  stress 4, 45, 49–50, 129
  velocity 49–50
  volume 59, 225
cross
  -fold joints 145, 151, 167, 169–170, 217, 241, 249, 251, 255, 260, 283, 293, 308, 318, 320
  fractures 140–141
crystallographic 33, 73–74, 95, 101, 117
cubic 56, 96
cup and cone 100, 116
curved branch 14, 16–17
curving 49, 123, 143–144, 149–150, 153, 167, 183, 219, 232, 275–277, 322
cusp 84–85, 93, 100
cuspate profile 162–164
cyclic
  torsion 77
  undulations 220
cylindrical
  specimens 202
  wall 122, 129

D piece 126
dead-weight 18
deflection rate 106
deformation
  dilational 176
  plastic 11, 12, 22, 25–27, 29, 73, 76, 95–96, 100–101
delayed failure 36, 42
density 47–48, 61, 82, 103, 114, 248
denudation 287, 292
determination
  of fracture stresses 243
  of paleostress on the joint surface 223, 229
  of stepping 178
devonian
  Catskill Delta 248, 297
  shales 205
  siltstones 144–145, 151, 159, 216, 240, 251, 309
diabase dikes 142

diameter of the bottle 123
diametrical
  split 126
  squeezing 120
diamond 83
diapiric 21
dilatancy 29
dilation 177, 278
dimples 33, 101
dip 199, 208, 219, 275, 302, 305, 308
discoid, s 140–142, 159, 161–164, 166, 170, 183, 185–187, 194, 215, 230, 259, 279
  conic 218
  embryonic 185
  symmetry of 187, 212
dislocation 7, 31, 33, 73, 262–263
displacement 7, 10, 24, 28, 120, 132, 177, 260, 263, 275–276
  constant 12, 18
dissociative chemisorption 34–35
divergent
  fan of joints 260
  plumes 203, 298
DM model 23–24
double cantilever 18, 30, 95
drilling 205–207
duality 50
ductile transition temperature 31
dune 84
dynamic 14, 20–21, 48–49, 51, 61, 187, 210, 323
  fatigue 46
  fracture toughness 48

earth sciences 20, 91
earth's crust 41, 226
earthquakes 46
echelon overlapping 172, 194, 234
edge
  crack 9, 11, 18
  dislocations 28, 32–33
  effects 105
effective
  pressure 41, 204
  principal stress 58, 240, 243
  radius 31
elastic
  energy 18, 21, 46
  fracture 1, 58
  microcracking 29
  strain energy 3, 5, 12
  stress concentration factor 3, 11
  waves 47, 156
electronic ray 192
ellipse 1, 3, 14, 91, 119, 137, 212
elliptical
  borehole 137
  cone 137
  crack 2, 16
  hole 1–3, 16
embryonic discoids 185
en echelon crack, s 17, 119, 134, 170–177, 179–181, 185–186, 190, 193–194, 199, 212, 215, 221, 226, 232–235, 237, 246, 248, 265, 285–286, 295–296, 300, 312–313, 318–319
  dimensions of 223, 232–233
  linkage 172, 203
endurance limit 44
energy
  activation 39
  balance concept 1, 3–4
  fracture surface 12, 51, 53, 108
  kinetic 21
  to-thickness ratio 190
  surface free 3, 12
Entrada Sandstone 204–205, 259, 274, 306, 309, 313
epeirogenic 293
equilibrium 4–7, 12, 18–19, 23, 39
erosion 267, 287, 289, 293, 296
eustatic 293
exfoliation
  in basalts 199
  in Chalk 199, 201
experimental 11, 14, 17, 26, 38, 44, 48–49, 52, 101, 109, 149, 164, 187, 192, 202, 235, 239–240, 322
explosive 47

face-centered cubic 100
factory roof 77–78
failure
  mode 52
  prediction 44–45
faint ridges 140, 142, 151
fan 61, 77, 143–144, 148, 156, 165–166, 190, 196, 224, 235, 300
fatigue 29, 36–39, 42–46, 53, 60, 77, 95–96, 99, 132–135, 199, 239, 246, 248, 317
  limit 36, 41–42, 237
  striations 69, 84–85, 96, 99, 119, 132
fault
  extremity 275–276
  -joint interaction 276, 318
  termination 275
fiber composite 52
finish 123
finite specimen 18
fissure 65, 80, 88, 197

fixed grips 7, 60
flat 25, 29, 65, 78, 85, 95, 101, 136–137, 139, 161–164, 170, 181, 183, 185–189, 199, 216, 218, 246, 279, 297, 308
flaw, s
  blunt 111
  depth 108
  deviating from a single plane 111
  initial flaw length 46, 103
  linear 38
  machining 110–111
  multiple 111
  point 38
  size 36, 38, 42, 108–110
  sub-critical 109
  surface 12, 108–109
flexural 'star' 129
flexure 50, 73, 76, 113, 114
flint 37, 67, 104, 140, 163, 166, 183, 185–187, 259, 267, 279, 282, 314
flow 25–26, 31, 52, 164, 190–192, 195–197, 205, 316
force
  crack extension 12, 17–18, 48, 60
  driving 12, 33, 53, 192, 217, 316
  perturbation of lines of 2
forking 88, 112, 114–115, 323
Fourier-transform 35
foyer declatement 65, 118
fracture, s 67
  bifurcations 278
  c-fracture 140, 142, 146, 171, 176
  clevage 95, 148
  concentric 126, 197, 262
  convergence 260
  coring-induced 205–206, 209
  cross fractures 140–141
  determination of fracture stresses 243
  diagenesis 120
  disc 206
  downward propagating 287
  ductile 24, 32, 95, 100–101
  elastic 1, 58
  feather-fracture 63, 142
  front 65, 79–82, 91, 93, 95, 150, 156, 180, 192–193, 195
  handling-induced 205, 209
  interaction 61, 203, 218, 245, 317–318, 320, 322, 324
  markings on mineral fillings 223
  mechanics 1, 9, 12, 17, 28, 34, 36, 39, 118, 134, 229, 238, 297, 322, 323
  parent 14, 29, 49–50, 76, 88, 113, 119, 155, 170, 172, 175–176, 183, 189, 209, 229–230, 286
  partial fracture planes 71
  petal centerline 206
  post-critical fracturing 229
  propagation is incremental 99
  satellite 284
  shear 28, 156, 203–204
  slanted 246
  stress 7, 41, 43, 49, 74, 89, 91, 101, 103–104, 106, 108, 116–117, 131–132, 193, 210, 229–230, 232
  surface energy 12, 51, 53, 108
fracture, s
  surface morphology 63, 149, 195, 202, 210, 234, 238, 293, 297
  terminal fracture velocity 47
  toughness 11, 18, 22, 24, 27, 48–49, 51–53, 55, 235
  velocity 1, 47, 49, 192–193, 223
  vertical fracture propagation 293
  width 309
Frauenbach Quarzite 182
free surface 29, 61, 70–71, 79, 153–154, 159, 217, 226
fringe zones 152–154, 156, 170, 241, 245
fringes 69, 73, 142, 148, 170–171, 174–176, 179, 185–187, 189–190, 211, 215, 226, 229, 233–234, 246–248, 255, 265, 267, 275, 282, 287, 296, 300, 313, 317
FSM 211–212, 215, 218–221, 223, 228–229, 237–239, 241, 243, 258, 282, 302
  overprinting 210, 221, 282
  size 213, 228

gas pore 96
gelatine 203
geologic formation 211
geometric
  constant 45, 108
  constraint 190
geometrical factor 12, 45, 49, 55
germanium 67
glacier 191–194
glass, es
  American Glass Research Inc. 121
  anomalous 41
  balls 55
  carbon 115
  ceramic 52–53, 101–102
  containers 121
  defects 67
  immiscible 51
  lappy 120
  lead 41
  molten 5, 63
  network 38
  normal 41

glass, es (cont.)
polyester 72, 235
pristine glass fibers 7
restrained 289
rod 8, 71, 101, 113
sheet 49, 93
silicate 22, 25, 31, 67
soda-lime 12, 26, 77, 108, 235, 322
tempered 63, 71, 104
toughened 106
grain
boundaries 51–52, 57–58, 76, 89, 100, 103, 109, 155, 230, 232
interlocked grain structure 52
resistance to cracking 51
size 42, 56–57, 76, 91, 109, 164
granite 45–46, 186, 195, 201, 229–232, 297
granodiorite 59
granular 22, 28–29, 140
graywacke 170, 177
Griffith
equation 228
energy balance 12, 19, 46
flaws 7
parabolic envelope 244
gull wings 92–93
gypsum 260, 284–285, 315
casts 166

hackle 48, 52, 66, 70, 74, 87–89, 91, 102–106, 110, 111, 114, 118–119, 142, 181, 194, 209, 216, 229, 231–232
coarse 87, 149
cuspate 103–104, 171, 180–183, 199, 201, 209, 215–216, 237, 274
healing 33
heel 121, 123–124, 126–130
height 73, 87–88, 130, 164, 169, 180, 201, 213, 228, 243, 260, 304, 308, 312–313
helices 130
hemi-spherical 293
hemicircular 160, 230, 308
herringbone 63, 65, 91, 144, 226
heterogeneous 13, 56, 59, 102, 108, 217, 316
hexagonal close-packed 95
hierarchial network 111
hole
elliptical 1–3, 16
Homalite-100 72, 88–90, 106, 235
horizontal
gradients 293
joints 149, 213, 297–298, 301, 305
split 124, 126–127
squeeze 129
surfaces 297, 299
Horsha Formation 245–246, 277, 285, 316
Huttonian gelogical cycle 238
hybrid 170, 245, 283
hydraulic 99, 135–137, 249, 282, 316
joints 249, 317
hydrocarbon 315
hydrogen
bond 35
embrittlement 99–100
flakes 134
hydroxyl 35, 38
hypabyssal 195, 230
hyperbolic 91

igneous 139, 142
impact 20–21, 53, 65, 70, 85, 87–88, 104, 119–120, 123–124, 127–129, 163, 166, 199, 201, 209
imperfections 95
inclined
joints 279, 301, 305
surfaces 301
inclusion 32, 51, 58, 67, 77, 86–87, 91, 93, 109–110, 185, 217
incubation 36, 58–59, 223
indentation 20–21, 96, 109, 322
indurated 248, 278
inertia 57, 108, 320
infrared spectroscopy 35
inhomogeneous 20, 240
Inmar Formation 197
intergranular 51–54, 57, 99, 103, 204
interlocking sutures 197
internal
crack 9, 11, 58, 76
friction 17
pressure 59, 74, 105, 113–114, 120–121, 123–124, 126–128, 131–132, 175–176
stresses 53, 63
intersection scarp 93, 119, 194–195
intraformational 262
ionic distance 48
iron 7, 95, 313, 314
isotensile 196–197
isothermal 196–197
isotope analysis 259
isotropic 7, 56, 58, 74, 105, 115, 151, 278, 289

J-integral concept 22
jellies 67
joint, s
arrest 61
burial jointing 240, 277, 313
columnar 157, 159, 195–199, 211, 217–218, 220, 296

columnar joint tiers 197
cross-fold 145, 151, 167, 169–170, 217, 241, 249, 251, 255, 260, 283, 293, 308, 318, 320
divergent fan of 260
early burial 241, 248
fillings 211
fringe 139, 153, 162, 171, 193, 237
horizontal 149, 213, 297–298, 301, 305
hydraulic 249, 317
incipience 58
inclined 279, 301, 305
intensity 313, 316
late burial jointing 245
length distribution 59, 223–224
macro-jointing 195
multi-layer 278, 283–287, 306, 308, 315
opening 278, 282, 298, 313–314
orthogonal joint sets 297
parent 173–174, 178–179, 183, 185–186, 188, 194, 229
periodic pore pressure jointing 199
plane 139–142, 151–154, 170, 172, 174–175, 179, 202, 237, 302, 315
post uplift jointing 301
radial columnar 196
release 285, 287, 289
ring 183
sequential formation of uplift 290
sets 210, 217, 219–220, 229, 232, 241, 243, 245–246, 265, 275, 279, 302
shape 308
strike 145, 217, 220, 241, 251, 254–255, 289, 308
syntectonic jointing 258–259, 275
systematic jointing 46
tectonic 249, 297, 317
type a 171
uplift 60–61, 201, 275, 279, 282, 285, 289–290, 292–293, 295–298, 300–302, 304–306, 308–310, 312–318, 320, 322
zone 274–275, 309
Jurassic 183, 197, 259, 274, 279, 295

knapping 37
knife edge spalls 209
knuckle 127

L/S 309
laboratory 12, 119, 202, 204, 246
lagging embayments 92, 143, 150
laminated structure 51
lattice 31, 33, 41, 95
layer
boundaries 58, 145, 149–150, 155, 157, 159, 167, 174, 189–191, 195, 212–213, 215–218, 226, 241, 245, 248, 255, 265, 267, 275, 298, 308, 312, 316
separation 294, 304
thickness 41, 190, 223–234, 255, 308
ledge 31
length 2, 5–7, 9, 11–12, 14, 16, 18, 22–26, 44–46, 48, 50, 54–56, 58–60, 78, 81, 86, 89, 95, 99–100, 102, 108–109, 112, 123, 130, 139, 159, 163, 169, 172, 175–178, 180, 193, 209–210, 213, 219, 223–226, 228–229, 232–233, 243, 246, 248–249, 265, 267, 274, 276, 278, 282, 304, 306, 308–309, 312, 318
critical 11, 96
critical flaw 103
initial flaw 46, 103
lightning 112
lignitic siltstones 146
limestone 30, 143, 150, 152, 166, 202–204, 209, 260, 274, 285, 308
concretion 163, 166, 199, 201, 312
line, s
arrest 81–82, 96, 119, 143, 151, 207–208, 212, 217, 228, 235, 249, 265, 274, 290, 296, 298
ultrasonic 86–87
liquid nitrogen temperature 36
lithographic 202
lithology 61, 145, 163, 201, 229, 320
lithomechanical 243, 248
load
loading 11–12, 14, 16–21, 30, 33, 36, 42, 47, 50, 54, 60–61, 71–72, 80, 87–88, 95, 103, 113–114, 135–137, 202, 243, 248, 251, 287, 322
vertical 120, 123–124, 127–128, 292
longitudinal 78, 121, 191, 193
looping 128
Lower Eocene 147, 151, 165–167, 169, 217, 219–220, 226–228, 241, 243–244, 246–247, 275, 278–279, 282–284, 287–288, 293, 299, 301, 304, 306, 309, 318, 320
LRT model 33
Luders 101
Luders' bands 204
lungs 111
lustrous 139

macrocrack 55–56
magnesium fluoride 113
Malm Zeta limestones 150
manufacturer 120, 123–124, 130
mapping 89
mark, s
beach 96, 108, 119, 132

mark, s (cont.)
  overprinting 285, 287–288
  ripple marks 63, 65–66, 77, 79–81, 84, 123, 127
marking, s
  combined 166
  contemporaneous 264
  counts of fracture 167
  giant 291
  induced by quarrying 209
  parabolic 90–91, 135
  that are in disharmony 199
  rib 63, 69, 143, 147, 159–163, 165–167, 169–170, 180, 185, 189, 211–215, 217–220, 223, 228–229, 235, 241, 244, 255, 261–262, 264, 282, 287–288, 296, 298, 300, 312
martensite 100
material, s
  bi-phase material 5, 13
  grainy heterogeneous materials 6, 17
  lithic materials 67
  science 240
  synthetic materials 142
  technical 63, 142, 155, 223, 237, 239–240
maximum 2–7, 12, 14, 16–17, 27, 29–30, 43, 45–50, 53–59, 61, 86–88, 91, 99, 114, 116, 129, 134–135, 172, 174, 176, 187–188, 190, 193, 211, 213, 225, 228, 232, 244, 246, 248–249, 275, 316, 323
measurable parameters 223
measurement of area 233
median 65, 152–155, 191–192, 195, 212, 226–267, 241, 304
Menuha Formation 213, 259–260, 308
Mereenie sandstone 295
mesoscale 317
metal 7, 12, 22, 24, 26–30, 36–37, 43, 67, 77–78, 84, 95–96, 99–100, 114, 119–120, 129, 132, 153–156, 205
  embrittlement 34
metallurgical 67
metamorphic 227
methanol 35–36
mica 7, 9, 73
microcline 232
microcrack 6–7, 13–14, 16, 22, 29–30, 32, 38, 50, 53–59, 76, 88–89, 151–152, 197, 230–232
  coalensence 30
microcrystals 67
microrupture 29
microstructure 42, 51–54, 56, 104, 151–152
microvoids 100
Middle Eocene 147, 151–152, 165, 173–174, 178, 214, 224–227, 232, 233–234, 245–248, 259, 275, 277–278, 282, 285–288, 293, 295, 297, 308, 313, 317–319
Mineralization 255, 259, 260, 263, 265, 282, 284, 309, 313–314
mirror, s
  asymmetrical 105, 108, 110
  bilateral asymmetric 106, 110
  boundary 50, 88, 105, 209
  cathedral 75
  intersecting 111
  overlapping 111
  radii 103
  tongue 74–75
Mishash Formation 199, 260–261
mist
  boundary 103, 131, 231
  zone 50, 82, 87–89, 91, 106
MLJ 282–284
Moab 273–274, 309
model
  atomistic slow fracture 35
mode, s
  combined 19
  failure 52
  loading 10
  mixed 14, 19, 61, 80, 103, 108, 110, 171, 223, 246
  of crack propagation 9
  opening 9, 16
  sliding 9
  tearing 9
molar volume 39
molecular 7, 35–36, 39, 80, 135
monotonically 33, 91
montmorillonite 260
Mor Formation 241, 246, 257, 316
multi
  -cuspate hackles 209
  -layer joints 278, 283–287, 306, 308, 315
multiple
  flaws 111
  sets 251
Muschelkalk 150
muscovite 7, 12, 33

nanometer 34–36
necking 33
Neogene 230, 284
  uplifts 292
Netser Formation 260
nodular 241, 283
nomenclature 118

non
permissible 19
-coplanar 71, 73, 78, 160, 232–233, 318–320, 322–323
-systematic 146, 237, 268
noncubic 56
nonlinear 22, 26, 30, 32
notch tip 30
nucleated 6, 32–33, 56

oblate 58
oblique 73, 80, 171, 176, 196, 202–203, 221–222, 288, 318
Obreimoff's experiment 7
obsidian 67
oil 55, 135, 316
Oligo-Miocene 265, 320
opening
crack 28, 36, 309
crack opening displacement 24, 26, 31–32
displacements 26
mode 9, 16
Ordovician 214
organic liquids (alkanes) 93
origin 10, 52, 65–68, 79, 81–82, 85, 87–88, 96, 99–100, 104, 106–109, 114, 119–120, 127–128, 132, 142–145, 149, 151, 153, 157, 159–163, 165, 167, 194, 196, 201–203, 205–207, 210, 295
orthogonality 71, 160, 191–192, 211, 219, 244, 263, 282, 297, 302, 310
oscillation 46, 80, 164, 167
overlapping
mirrors 111
of fracture markings 221, 300
overshoot 48
oxygen 33–35

paleostress 46, 187, 189, 190–191, 228–230, 237, 251
Paleozoic 164, 213, 259, 267, 291–293, 298, 300
parabolic
-like Wallner lines 82
markings 90–91, 135
Mohr envelopes 155
undulations 160
parent
crack 14, 112, 171–172, 178
fracture 14, 29, 49–50, 76, 88, 113, 119, 155, 170, 172, 175–176, 183, 189, 209, 229–230, 286
joint 173–174, 178–179, 183, 185–186, 188, 194, 229
percolate 60, 195, 226
percussions 67
periodicity 210, 219–220, 223–224, 255, 267
Permian 249
permissible 19, 260
perpendicular 9–10, 16–17, 24, 28, 70–72, 79, 100, 129, 139, 145, 155–156, 159, 177, 197, 211, 320
Perspex 21, 136
petites ondulations conjugees 65
petroleum 293
pH 39
phosphorite 199
pigmentation 279
pitchfork 112–113
plastic
deformation 11–12, 22, 25, 29, 73, 76, 95–96, 100–101
limit 101
stress 29
zone 1, 22–27, 29–30, 77, 246
plateau 40–41, 47–48
Pleistocene 149, 315
plume, s
along the distal direction 183
B1 plume 144
bifurcation 143–144, 150, 183
bilateral 156–157, 159, 179, 201, 212, 224–226
chaotic 145–146
chepy-type 265
coarse straight 151, 212
combinations 145–146
delicate straight 152, 167
divergent 203, 298
end-member 144, 167
Incremental plume growth 220
length 223–225
periodicity 223–224, 265
spiral 143–144, 156, 273–274
S-type 144–145, 167, 223, 249
unilateral 159, 225
vertical 145, 212, 295, 310, 312
vicinal 145–146, 268, 282, 301
waviness 223, 227
wavy 145, 202, 215, 217, 228, 273
plumose 140, 142, 146
PMMA 14, 16–17, 135
polycrystalline 1, 41, 51–53, 57, 73, 75–76, 83, 91, 101, 103–104, 109
polymers 22, 48, 91, 114, 234
polymethylmetacrylate 16
porcellanites 199
pores 109–111, 115
power-law 59
practical 12, 35, 38, 51, 237

pre-split blasting 210
Precambrian 227, 301
prehistoric 67, 166
pressure
  effective 41, 204
  fluid 58, 60–61, 71, 135, 137, 204, 249, 255, 317
  internal 59, 74, 105, 113–114, 120–121, 123–124, 126–128, 131–132, 175–176
  negative 121
  overburden 58, 243, 248, 293, 298, 309, 316
  partial 41
  periodic pore pressure jointing 199
  Positive 121
prismatic columns 142
process zone 58
profile 84–86, 116, 162, 164, 177, 196, 226
prolate spheroid 58
proof testing 45
propagation
  direction of 79, 129, 150, 286, 312
  resistance of the 11
  stable crack 17
prototypes 31–32
pure
  shear 78
  torsion 114–115
pyrite nodule 206

qualitative characterization 211, 229
quantitative characterization 223
quarrying 202, 210, 301
quartz 67, 73–74, 83, 86, 151, 197–98, 203, 231–232, 260
quasi
  brittle 67
  -ellipse 212
  -spiral 161
  -symmetric 211

radial
  barbs 190
  columnar joints 196
  crack 66, 109, 129
  scars 102, 108
  stress component 10
  striae 66–68, 74–75, 77, 167, 185, 230–232, 299–300
radius
  crack 45–46, 134
  effective 31
  of circular plastic zone 22
  of curvature 2–3, 16, 23, 39, 193
  tip 43
ratio of 16, 27, 42, 46, 55, 78, 105–106, 109–111, 114, 122, 176, 189, 229, 234, 243, 246, 267, 274, 289, 309, 317
  aspect 223–224, 228, 243, 308
  energy-to-thickness 190
  Poisson 156, 186–187, 243
Rayleigh surface 47
reamy 120
rebranching 113, 323
rectangular 10, 49–50, 63, 65, 69, 79, 84, 105, 113, 308
relief 140, 144, 151, 211, 218, 275, 313
resin 67
Rhine Graben 264
rhythmic 60, 143–144, 148, 150, 157, 167, 179–180, 189, 218, 224, 235, 239, 255, 258, 265, 310
rib
  arcuate rib markings 160–162, 164
  concentric rib markings 144, 160–161, 185, 187, 189, 212–214, 235, 258, 260, 287, 295
  marking 63, 69, 143, 147, 159–163, 165–167, 169–170, 180, 185, 189, 211–215, 217–220, 223, 228–229, 235, 241, 244, 255, 261–262, 264, 282, 287–288, 296, 298, 300, 312
richterite 53
ridges 91, 95, 101, 103–104, 142, 144, 151–153, 159–160, 167, 181, 185, 197, 209, 234
rigid 6, 17, 21, 129
ring
  crack 21
  joints 183
river patterns 68, 73, 78, 119
rock, s
  blasting 149, 181, 209, 221, 301
  competent 260
  fabric 287, 289
  slabs 202
  splintering 277
root zone 76–77, 172, 174–175, 180, 183, 194, 233–234
rotation 19, 71, 81, 172, 175–176, 179–180, 188–189, 246–247, 258, 275, 320, 322
rubber 67

S.A.E. 4130 95
S.A.E. 4340 100
saddle 123
salt 275
sandstone 30, 67, 159, 164, 197–198, 203–204, 213, 230, 249, 259, 274–275, 282, 291–293, 298, 300, 308, 312

Santonian 258–260, 262–263, 279, 302, 308, 314–315, 318, 320
sapphire 67
scale factor 58
scalloped 140, 209
scanning electron microscopy 89
Schmidt Hammer Concrete Test 248
scratches 67, 129
screw dislocations 33, 73–74
scuffs 67
secondary 14, 17, 21, 29, 33, 49, 73–74, 76, 82, 92, 105, 111, 114–115, 128, 172, 177–178, 267, 313
sedimentary 139, 153–154, 157, 189, 195, 212, 217–218, 229, 238, 245, 255, 289
SEM 78, 95, 99, 107, 110, 117, 204
semi
  circular 3, 84, 106, 110, 130, 134, 162, 232
  elliptical 84, 106
  infinite specimen 18
Senonian 163, 167, 213, 220, 264, 296, 298
sequence 2, 63, 72, 176–178, 194, 196, 205, 211, 217–218, 220–221, 241, 245, 260, 262–263, 265, 283, 290, 293, 314, 323
shales 61, 145, 217, 249, 251, 254, 259, 274–275, 293, 296
shape 1, 3, 10–11, 14, 17, 20, 22–24, 31, 36, 39, 42, 49–50, 58, 67, 71, 73, 80–81, 91–93, 101, 103, 105–106, 108–109, 118, 120–121, 123, 127, 129–130, 139, 142, 144, 147, 163, 165, 167, 174, 183, 185, 188–190, 199, 214, 218, 237, 249, 264, 267, 279, 282, 295, 297, 308–309
  process 189
shark's fin 123, 130
shear
  fracture 28, 156, 203–204
  lip 133, 135, 246
  stress 10, 27, 29, 61, 69, 101, 113, 156, 175–176, 180, 187–188, 203, 215, 320, 322
  zones 171
sheets 33, 149, 213, 282
shelf 259
shielding effect 287
shingles 69, 178
shoulder 122, 124, 127–128, 130, 139–142, 146, 153–154, 172–175, 177–178, 186, 194, 234, 286
SiC fiber 102
side walls 126–127
silanol 35
silica 26, 33–36, 38–39, 41–42, 54, 89, 130, 235
silicon 31, 34–35, 73, 104
siloxane bond 38
siltstone 61, 143, 148, 150, 167, 213, 215, 217–218, 221, 224, 248–249, 251, 254, 293
single crystals 30, 51, 67, 73, 76, 83, 95, 101, 111
sintered 109–110
sinuous 142, 145
size 6, 23–27, 31, 33, 35–36, 45, 49, 51, 54, 56–57, 73, 89, 95, 101, 103, 105–106, 109–111, 114, 118, 137, 139–140, 151, 155, 163, 174, 177, 180, 183, 185, 194, 199, 203, 210–211, 220, 229, 232, 240, 246, 264, 274–275, 279, 282, 285, 298, 312
slabs 23–24, 202–203
slates 183, 230
slip bands 25, 101
slit 10, 14, 16–17, 22–23, 31–32
SLJ 283–284
soil 279, 284, 309, 315
Solenhofen limestone 204
spatial 188, 191, 298, 302
spherical 20–21, 87, 91, 115
spontaneous growth 11
squashed ball 124, 127
stabilized ZrO_2 107
static fatigue limit 42
steamed 37
steatite 104
steel, s 21, 23, 31, 48, 67, 95–96, 99–101, 114, 134, 136, 148, 152, 190, 192, 237, 248
  AISI 302 96
  AISI 4340 77–78
  austenitic stainless 99
steepnesses 163
stepping 134, 178, 182, 185, 275–276
steps 19, 34, 71–75, 77–79, 89, 91, 107, 119, 141, 171–172, 174, 176–178, 180, 188, 193–194, 203, 208, 210, 233, 267
  cleavage 68, 73, 75–76, 78, 119
stiffness 18, 60
stones 67, 120
straight Wallner lines 83
strain
  anti-plane 10
  critical strain energy release rate 12–13
  elastic strain energy 3, 5, 12
  energy release rate 14, 114
  figures 101, 205
  high strain rate 31
  plane 10–11, 14, 18, 23, 25–29, 58, 60, 246, 297
  point 130
  tensile 24, 296
  uniaxial tensile yield 24

strength 5, 7–9, 17, 21–23, 27, 29, 31, 33, 36–37, 42, 53–56, 58, 89, 95, 99–100, 104, 134–135, 155–156, 189, 239, 243, 245, 316, 320
  inert 42
  tensile 99
  theoretical 7, 36
  theoretical cohesive 2, 7
stress, es
  alternating 36, 43
  axial 122–123
  circumferential 114, 116, 121, 123, 128
  concentration 1–3, 6, 16, 29–30, 36, 53, 196, 217
  Concentration Factor 1, 58
  constant stress rate loading 45
  contemporary tectonic stress 287, 289
  corrosion 36–41, 46, 53, 99, 105, 108, 132, 235
  critical 4, 45, 49–50, 129
  crustal 292
  difference 29, 155–157, 245, 258, 320
  dilatational 47
  displacement curve 7
  determination of fracture 243
  effective principal 58, 240, 243
  elastic stress concentration factor 3, 11
  field 2, 10, 20, 33, 59, 61, 79, 102, 114, 176, 188–189, 262, 275, 279, 320
  fracture 7, 41, 43, 49, 74, 89, 91, 101, 103–104, 106, 108, 116–117, 131–132, 193, 209, 229–230, 232
  hinge 128
  hoop tensile 99
  intensity at crack branching 50
  internal 53, 63
  least effective principal 58
  local 47, 178, 297, 302
  localized flexural 128
  microstresses 56
  normal 20, 175–176
  oscillating 46
  pair of 69
  paleo-principal 187, 237
  plane 3, 11, 13–13, 18, 22–29, 246
  planes of maximum shear 24, 28–29, 188
  plastic 29
  radial stress component 10
  remote 1, 11, 26, 175, 179
  shear 10, 27, 29, 61, 69, 101, 113, 156, 175–176, 180, 187–188, 203, 215, 320, 322
  skin 130
  state of crustal 310
  sub-critical stress intensity 229
  tangential 2, 10, 14, 134
  thermal 77, 130, 195–196, 199, 201, 289, 292–293
  vertical gradient in 293
  waves 49, 81, 106, 137, 155
  yield 11, 22, 26, 29
striae 65, 68–75, 77–79, 89, 95, 118–119, 130, 144, 149, 160, 171, 190, 215, 234–235, 286, 291
strike 145, 179, 206, 217–219, 241, 244, 251, 254, 259–260, 263, 265, 274–276, 278–279, 283–285, 293, 302, 304, 306, 308, 310, 320, 324
sub-critical 39, 41, 61, 157, 190
  flaws 109
  stress intensity 229
sub-recent 301, 315
subcritical 1, 36, 39–42, 44, 46, 48, 57
supercritical 1, 40, 46, 181, 209
superposition 71, 147, 166
symmetrical lances 71
symmetry 31, 65, 159, 187–189, 201–202, 211, 237, 258, 297, 322
syncline 151, 178, 259, 278, 316
Syrian Arc 259, 314

tectonics 179, 239–240, 246, 279, 290, 301
tectonofractography 91, 239, 323
tectonophysics 171, 240
temperature 5, 31, 36–37, 39, 41–42, 45, 47, 52–53, 73, 95, 99–100, 104–105, 121, 130, 132, 135, 197, 289
tensile
  strain 24, 297
  strength 99
tension 1–2, 4–6, 13–14, 16–19, 27, 29, 39, 49–50, 55–56, 58, 60–61, 65, 69–71, 81, 84, 95, 100–101, 103, 105–106, 111–116, 129–130, 134, 156, 167, 172, 191, 202, 205–206, 226, 240, 245–246, 292, 296–297, 316–317, 322–323
termination 51, 210, 217, 319
terminology 89, 118–119, 149, 234, 289
tertiary
  bridges 172
  en echelon 177–178
  -Quaternary 259
tetragonal 100
tetrahedron 34
texture 63, 65, 67, 79–80, 92–93, 114, 117, 132, 192, 201, 232
theory
  reaction-rate theory 39
  rebound theory 46
  static arrest theory 51

theoretical
  cohesive strength 2, 7
  stength 7, 36
thermal
  -elastic contraction 287
  expansion anisotropy 56
  metamorphism 197
  shock 53, 120, 123–124, 127, 130–131
  stress 77, 130, 195–196, 199, 201, 289, 292–293
three point bend 115
three regions 40, 45
thumb-nail 96, 99
tier 195, 199
tilt 19, 93, 259–260, 263, 279, 298, 305
time to failure 1, 37, 42, 45–46
tip
  radius 43
  sharpening 42
  -to-plane linkages 176
  -to-tip 176
titanium 95
transformation-toughened 114
transgranular 51–52, 54, 57, 77, 91, 99, 104
transition lines 93
transport 39, 128, 315–316
trapezoid 123
Tresca criterion 27, 29
Triassic 150
tributaries 76
tridymite 33–34
Tsor'a Formation 297
tungsten 83
Turonian 166, 259–260
Type b 171

ultrasonic
  crack modulations 86
  lines 86–87
  waves 86
undercutting 129
undulations 65–66, 73, 77, 79, 86–87, 102, 108, 119, 129, 135–136, 149, 159, 161–163, 177, 190, 192, 199, 211–212, 215, 218, 221–222, 237, 239, 255, 260, 267, 291–292, 298, 300, 310, 312, 313
  parabolic 160
uniaxial
  compression 2, 14, 16, 202, 248
  tensile yield strain 24
universal fatigue curve 37, 39
unloading 14, 17, 206, 246, 287, 289, 292–293, 297, 316–317
unstable growth 11, 18
uplifting 238, 274, 317
upper Middle Chalk 267
Uppper Old Red 197

vacancies 31
vacuum 7, 38, 41
vein networks 259
velocity, es
  constant 82
  critical 49–50
  fracture 1, 47, 49, 192–193, 223
  overshoot 40, 48
  terminal fracture 47
  transitional 11
vertical
  fracture propagation 293
  gradient in stress 292–293
  load 120, 123–124, 127–128, 292
  plumes 145, 212, 295, 310, 312
  split 123–124, 127–128
  wall 124, 126–127, 130
virgin 33–34
viscous 21, 52
vitreous carbon 67
void 29, 31, 90
volcanic 54, 145, 195

Wallner lines 65–66, 79, 81–84
water 33–39, 41–42, 60, 84–84, 93, 95, 130, 164, 167, 189, 195–197, 202, 226, 235, 243, 309, 316
  carbonated 121
  ground 289
  ionized 39
  meteoric 259, 263
wave, s
  elastic 47, 156
  fractography 87
  lengths of the rib sections 178
  transverse 86
  ultrasonic 86
wedge 7, 30–32, 34, 60, 93, 95, 106, 114–116, 119, 142, 194, 202, 226
well 4–5, 11, 14, 20–21, 43, 52, 55, 60–61, 65, 71, 73, 77, 82, 88, 100, 103, 110–111, 118, 123, 132, 134, 140, 142, 146, 148, 150, 159, 170–172, 177, 189–190, 197, 203, 208, 210, 230, 232–233, 239–240, 246, 262, 264–265, 268, 275, 284–285, 289, 297, 304–305, 312–313, 315, 323
Wellenkalk 150
wet 93, 119, 168, 203
width
  fracture 309
  of the mist region 89, 109
wind 84

work-hardening 25
workability 37

Young's modulus 12, 18, 47, 50, 55, 108, 230, 289

zone, s
  cracked 6
  crack tip 22
  damage 1, 6, 12, 22, 30
  mist 50, 82, 87–89, 91, 106
  plastic 1, 22–27, 29–30, 77, 246
  primary 33
  radius of circular plastic 22
  shear 171
  tearing 65, 118
zig-zag 51, 73
zirconate titanate 42